The View from Vermont

THE VIEW FROM VERMONT

Tourism and the Making of an American Rural Landscape

BLAKE HARRISON

University of Vermont Press
Burlington, Vermont

PUBLISHED BY UNIVERSITY PRESS OF NEW ENGLAND
HANOVER AND LONDON

University of Vermont Press
Published by University Press of New England,
One Court Street, Lebanon, NH 03766
www.upne.com
© 2006 by University of Vermont Press
Printed in the United States of America

5 4 3 2 1

Library of Congress Cataloging-in-Publication Data

Harrison, Blake A., 1970–
The view from Vermont : tourism and the making of
an American rural landscape / Blake Harrison.
p. cm.
Includes bibliographical references and index.
ISBN-13: 978–1–58465–566–4 (cloth : alk. paper)
ISBN-10: 1–58465–566–6 (cloth : alk. paper)
ISBN-13: 978–1–58465–591–6 (pbk. : alk. paper)
ISBN-10: 1–58465–591–7 (pbk. : alk. paper)
1. Tourism—Vermont.   2. Tourism—Social aspects—Vermont.
3. Vermont—Social life and customs.   I. Title.
G155.U6H245 2006
338.4'791743–dc22   2006023051

 University Press of New England is a member of the Green Press Initiative. The paper used in this book meets their minimum requirement for recycled paper.

For Rebekah and Dahlia

Contents

Illustrations

Preface

Like millions of other Americans, I first encountered Vermont as a tourist. Not long after I was able to drive, I began visiting the state, searching its back roads and small villages for scenes that felt distinctly "Vermontish" and marveling at what seemed to be the dramatic contrast between Vermont's landscape and that of my home right next door in northern New York. During my travels, I forged encounters with Vermont that I thought were quite special, which, of course, they were. They were not, however, unique. For over a century visitors have traveled to Vermont in search of some of the same icons and affirmations about rural life that I was after: covered bridges, old stone walls, flaming red maple trees, and village greens seemed to embody the sentimentality from which I, and so many others, approached the state.

In the early 1990s I crossed over into Vermont not as a tourist but as a resident-to-be. I took a job at a ski resort and moved into an early-nineteenth-century farmhouse long abandoned by its original farm inhabitants. Living in Vermont over the ensuing years gave me a different set of perspectives on the state — perspectives based on work as much as leisure; on modern ski development as much as sentimentalized farm scenes; and on relationships with others whose old farms had also become still, often later than mine but for similar reasons nevertheless. I found myself discovering a Vermont that was entirely new to me, a place of work, social tension, and even dynamic change. And again, my experiences were not unique.

The last farmers to inhabit my old house left Vermont sometime in the 1860s, moving west in search of farmland more promising than the stony hillside out back. Following their departure a minister from Massachusetts with a passion for trout fishing purchased the hundred-plus-acre farm, for roughly ninety dollars. From that point on the house served mainly as a summer residence for him and his descendents. In the decades to come thousands of such properties in Vermont made a similar transition from working farms to leisure getaways. A new class of seasonal settlers had discovered new kinds of opportunities in Vermont, and as they did, they reshaped and redefined the state's rural landscape for themselves, for those who lived there, and for those who followed. Other types of visitors did much the same, only each new group brought different patterns of land use to the state, different ways of defining its rural landscape, and different ways of interacting with those who lived there.

During my time in Vermont, I joined these visitors and residents in what I now see was a larger historical progression — one defined by an ongoing construction of landscape and identity in rural Vermont. Ours was a landscape produced through diverse and often conflicting opinions about how to use and define rural space in the wake of tourist development. Ours was a landscape defined by its simultaneous associations with old and new, stability and change — a landscape where form and identity were continually "reworked" through the changing contexts of rural leisure and rural work that have marked its passage through time. *The View from Vermont* explores the evolution of that landscape.

Acknowledgments

This book began as a dissertation in 1999 while I was a graduate student in the Department of Geography at the University of Wisconsin–Madison. Since that time many people have contributed to its completion. For starters, there are those who read and offered advice on my work. Reading other people's work takes time and effort, and I am grateful to the readers who made room in their busy schedules to offer advice. Chief among these is Bill Cronon, who advised this project when it was a dissertation and who has continued to lend support and advice years after I moved on from Madison. Bill worked with me every step of the way by offering thorough and rigorous critiques. I am truly grateful for his support, fairness, inspiration, and high expectations. Perhaps above all else, Bill continually reminded me that even big ideas and big arguments mean little if they are not stated in clear and accessible prose. I have tried my very best to abide by that wisdom on every page that follows.

In addition, I would like to thank Arnold Alanen, David Bernstein, Alan Berolzheimer, Dawn Biehler, Mark Cantrell, Mary Curran, Ron Davidson, Jeffrey Sasha Davis, Greg Downey, Ken Foote, Josh Hagen, Gareth E. John, Matthew Kurtz, Bob Ostergren, Tom Robertson, David Robinson, Richard Schein, Michael Sherman, Matt Turner, Tom Vale, Bob Wilson, David

Wrobel, and Bill Wyckoff, all of whom read and commented on material that, in one way or another, found its way into this book. Among these, I would like to extend a special thanks to Richard Schein, who kindly lent advice and encouragement on a number of different occasions. Finally, there are those who I may never meet — anonymous reviewers who offered advice on the book proposal, the book manuscript, and a handful of journal articles covering related material. Their thoughts have mattered a great deal to me.

I would also like to thank Richard Pult, the book's editor at the University Press of New England, for his encouragement, advice, and candor. The support and efficiency that he and Phyllis Deutsch offered have reminded me time and again of how fortunate I am to have found a home, for this book with UPNE. I owe them, as well as a great many others at the press, a debt of gratitude for guiding me through the publishing process.

My ideas have also benefited from valued conversations about landscape, tourism, and environmental history with Mark Adams, Josh Becker, Eric Carter, Eric Compass, Zoltan Grossman, Steven Hoelscher, John Isom, Eric Olmanson, Mike Yochim, Yolonda Youngs, and all of the members of Bill Cronon's weekly breakfast meetings. David Bernstein, Ryan Galt, and Chris Edwards also offered valuable technical advice. Thanks as well to John Mack Faragher and Jay Gitlin of the Howard R. Lamar Center for the Study of Frontiers and Borders at Yale University for opening their community to me during the year in which I finished revising my dissertation into a book manuscript. That year would have been a very lonely time intellectually were it not for the Lamar Center.

Parts of chapters 2 and 5 appeared in earlier form in the *Journal of Historical Geography* and *Vermont History*, respectively. I am thankful for the permission granted by Elsevier and the Vermont Historical Society to use that material here.

I received financial assistance from the University of Wisconsin Graduate School, the Historical Geography Specialty Group of the Association of American Geographers, the Vermont Historical Society, and the New England Ski Museum. Their support helped considerably with research expenses. I am also extremely grateful for the logistical support I received from librarians and curators in Wisconsin and New England. I would like to thank the staffs of the following institutions: the Geography Library and the Inter-Library Loan Department at the University of Wisconsin–Madison, the Vermont Historical Society, the Vermont Folklife Center, the Office of the Vermont Secretary of State, the New England Ski Museum, Special Collections at the University of Vermont's Bailey-Howe Library, *Vermont*

Life magazine, the Vermont Department of Libraries, the Vermont Department of Tourism and Marketing, and all of the local historical societies, libraries, museums, and town offices who opened their doors to me. Finally, thanks go to those who aided me in gathering and securing permissions to use the illustrations in this book. My work benefited directly from theirs.

Many others offered support along the way, ranging from food and shelter to an encouraging word. I would like to thank the Harrison, Comfort, and Irwin families for their support and ongoing interest in what I know has seemed like a never-ending project. In particular, I would like to thank my father, Derek Harrison, for being both a fan and a valued commentator, and my mother, Judith Comfort, for always reminding me that life is full of many joys — of which writing a book is only one. Special thanks also go to Doug and Evelyne Skopp, Jeannie Kelly, John Eisenhardt, Will Kelly, Dr. George Humphreys, John Moravek, Matt Collins, Jen Mallet, and the family of Mary Canales, Richard Spindler, and Rosa. Thanks also to Jon Abel and Wally Lowe for teaching me how to repair ski lifts, plus a whole lot more about Vermont. Finally, I want to thank Hildegarde H. von Laue, whose house inspired this project and whose bottomless kindness has enriched my entire life. I am very sorry she cannot see the final product.

Only one person has played the roles of critic, supporter, and librarian all at the same time — my wife and best friend, Rebekah Irwin. She and I met seven years ago, just as I was starting to work on this project, and I cannot imagine that the history of this book's creation would have been as bright as it was without her. I can easily imagine a bright future without the book, however. For now, as Rebekah and I raise our daughter, Dahlia, I find new reasons each day to be thankful for my incredibly good fortune.

The View from Vermont

FIGURE I.1. Vermont and the northeastern United States.

INTRODUCTION

Tourism and the Reworking of Rural Vermont

Most of the people I have talked to over the years about this book have vivid impressions of Vermont. Whether they have ever been to Vermont or not, they, like millions of other Americans, have come to expect certain things from this small New England state. Pastoral views segmented by stone walls and patches of forest; charming white villages set against a rolling mountain backdrop; opportunities for outdoor recreation such as hiking and skiing; slow drives on country roads lined with maple trees; a quiet, dignified people whose days are long but whose work is honest: these are the kinds of images that have traditionally informed popular impressions of Vermont, and they are the kinds of images that have motivated generations of travelers to visit the state.

Taken as a whole, this "typical" Vermont scene has given the state an enduring reputation defined, above all else, by its associations with *rural land and life*. Again, the people I have talked to about my interests would all readily agree that the state is an essentially rural rather than urban place, and most would deploy fairly similar sets of icons and imagery to explain why they think this is so. But what do we mean when we use the term "rural" to describe Vermont? Why is it that Vermont and the category "rural" are so often conflated in the popular mind? How did this relationship form, and why did it become so appealing, so persistent?

One way to approach such questions is through statistics. With a long history of farming, no major urban areas to speak of, and a contemporary population of just over half a million people, Vermont consistently comes in as one of the United States Census Bureau's most "rural" of states. But statistical categories about rural communities can easily mislead, and numbers alone can never fully explain the social and cultural circumstances in which Vermont's rural reputation has evolved.[1] For that we also need to explore the values and ideas that people from Vermont and elsewhere have used to define rural landscapes. We need to consider how those values and ideas were translated into action.

A second way to approach questions about Vermont's rural reputation is to turn to the state's physical landscape itself. With its attractive mix of farm, forest, and village, one might say that Vermont simply *looks* as rural as rural can be. But, of course, not all of Vermont's 9,600 square miles looks quite like the pictures found in coffee table books or scenic calendars. The central spine of the state's Green Mountains looks and feels very different from the lowlands of Lake Champlain, for instance, and the forested expanses of northern Vermont can at times feel only remotely related to its settled agricultural valleys. Family farms have long been on the decline statewide, recreational home lots and even suburban developments have long been on the rise, neon signs vie with handmade ones along the state's roadsides, and most Vermonters get their produce from grocery stores rather than their backyards. As much as we would like to point to a picture-perfect scene of white-steepled villages and farms fading to a mountainous backdrop and say, "*That* is rural Vermont," this, again, is too simple. For whom, after all, does this conception of the rural ring true? Whose stories does it tell, and whose does it leave out?

Rather than turn to statistics or to the physical appearance of the landscape, *The View from Vermont* explores questions about the category "rural" and its associations with the state through the context of tourism. Naturally, the history of rural communities in Vermont and elsewhere is shaped by causal forces other than tourism. But for over a century tourism has exerted tremendous influence over land use, identity, and social relations in rural communities from Vermont to Montana to California to Maine, and the extent of its reach into all aspects of rural land and life has been remarkably long. As a cultural and historical geographer, my approach to tourism's history is influenced by my discipline's broader concerns with the production of landscape and identity. I am interested in the ways in which people transform and make meaning for the landscapes they inhabit, and how, in the process, they transform and make meaning for themselves and others. I use this book, then, to explore tourism's influence on landscape and identity in rural Vermont from the late nineteenth century to the late twentieth century. When we consider that influence — when we view the historical geography of rural tourism from the context of Vermont — we find ourselves contemplating stories that speak more broadly to the making and meaning of American rural landscapes. That starting point defines my rationale for writing *The View from Vermont*.

The late nineteenth century, Hal Barron explains, was a time when the lives of rural northerners were transformed by political realignments, the emergence of a new corporate culture, and the rise of mass consumerism.[2] Tourism emerged as part of this mix as well, transforming rural places in the eyes of visitors and of the residents who catered to them. As growing numbers of twentieth-century Americans made their homes in cities and suburbs, tourism and tourist-directed advertising became powerful contexts — and in some cases, the *only* contexts — through which many encountered and defined rural landscapes. That process taught rural residents to think about and to use their surroundings in new ways as well. Generations of rural residents have tried to capitalize on tourism's economic potential by transforming the rural landscape according to the demands of visitors, only to find themselves struggling to retain control over that landscape as their power to define its future eroded.

Here I explore trends such as these by drawing on the work of geographers and historians of tourism, many of whom examine fundamental questions about tourism's impacts on the material landscape and on the group- and place-based identities that are bound up in the landscapes we create.[3] I do so with two objectives in mind. First, each of the book's chapters examines a range of social and cultural discourses through which visitors and rural residents have made sense of rural tourism in Vermont.[4] What these chapters suggest by doing so is that the histories of seemingly quiet, uncontrived rural tourist communities are often more diverse and contentious than we might expect. Second, and more importantly, each chapter places those histories within an analytical framework defined by its focus on work-leisure relations. I argue that tourism has historically had profound consequences for the nature of work-leisure relations in rural communities like those in Vermont, and that, in turn, the act of trying to define work, leisure, and the relationship between them in the wake of tourist development has had significant consequences for the ways in which generations of Americans have encountered rural people and place. I call this process — this negotiation of landscape and identity according to the context of work-leisure relations — the "reworking" of rural Vermont.

When historians and geographers talk about the appeal of rural landscapes in American culture, they often begin by talking about differences between the rural and the urban. For over two centuries, the geographer Michael Bunce notes, Americans have championed the countryside as a "symbolic antithesis of the city; a place for reconnecting to natural processes and ancestral roots."[5] That sentiment emerged with particular strength during the last decades of the nineteenth century, as the pace of urban-industrial growth accelerated throughout much of the northeastern United States and as economic, cultural, and social differences between rural and urban places grew increasingly pronounced. Many Americans responded to such changes by embracing a pastoral, Jeffersonian ideal through which rural land and life emerged as hallmarks of national identity.[6] Generations of artists and literary critics, in particular, critiqued the city in favor of romantic notions about farm and village life, helping to establish the rural landscape as one of the nation's most enduring and cherished cultural icons.[7]

The roots of rural tourism lie in real and perceived differences such as these. According to the sociologist John Urry, tourism represents a fundamental search on the part of travelers for places, people, and experiences different from those that define their everyday lives.[8] Generations of travelers from American cities and suburbs have performed that search for difference by visiting rural places from New England to the Rocky Mountains. There they have sought rest and outdoor recreation, all set, importantly, in a cultural landscape that, for many, seems to embody some of the very best of American values.

Scholars typically deploy a number of assumptions to explain tourist-based encounters with rural places, shaping how we think about the historical motivations for rural tourist travel. First, we often equate rural tourism with a search for expressions of *timelessness* or *stability*. Years of sentimental representations in tourist promotions and the popular media, for instance, have emphasized New England's apparent continuities with the nation's rural past, turning the region's pastoral scenes into symbols of nostalgia capable of attracting visitors to their midst.[9] Second, we often equate rural tourism with a search for *organic, uncontrived* landscapes. Tourists, from this perspective, typically view the rural landscape as a genuine, natural product of hardworking, independent residents, rather than something of conscious or manipulative design. Finally, we often equate rural tourism with a search for expressions of *harmony*, both in terms of rural society and in terms of people-environment relations. Seemingly less troubled by the complexities of urban life, the rural landscape in this view becomes a place of balance, a place of accord.

The popularity of sentiments like these has proved astoundingly resilient over the past century, perhaps nowhere more so than in Vermont, where

FIGURE I.2. Pastoral scene in Corinth, Vermont, no date. Promotional photography such as this often contains traditionalized and seemingly timeless icons of rural Vermont: farms, fields, and forests circling a small hamlet complete with a white church steeple. Photo by Vermont Travel Division. Courtesy Vermont Historical Society.

they have consistently informed the kinds of language and imagery that Vermonters and their visitors turn to when they talk about the state. Vermont's church steeples, town commons, covered bridges, and stone walls, for instance, have all become icons of the state's appealing sense of difference — one that, in turn, has been marketed to generations of visitors from nearby cities such as New York, Boston, Hartford, Providence, or Springfield. Some of these visitors have come to Vermont to stay in resort hotels, on family farms, or in their own private vacation homes. They have come to hike, to fish, or to ski. They have come to drive the state's backcountry roads, taking in the scenery as it passes by their windshields. But no matter what the visitor's agenda, they and the promoters who have catered to them have consistently defined Vermont as an anti-modern foil to the challenges of urban, industrial America.[10] As one mid-twentieth-century promoter explained, Vermonters "have held fast to the ways of their fathers, and perhaps in doing so have preserved on their small farms and in their quiet villages something valuable that might otherwise be lost in the larger pattern

of American life."[11] When viewed in this light — as a cultural home to the
best of the nation's rural traditions — Vermont seemed a very different and
very attractive place indeed.

Of course, rural Vermont is different in many ways from, say, rural New
Mexico, rural North Dakota, or even nearby rural New Hampshire. This is
an important point to keep track of. It reminds us that rural landscapes can
differ dramatically in geographic form, in historical experience, and in popular meaning. It reminds us that the particulars of Vermont's historical geography have shaped its experience along distinct lines: that factors such as
the state's position relative to major urban centers, its predominantly Anglo
heritage, and its natural environment matter in what can be very specific
ways.[12] Indeed, there is an undeniable mystique to the name Vermont — a
mystique that, in this case, operates at the state scale, but one that has also
made Vermont a powerful and enduring symbol of rural exceptionalism,
both on regional and on national scales. That fact, I believe, makes Vermont an opportune place to explore tourism's hand in shaping the contours
of rural exceptionalism through time. In this sense, Vermont is unique in
many ways, but its uniqueness is often only a matter of degree. Idealized
conceptions of rural land and rural people are not confined to Vermont;
they spill easily over political boundaries, flooding rural communities across
New England and beyond. Part of this book's job is to explore tourism's
role in their persistence, but part of its job, too, is to uncover alternate and
often contentious readings of the rural tourist landscape. To do so, we need
to begin from what may seem to be an oddly contradictory vantage point
by thinking about rural landscapes not as being entirely different or entirely unique but rather *very much like* other kinds of landscapes. We need
to begin, that is, by thinking about rural landscapes as a geographer might,
searching for answers not only about what they are or how they look but
also about what they do, about who and what they serve.

On one level, I use the term "landscape" in this book to refer to the tangible, visual scene — the trees, farms, white steeples, covered bridges, resorts,
roads, and so on that constitute physical space in Vermont. On another
level, I define landscape according to the ideas that visitors and residents
have carried around with them in their heads. Here I follow a long tradition in cultural geography that maintains that landscapes offer visual clues
about the values and aspirations of those who create them.[13] In this book,
for instance, we might "read" the landscape of rural Vermont for what it reveals about cultural attitudes toward concepts such as travel, nature, land
use, outdoor recreation, rural work, and rural leisure.

But my definition of landscape does not stop there, for in addition to *re-*

flecting insights about people and places back to their observers, landscapes also help *to shape and to reproduce* the social and cultural discourses that we use to make sense of our surroundings, our neighbors, and ourselves. "As a material component of a particular discourse or set of intersecting discourses," the geographer Richard Schein has argued, "'the cultural landscape' at once captures the intent and ideology of the discourse as a whole and is a constitutive part of its ongoing development and reinforcement."[14] From this perspective, landscapes play a part in the ongoing production and reproduction of group- and place-based identities, of social and cultural values, and ultimately of the power relations inherent in any society. They do so by naturalizing the ideologies of one group over another, creating expectations about what is "normal" through their physical structure and symbolic representation. In many respects, we learn to define social power by the ability of dominant groups to express their understanding of the world in visual form. And conversely we learn to challenge that power by manipulating the landscape according to our own views. When one controls the landscape, one controls much about how we identify and use our world.[15]

For over a century, a range of social groups have claimed a stake in defining landscape and identity in rural Vermont: vacation home owners, outdoor enthusiasts, private and state-sponsored promoters, farmers, politicians, and resort developers have all transformed their surroundings in ways that reflected and (ideally) reproduced their perspectives about what constituted an "authentic" rural landscape. But what, if anything, did it mean to be "authentically rural" in a place invested in by so many different groups? What did it mean to be "authentically rural" in a place so profoundly shaped by tourism? Geographers of tourism such as Dydia DeLyser urge us to recognize authenticity as a socially constructed and contested category. No matter how powerful a dominant group's conception of authenticity may be in any given tourist context, DeLyser notes, others will always hold competing interpretations of just what is and is not authentic about a place or its people.[16]

Competing interpretations of authenticity — like any social and cultural discourse through which we approach our world — are not always so easy to discern in the landscape itself. As Don Mitchell has demonstrated, landscapes can be cast in ways that mask the social, economic, and political complexities inherent in their production, naturalizing within themselves a sense of tranquility, a sense of uncomplicated, even innocent inevitability.[17] This holds particularly true for tourist landscapes, which travelers often view uncritically as the authentic reflections of reality they often purport to be.[18] This also holds true for rural landscapes, which, as Raymond Williams has noted in the case of England, have long hidden the labor and social tensions embedded in their production behind an attractive facade of social harmony and scenic beauty.[19]

What this all means, on a fundamental level, is that we should avoid taking rural landscapes at face value, for there is always more at work in them than meets the eye. It means that we should avoid assuming that traditionalized impressions of timelessness, organicism, and harmony tell the whole story. Those impressions have become naturalized in the symbolic and material spaces of Vermont's tourist landscape, reproduced by those who have shepherded landscape and identity along sentimental, romantic lines. They have become a dominant means by which tourists and many who study tourists make sense of Vermont. But despite the power of such ideas, Vermont's tourist landscapes have been implicated in the production of other, less traditional, and less obvious discourses about the rural as well. Since the late nineteenth century Vermonters and their visitors have approached the state's rural tourist landscapes through terms other than nostalgia and timelessness alone; they have approached them simultaneously through a collection of progressive, modernist, and reformist perspectives as well. When seen through the lens of tourism, rural Vermont has been more than a place of organic, uncontrived authenticity; it has been a place where people have eagerly transformed their surroundings to accommodate some types of development, even as they struggled to prevent others. It has been more than a place defined by impressions of social and natural harmony; it has been defined as well by problems such as prejudice, social exclusion, and environmental degradation.

For these reasons, I do not rely exclusively in this book on dominant, naturalized perspectives about rural places and rural people to explain the historical geography of tourism in Vermont.[20] Rather, I add to these a diversity of alternate and often competing discourses about travel, recreation, land use, rural culture, and social relations. I do so not to prove that what we know about the history of rural tourism is somehow all wrong, or that traditional perspectives about rural scenery and rural life can be trumped cleverly by others that are more accurate. Rather, I do so because I think that by considering traditional perspectives, non-traditional perspectives, and the points of overlap that developed between them we challenge ourselves to develop a more critical picture of rural tourism, whether in Vermont or elsewhere. Seemingly contradictory discourses about rural tourism were less at odds than in dialogue among each other. And what emerged from that dialogue was not a single, definitive take on rural Vermont, but many takes.

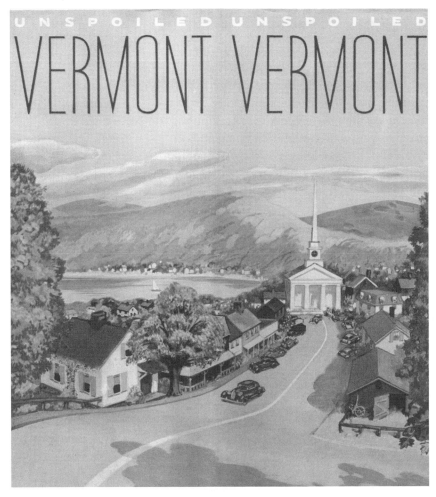

FIGURE I.3. Old and new in unspoiled Vermont, c. 1940. Tourist promoters often encouraged visitors to think of Vermont not merely as a throwback to an earlier time but as both a traditional and a modern, comfortable place. Here one would encounter "unspoiled" and old-fashioned rural beauty as well as modern roads and other amenities. Courtesy Vermont Historical Society.

Understanding Vermont's encounter with tourism in this way is a valuable exercise, not merely because it draws attention to complicated-sounding ideas like "alternate and often competing discourses," but because those ideas are part of a critical perspective that focuses attention on a more fundamental fact about tourism: Like any land use, tourism has very real consequences for how we define and interact with other places and other people. This holds true whether we are the visitors or the visited, and it often holds true in regrettable ways. As Hal Rothman notes in his influential book *Devil's Bargains*, tourism is not an "inherently bad" choice for communities and individuals seeking economic renewal, but it is a choice that, more often than not, trades the "souls" of places on the false promise of economic gain, ultimately turning those places into hollow "caricatures of their original identities." This, Rothman argues, is the devil's bargain of tourism: "The inherent problem of communities that succeed in attracting so many people," he writes, "is that their very presence destroys the cultural and environmental amenities that made the place special."[21] In Vermont, tourism has altered the look, feel, and social structure of the state, rescripting place, disenfranchising local residents, and crowding the landscape with vacation homes, shopping centers, and sprawling resorts. Enticed by tourism's promise of economic gain, Vermonters often ended up watching their state become something unfamiliar, even unwanted.

Conceptualizing tourism as a devil's bargain reminds us of how important it is to explore and critique tourism's fundamental power to transform. It reminds us to analyze tourism's unintended consequences, to identify those who lose and those who gain because of it. But at the same time, we need to be careful, for if we let ourselves view tourism too critically as a destructive force, we risk essentializing (or even romanticizing) pre-tourist landscapes. Tourism undoubtedly changes the identities of places, but how should we define the essential baseline for a particular place's identity, and how should we evaluate the value of that baseline? For example, can we point to a time and place in Vermont and call that the state's "true," pre-tourist identity? Can we assume that landscape and identity in Vermont's rural communities were fixed, agreed upon, or even all that beneficent before tourists arrived and changed them into something else entirely?

Some tourist scholars urge us to avoid making such assumptions, and to avoid thinking of tourism in terms of stark divides such as authentic or inauthentic, true or false, winner or loser. The culture of tourism presented to visitors, the anthropologist Edward Bruner notes, is not merely a mock-up of an original soul but rather something unique unto itself — something that is good for some and bad for others, but something that is always constructed according to social, political, and economic contexts that themselves never stand still.[22] This assertion has parallels in the study of rural landscapes as well. Scholars of critical rural studies remind us that landscape and identity in rural communities are never static, regardless of the mythical (and often tourist-inspired) images that encourage us to see them as such.[23]

The View from Vermont follows this logic. My hopes are that we think about rural landscapes and rural identities as always being open for debate, always defined by change, always constructed socially according to categories such as gender, race, ethnicity, and class, and that we think of tourism as one of a number of important forces that make this happen. Places certainly have dominant identities that define them at any given time, but that dominance is itself dependent on the ability of one group or another to define and control it — often to the detriment of others. And place identities are certainly altered by their encounter with tourism, yet that does not mean that pre-tourist identities were necessarily fixed, accepted by all, or even more just and equitable than those created by tourism. Nor does it mean they would re-emerge and stabilize over time if the effects of tourism were washed away. Despite Stowe, Vermont's extensive resort development, for instance, this well-known ski town is still considered a national rural icon. Stowe is certainly not the same kind of rural place that it was when tourists first started visiting it over a century ago, but that does not mean that the pre-tourist Stowe of the mid-nineteenth century was the definitive embodiment of rural culture. Nor does it mean that Stowe would look and feel the same today were it not for tourism.

Just because identities are hard to define, of course, should not imply that "anything goes" in terms of tourist development. Nor should it absolve us from taking a hard look at its consequences. Tourism contributes more often than it should to social and environmental outcomes that are both unjust and unsustainable. I will not avoid these. But what interests me most in this book are the social and cultural politics that generate those outcomes — politics that are themselves a consequence of tourism. To give those politics coherence, we need to frame a century of tourism according to a definable narrative — one that I hope will suggest connections to other rural places while at the same time prompting us to reflect critically on our own relationships to rural tourism, whether as travelers, rural residents, researchers, or students.

I define that narrative according to a framework of work-leisure relations. For over a century, tourism has forced rural work and rural leisure into continual contact with one another, such that the meanings, practices, and spaces associated with each category have informed one another to the point of becoming mutually constitutive. Tourism blurred the boundaries between work and leisure in rural places, making each inseparable from the other. This process has been persistent yet not uniform, either in its mechanics or its outcomes. It has played out differently depending on the people, places, and activities in question, and on the social and cultural discourses (traditional and non-traditional) that different historical actors deployed in their encounters with rural tourism. Nonetheless, the interaction between work and leisure that follows in the wake of tourist development has always been there, providing a framework through which visitors and residents have constructed landscape and identity in rural communities nationwide. I think of this construction — as guided by the context of work-leisure relations — as the "reworking" of the rural landscape.

As we might expect, work-leisure relations are important in *any* tourism context, whether rural, urban, or wild. As Rothman notes, community leaders throughout the American West embraced tourism as a solution to declining job opportunities. In time, however, tourism's new workers and new work regimes only compound the inevitable transformation of place that follows on the heels of tourism.[24] Likewise, generations of Vermonters have turned to tourism as an alternative to the state's declining agricultural fortunes. But rather than finding salvation, many instead found themselves in low-paying, seasonal positions with little hope of advancement. The historian Cindy Aron also deploys work-leisure relations as a means for framing her history of vacationing in the United States. In her book *Working at Play*, Aron argues that growing opportunities for leisure travel forced middle-class Americans to reconcile their collective desire for vacations with their class-based commitment to hard work. Their efforts to do so, Aron asserts, suggests how important tourism was to the evolution of middle-class identity.[25]

But what makes work-leisure relations particularly important to the context of *rural* tourism? The answer to that question has everything to do with the nature of the rural landscape itself. Some scholars encourage us to think of tourism as being defined by a division or opposition between work and leisure. John Urry has written, "Tourism is a leisure activity which presupposes its opposite, namely regulated and organised work. It is one manifestation of how work and leisure are organised as separate and regulated spheres of social practice in 'modern' societies."[26] The historian Paul Sutter makes a similar point with reference to wilderness travel. Examining the automobile's role in the history of the American wilderness movement, Sutter argues that the wilderness seeker's preference for places untouched by human labor points toward a broader segregation of working and leisure spaces that is characteristic of twentieth-century American culture.[27]

When viewed in the context of wilderness (as in the case of Sutter) or in the context of the traveler's daily working routines (as in the case of Urry), arguments about the segregation of work and leisure make a great deal of sense. After all, when we go on vacation we like to "leave our work behind." And if we were to go backpacking on our vacation, we expect to leave *everybody else's* work behind as well. But what if we went on a vacation that took us to and through rural communities? What would our relationship to work look like then — not merely our own work back home, that is, but the work of others?

When tourists visit rural places, I would argue, they set in motion a process defined not by a dichotomy between work and leisure but by their overlap. Rural landscapes are lived-in places whose spaces and identities are linked in important ways to the work of rural people, to the visible expressions their labor leaves on the land. On the one hand, this link applies to rural residents for whom work — visceral, physical toil — has always shaped their encounters with landscape.[28] But it also applies to visitors, although typically in less visceral and more romantic ways. When vacationers travel to rural places, whether for a weekend drive or a longer stay, they leave their own work behind in order to visit a place shaped by the work of others — work that is typically agricultural but at times industrial as well. That is not to say that visitors necessarily develop a well-informed understanding of the realities of rural labor. But it is to say that the very act of vacationing integrates the spaces and experiences of rural work into the tourist trade, redefining them according to the social and cultural discourses guiding that trade in any given context. That integration has made the meanings, spaces, and practices of leisure inextricable from those of rural work. This continual reconfiguration of work, leisure, and the relationship between them, I believe, is what gives historical continuity to the rural tourist experience.

That has never been a straightforward or easy process. New work-leisure relations always brought new challenges and new debates to Vermont, both between and among visitors and residents. Opponents of tourism argued for decades, for instance, that a leisure economy worked at cross-purposes to the state's "proper" rural economy and culture. Vacationers set a bad example for young Vermonters, they argued, undermining the state's reputation for hard work as well as its entire agricultural way of life. For their part, visitors often worried that the day-to-day realities of rural work threatened their conception of a properly ordered rural landscape. Some complained when farms appeared less than tidy, for instance, while others opposed billboards as a blight on the scenic landscape and a threat to the value of their summer homes. But for many Vermonters, farmyards were working spaces that sometimes had to be less than tidy. And for others, billboards were a very legitimate and necessary part of the state's working economy — one, in fact, that helped tourists locate the services they needed.

In many respects, such differences of opinion echo a common divide in how we think about rural tourism — a divide defined by tensions between landscapes of resource *production* and landscapes of resource *consump-*

FIGURE 1.4. Tapping a maple tree for sap, 1947. Depictions of rural work such as this cover photograph for *Vermont Life* magazine have long been common in the popular media. Visitors to Vermont often romanticized rural work and rural people, drawing their work and their traditions into a consumptive tourist trade. Photo by Warren W. Dexter. © and courtesy of *Vermont Life* magazine.

tion.[29] Rural Vermonters who were accustomed to using their land for producing resource commodities, this model suggests, often came into conflict with tourists accustomed to consuming scenery as an aesthetic commodity of the tourist trade. But differences of opinion developed between different kinds of tourists, and they developed between rural Vermonters of different minds as well, suggesting that more was at work than conflicts between economies of production (on the local's side) and consumption (on the tourist's). As we shall see, visitors espousing different brands of leisure vied for control over the recreational landscape, as did rural residents for whom some kinds of leisure felt more appropriate and more valuable to working Vermonters than others.

No matter how we frame them, discussions about work and leisure consistently guided the construction of landscape and identity in Vermont; that is, discussion about work and leisure guided the reworking of rural Vermont. Each of the following chapters traces that process through a combined topical and chronological sequence. Chapter 1 explores the role that both nostalgia and progress played in defining and binding together the state's turn-of-the-century economies of tourism, agriculture, and industry. Chapter 2 examines the sale of abandoned farms as summer homes and the turn-of-the-century reform agendas associated with this transfer of property from work to leisure. Chapter 3 highlights the language of accessibility that guided the spatial expansion of tourism in the early decades of the nineteenth century, noting the debates that emerged about different forms of accessibility and about the wisdom of allowing leisure land uses to spread statewide. Chapter 4 looks at the mid-century expansion of tourism beyond summer alone, pointing out the ways in which traditional, seasonal patterns of rural work and leisure were transformed by their absorption into the tourist economy. Chapter 5 focuses on skiing, exploring the ways in which modern technological ski landscapes redefined the pace of leisure for visitors and the face of work for residents, many of whom were displaced by changes in the dairy economy. Chapter 6 discusses the relationship between tourism, scenic preservation, and environmentalism, arguing that by the 1960s and 1970s, Vermonters looked increasingly to planning and legislation as means for balancing the competing demands of work and leisure.

Each of these chapters is connected to the other by a common thread: each describes the creation of what I think of as a "middle landscape," a construct specific in its associations in this book with tourism and the reworking of rural Vermont but general in its associations with American culture. This middle landscape is what we encounter when we ponder the view from Vermont.

Generations of geographers, historians, and other social commentators have used phrases like "middle landscape" or "middle ground" to describe rural places. Typically, this idea of the rural-as-middle-landscape is meant to suggest its favored (and often favorable) positioning between the urban and the wild, between nature and culture. Rural places are domesticated, but they do not have the same problems of crowding and scenic blight that domestication brings to the city. They are places where one can be in touch with nature without losing touch with the comforts and familiarity of culture. By this kind of logic, rural people in places like Vermont may seem to live a more balanced life — one in which they have resisted the crass materialism and excess of urban and suburban lifestyles.

It would certainly not be wrong to argue that romanticized conceptions of the rural middle landscape have contributed to Vermont's enduring success as a tourist destination. We might even get away with leaving it at that. But I believe it is worth thinking about the rural middle landscape in a slightly different way — one that is more symbolic than material, perhaps, but one whose power does not exist independently of the very physical ways we treat each other and the world around us.[30] I do not think of rural landscapes as middle landscapes merely because they combine aspects of urban and wild, nature and culture, old and new, tradition and progress, or even work and leisure — although, as we shall see in this book, they do all that, all the time. Instead, I think of rural landscapes as middle landscapes because they are where the people who act on such things as nature, culture, work, or leisure come together and collide with one another in their efforts to embed their beliefs in the landscape itself. Rural landscapes, *like all landscapes*, are poised squarely in the "middle" of the complex social relations of which they are both a product and a part. They are the ground on which and through which we negotiate our place in the world. And for me, that is why they matter. For on one level this is a book about Vermont, about tourism, and about rural places. But on another level it is a book about the kinds of fundamental social struggles and power dynamics that I believe are at the core of American culture. In my view, thinking of the rural as a middle landscape helps us focus on that core, for it is in this middle ground that struggles to control landscape and identity become struggles to control one another. It is in this middle ground that the reworking of rural Vermont unfolds.

CHAPTER 1

Two Worlds of Work

Nostalgia and Progress in Turn-of-the-Century Vermont

As seen from the mountain summits, there is the cultivated farm, the country hamlet, with its single church spire and country store; the manufacturing town, with the smoke from its many chimneys; the ponds and lakes nestled among the hills and mountains; the stretch of river glistening in the sunlight, and a grand panorama of hills, mountains, and valleys extending in all directions, combining to produce a scene of rare beauty and interest, which it is an ever increasing delight to study.[1]

The visitor to Barre debarks from the train at a convenient Depot in the center of the Village, and is immediately impressed with the magnitude of the granite industry. On the opposite side of the track from the depot, he views scores of granite cutting establishments, towering above which are huge derricks; and great piles of enormous blocks of beautiful granite ready for the chisel of the sculptor are everywhere seen.[2]

The passages quoted above capture the spirit of two different kinds of landscapes encountered by visitors to Vermont in the late nineteenth and early twentieth centuries. The first, written in 1892, paints an essentially pastoral scene whose picturesque, romantic tone would have been familiar to many tourists at the time. Growing numbers of nostalgic urban travelers were traveling to Vermont by the late nineteenth century, and sentimental depictions of farms and rural hamlets framed by rolling hills were becoming more common in the pages of the state's promotions all the time. To its visitors, Vermont's rural landscape seemed somehow more traditional and stable than the places they left behind — an impression captured best, it seemed, among scenes like that above.

The second passage, written in 1887, captures a different kind of Vermont, where enthusiasm for bustling industrial development of the kind seen in Barre seemed worlds apart from the state's quiet, pastoral reputation. Although industrial scenes such as these were less prevalent in the tourist literature than their agrarian counterparts, they nevertheless factored into the creation of Vermont's emerging tourist identity. Turn-of-the-century Vermont did not have large industrial cities akin to those in southern New England, but it did have its share of manufacturing villages and moderate-sized industrial towns, where lumber milling, stone working, and even textile production all took place. In such places, the Barre passage suggests, visitors to Vermont could witness the bustle and budding promise of the state's industrial and commercial future. In such places visitors could find a state that was not a throwback to an earlier time but instead a decidedly progressive place.[3] By using industrial development as a theme in tourist literature, Vermont's promoters connected industry and tourism by a common language and common pursuit of progressive economic growth. They added depth to a tourist economy known widely for its associations with agricultural landscapes, and they tied their state to a larger embrace of material progress and economic development that was so central to American culture between the 1870s and the 1910s.

This chapter explores two different worlds of work — one agricultural, the other industrial — as a framework for narrating the production of the nostalgic and progressive discourses that informed passages like the two above. First, we encounter a largely familiar chapter in rural tourist history in which promoters, town boosters, and visitors equated farming and the experience of farm landscapes with nostalgic impressions of rural people and place. Even as Gilded Age and Progressive Era Americans celebrated the progress of American commerce, industry, and urban development, they also recognized that progress had its costs. Tourist promoters played on this ambivalence by encouraging American travelers to think of rural places as offering a timeless and refreshing antidote to the changes associated with turn-of-the-century industrial capitalism. In turn, tourists mapped the experience of leisure travel onto Vermont's working farm landscapes, reworking the form and meaning of those landscapes according to new conceptions of rural space and implicating them in the production of a rural nostalgia to which, notably, not all Vermonters were willing to ascribe.

In the chapter's second half, we encounter a less familiar story about rural tourism — one in which promoters and travelers forged complex links between tourist and industrial economies in rural communities, merging the two according to an embrace of modernity and progressive economic development.[4] Although many turn-of-the-century Vermonters bemoaned their state's lack of clear economic progress (about which I say more in chapter 2), many entrepreneurs, boosters, reformers, and politicians were also quite enthusiastic about the state's potential. Neither a national economic depression in the mid-1890s nor a persistent out-migration of rural residents from hundreds of farm communities statewide dampened their enthusiasm entirely; new industries, investments in rail lines, and budding urban centers, optimists maintained, all pointed to a bright future for Vermont.

Optimists envisioned that future as one where tourism and industrial development coexisted with one another, each contributing to economic growth and town development without compromising the state's scenic beauty and without undermining the faith in progress to which so many Vermonters and so many of their visitors subscribed. Town boosters often advocated tourism in the same breath as industrial development, hoping to diversify local economies and lure visitors to Vermont for whom evidence of material progress, rather than its absence, actually factored quite deeply into their travel decisions. Indeed, the middle class's embrace of progress proved to be a crucial part of tourist culture in late-nineteenth-century New England. In his recent book on Maine's Poland Spring resort, David Richards argues that post–Civil War summer resorts were products of more than escapist tendencies among urban travelers. Rather, they were "constructive" cultural products, born of summer visitors' desires to connect with quiet natural charms while still maintaining their "urban conceptions of status, consumption, health, nature, leisure, and culture." Proprietors sought to accommodate their desires, Richards argues, by creating a literal "middle landscape" in which "the best virtues of a nostalgic rural agrarian past" were merged with "the modern urban industrial present."[5]

Visitors to Vermont, we shall see here, proved equally complex and equally capable of demanding urban amenities and rustic rural charm simultaneously. But what we will also see is the degree to which such demands overlapped with the goals and aspirations of industrial developers and town builders within Vermont itself—a historical pattern that challenges us to think in new ways about the aims and consequences of rural tourism's early manifestations. I explore this in detail by focusing on a case study from the Deerfield River Region of south-central Vermont. Here promoters, town leaders, and the visitors they courted acted within a rural landscape shaped simultaneously by industrial and tourist development. Just as these same sets of actors were capable of reproducing nostalgic sentiments by linking leisure to agricultural life, so too were they capable of reproducing progressive sentiments by linking leisure to industrial aspirations in places like south-central Vermont.

Little consensus existed in Vermont about how to go about creating the state's turn-of-the-century tourist landscape. Whether one linked tourism to nostalgia or progress, to farms or industrializing communities, most Vermonters (and many visitors as well) viewed it as a mixed bag. They did so with good reason: The new places and the new kinds of land uses that emerged from Vermont's early engagements with tourism did not necessarily yield an attractive, beneficial middle landscape reconciling past and present, urban and rural, farm and factory. Rather, the end result was a landscape positioned in a contentious middle ground between the competing demands of those who participated in its creation. From the late nineteenth century forward, those demands would revolve around an enduring primary question: How, in the wake of tourist development, does one define the meanings and practices associated with rural work and rural leisure? Visitors and residents were now beginning the process of negotiating work-leisure relations both between and among themselves — a process that hinged on their ability to control the reworking of rural Vermont.

Searching for a Tourist Identity

Vermont's earliest tourist economy centered on a handful of mineral-spring resorts, where mid-nineteenth-century travelers drank mineral-rich spring water and took part in elaborate "water cures" in the hopes of improving or maintaining their health. Although Vermont's mineral-spring resorts never earned the same renown as places like Virginia's White Sulphur Springs or New York's Saratoga Springs, the state managed to make a modest name for itself among antebellum travelers, particularly as early railroad lines extended into Vermont and as opportunities for leisure travel increased among the nation's emerging middle class. Roughly a dozen Vermont communities, including Manchester, Alburg, Highgate Springs, and Clarendon Springs, boasted mineral-spring resorts by mid century.[6]

The popularity of such places was fading by the 1860s, however, as tastes changed among middle-class American tourists, both in Vermont and beyond. As the Vermont historian J. Kevin Graffagnino has argued, travelers in the second half of the nineteenth century increasingly "valued impressive scenery, a variety of amusements and activities, elegant accommodations, proximity to other attractions, and convenient travel connections as highly as they did the supposed medicinal properties of a particular spring."[7] Some mineral-spring resort owners in Vermont adapted as best they could to the changing times by placing a greater emphasis on scenery and "delightful drives" rather than the "curative powers" of their waters alone.[8] But many ultimately failed to make a transition away from their reputation as mineral-spring resorts, and Vermont's standing among post–Civil War tourists continued to lag behind that of nearby states.

This was most obvious to Vermonters when they compared their state to neighboring New Hampshire, where the rugged White Mountains had become one of the nation's most popular tourist destinations. Aggressive promotions, ambitious entrepreneurs, new rail lines, and the construction of elegant new hotels all fueled the region's success. But it was its associations with literary and artistic styles popular among mid-century American travelers that gave the White Mountains a truly salable tourist identity.[9] Drawing on established traditions among European Romantics, America's mid-nineteenth-century writers and painters constructed powerful aesthetic conventions through which tourists learned to comprehend and bestow meaning upon the natural world. The most popular of these for mid-century tourists was the "sublime." More than a mere descriptor, the sublime conflated nature with Christianity and with powerful emotions of awe and terror. To stand in the presence of sublime scenery was to confront an unspeakable sense of one's own humility in the presence of God.[10] Between the 1850s and the 1870s, New Hampshire's White Mountains were one of the nation's most popular expressions of the sublime. Although southern New Hampshire was fast becoming a hub of urban and industrial development, the White Mountains to the north were not. Here the traveler could encounter a seemingly incomprehensible "trackless wilderness" of lofty, mist-shrouded mountains and dark, impenetrable forests. Here one could gaze on astoundingly popular natural curiosities such as the "Flume" and the "Old Man of the Mountain" — a rock-like visage that filled visitors with a "thrill of half surprise and half fear."[11]

In an effort to capture a share of the nation's growing tourist trade, Vermont's promoters tried hard at first to define their state's Green Mountains in terms of the sublime. Travelers were encouraged to visit places like Lake Willoughby, for instance, where dramatic cliffs, dense forests, and the serenity of reflective water combined to make the area one of the state's best expressions of sublime nature. Others were encouraged to visit Mount Mansfield, Vermont's highest peak, whose long and jagged profile looked from a distance like an upturned face, complete with nose, chin, and forehead. The mountain was crowned with a summit hotel in the 1860s from which visitors could explore Vermont's own expressions of the sublime, including the "Rock of Terror" and the "Old Woman of the Mountains."[12]

Despite the modest appeal of places like Mansfield and Willoughby, they never matched the success of comparable resorts in New Hampshire's White Mountains.[13] One explanation for this was natural. Compared to the higher and more rugged White Mountains, Vermont's Green Mountains seemed to many travelers more like a narrow chain of hills than towering seats of the sublime. Another explanation was cultural. Generations of Vermont's pioneer settlers had transformed their state's landscape dramatically in the service of agricultural and industrial work. Once covered almost entirely with mature forests of spruce, fir, hemlock, maple, and beech, nearly 70 percent of Vermont had been cleared by the 1870s for fuel and lumber, cropland, and pasturage that often swept high up the state's steep hillsides.[14] By the last decades of the nineteenth century, then, Vermont's landscape felt far more rural than sublime—a feeling which some mid-nineteenth-century tourist guidebooks described in decidedly plain and unenthusiastic terms. At best, parts of Vermont had "the appearance of a good agricultural region," and at worst they were "void of interest."[15] Vermont's lasting success among tourists would have to wait until a time when popular conceptions of leisure matched more closely the state's landscapes of work.

rural than Sublime

Nostalgia and Farm Boarding in Rural Vermont

That time arrived during the last two decades of the nineteenth century, as the sublime's popularity among travelers weakened and the popularity of another scenic convention—the "picturesque"—grew. Compared to the sublime's emphasis on wild, rugged, and mystifying places, the picturesque emphasized a wider range of natural and cultural landscapes. As the art historian Barbara Novak has argued, clearings in the forest, small but prosperous villages, and even the smoke of an oncoming train all fit into the scope of the picturesque. Scenes such as these increasingly found their way into art and literature, inspiring a sense of pride in America's civilizing mission in the wilderness and boosting the scenic value and cultural importance of the nation's rural landscapes.[16] Both the sublime and the picturesque had existed simultaneously during the middle decades of the nineteenth century, but the growing influence of the picturesque toward the end of the century made possible a new emphasis among travelers on rural landscapes such as those found in Vermont. Such places seemed to embody the aesthetics of what Leo Marx has called a "pastoral ideal"—a celebration

of a "chaste, uncomplicated land of rural virtue." For many middle-class Americans, that ideal lay at the heart of their nation's cultural identity.[17]

The popularity of picturesque rural scenery owed much to the nation's emerging middle class, whose members were often ambivalent about the scale and pace of change that accompanied late-nineteenth-century industrialization, immigration, and urbanization. While the middle class championed industrial capitalism and the material progress they associated with it, they also responded to the uncertainty it brought by embracing popular expressions of antimodernism.[18] Rural tourism became an outgrowth of that antimodern sentiment, particularly as growing numbers of urban residents — some of whom had grown up on farms — now found themselves distanced from rural places and rural people. Consequently, the historian Dona Brown has argued, many turned to leisure travel as a means for connecting to the nation's ancestral rural roots. Rural New England's farms and agricultural hamlets captured the imagination of these travelers like no other region. This was especially true for the northern New England states of Maine, New Hampshire, and Vermont, where the relative pace of change was slower and where the persistence of farm and village life seemed to offer a refreshing contrast to New England's urban core to the south. Popular writers and artists garnished northern New England with sentimental visions of a pre-urban, pre-immigrant, pre-industrial past, naturalizing a nostalgic image of the region in the tourist's mind.[19] In Brown's words, tourists from cities across the northeast increasingly viewed New England "as though it were a kind of museum, a storehouse for a whole collection of old-fashioned ways of being, of old-fashioned values and beliefs."[20]

Vermont's working farm landscape — much of which by this time was geared toward the production of milk, cheese, and butter for regional markets — became a crucial link in the construction of this nostalgic discourse and the tourist identity it helped to create. Together, the state's picturesque beauty, attractive sense of timelessness, and calming impression of harmony gave Vermont a popular identity among travelers, turning its rural landscape into a valuable commodity on the tourist market. Associations between Vermont and rural tranquility within the regional tourist culture were captured in passages such as the following from 1888, in which readers were invited to contemplate the view from Vermont's Mount Mansfield:

The scene that is spread out before the eye on either of the summits is gorgeous and beautiful in the extreme. It differs from the views to be had at the White Mountains, as there in whatever direction you look nothing is to be seen but rugged mountain tops, with occasional valleys between them; while here, besides a similar view towards the east, there is spread out before you on the west the level, fertile land of Western Vermont, diversified by pretty hills, bordered by the silver waters of Lake Champlain, with the deep blue Adirondack Hills in the far distance beyond. This view is singularly attractive. You see the farm houses clustering into vil-lages; you can follow the courses of the winding trout-laden streams among the hills and forests; you can see the dark green of the waving grain, and can almost distin-guish the farmers at their toil. It is the wildness that is chiefly impressed upon the mind at most mountain tops, but here, while there is enough that is wild and romantic in the scenery, there is another and different sensation. The beauty of the landscape, the feeling that all this pleasant land is filled with life and intelligence, together with the suggestion that the distant Adirondacks contain within their dark recesses an undiscovered world of loveliness, combine to chain one irresistibly to the spot.[21]

Passages such as this suggested to visitors that Vermont's rural landscape embodied very real, tangible expressions of America's celebrated agrarian roots. Here the nostalgic traveler could surround themselves with pictur-esque rural scenery and in so doing connect to a way of life seemingly lost to so much of the rest of the nation. Settled rather than wild, pretty rather than rugged, picturesque rather than sublime, Vermont's scenery was grow-ing more popular *because of* rather than *in spite of* its humanized feel. In the process, those who promoted and those who visited Vermont began to rework the physical and symbolic spaces of its rural landscape, transform-ing them according to cultural dictates that were geared toward leisure rather than work alone.

Northern New England's turn-of-the-century farm-boarding economy became one of the best expressions of this process. As Dona Brown has shown, many middle-class tourists during the 1890s and 1900s chose to spend their vacations on working farms, often for weeks on end. Although travel-ers in isolated parts of Vermont had long turned to rural residents for ac-commodations, farm boarding sprung less from convenience and more from a nostalgic desire to connect with what travelers hoped would be an authentic rural experience. According to Brown, such visitors "looked to the northern farm in itself — not to scenery or to fresh air alone, but to a pastoral, nostalgic vision of rural life — for their fulfillment."[22] Beginning in the 1890s, the Vermont State Board of Agriculture (VSBA), railroad companies, and private development associations provided prospective visi-tors with pricing and location information for scores of farm families state-wide who were now opening their homes to as many as a dozen visitors at a time. For anywhere from five to ten dollars a week, travelers could buy the privilege to sleep, eat, and even do a little work on an actual Vermont farm.[23] Some of these visitors had grown up on farms, some had not. But all embraced farm landscapes as expressions of a seemingly more wholesome, pre-industrial American past. Even the smell of country life and country people, one Vermont visitor went so far as to suggest, recalled "many happy memories of childhood days on the old farm."[24]

The farm boarders' nostalgic interpretation of rural life involved more than sentiment, it involved an intentional, leisure-based encounter with rural work. For them, vacations that incorporated wholesome encounters with traditional farm work would lead to personal rest and rejuvenation, leaving them better prepared to carry on with their own work back home. As one writer put it, the visitor to Vermont was sure to find "a paradise of rest and health, of wholesome relaxation and recreation, a dreamland of revel in the true luxury of purposeful idling."[25] The Central Vermont Railroad added that a visit to Vermont was "a matter of the utmost importance in its bearing upon the physical, intellectual and moral well-being of [travelers] and their families."[26] Exposure to the state's rural landscape reinforced the visitor's sense of "well-being" and "purposeful idling," particularly when visitors compared rural landscapes to the urban landscapes from which so many of them came. Viewed from an elevated distance, for instance, Vermont's farm landscape was praised for having a calming, restful effect on the weary traveler — an effect far different from what one might feel gazing upon urban scenes back home. Even on the ground — even out in the farmyard itself — visitors chipped in to help with farm chores, mapping their pursuit of leisure onto the working routines of residents. The work they performed as part of their vacation in Vermont was not done to earn an income; this was leisurely work that they *paid* to perform in the hope of being better prepared to succeed back home.

That hope was based on an encounter with a rural landscape that for residents had everything to do with work. Yet for Vermont's farm families, of course, that work had far less to do with leisure and restfulness than it did for their visitors. Nonetheless, farm vacationing forced residents to rethink and redefine farm work in terms of leisure, heralding a shift in how many Vermonters understood the working farm landscape. Vermonters now had to learn what it meant to be visited, to have even their most mundane daily activities scrutinized and celebrated by tourists who must have been just as curious to Vermonters as they were to tourists. They had to redefine the identity of rural work in order to cater to the expectations of leisured guests. They had to rework the landscape of their family farms, transforming their physical spaces and conceptual meanings according to the dictates of tourism rather than farming alone.

Farm boarding offered Vermont's farmers a new opportunity to generate income without having to leave the farm. That was certainly a blessing to many families, but it was one that brought with it an additional burden of labor. Farmwomen bore the brunt of that burden; on top of their normal daily routines they now assumed the tasks of cooking and cleaning for as many as a dozen guests at a time. Farm families also found themselves having to transform their homes and farmyards to fit visitors' expectations. It was not enough to merely "be themselves," for the reality of farm life, it seemed, was often too messy and complicated for the sentimental visitor. Rather, families now had to rework their domestic surroundings in order to reap the benefits of the tourist trade.

Formal and informal networks of advice were in place to help them accomplish this, including advice from state agricultural officials eager to establish certain standards of behavior and presentation among those who participated in the farm-boarding economy. George Perry of the Vermont State Horticultural Society, for one, advised farm families on issues including food preparation, entertainment, and arrangement of the house and grounds. According to Perry, the host's farmhouse had "better be old." Nothing about the house should look fancy or pretentious; instead it should be clean, comfortable, and "open for inspection." Outside the house, Perry urged farmers to cultivate a landscape that offered recreational activities and that fit stylized, leisure-based notions of rural life. Farmers were encouraged to plant apple orchards, not so much for their fruit but for their charm and for the opportunities they offered to sling hammocks from tree to tree. Rather than manage the family's woodlot for production alone, Perry suggested, part of that lot should be allowed to grow wild, thereby assuming a more natural feel. A lazy, tame horse for riding was a good thing to keep around the farm as well, he advised, even if that meant feeding an animal that might otherwise have been useless to working farmers. Perry even suggested that pigpens and henhouses be removed to a comfortable distance from the house so as not to offend visitors.[27]

The fact that Vermonters were being told to modify their surroundings suggests the degree of complexity that was emerging at the confluence of tourism and farm life. Farm boarding demanded new ways of thinking and using the landscape in rural Vermont. It demanded that residents learn to perform a new genre of work that was both unfamiliar and surely not nostalgic. The expansion of Vermont's leisure economy into the homes and yards of farm families introduced a new level of complexity to the nature of rural work and to the state's traditional social structures. Farm boarding may have been based on nostalgia, but, in fact, it represented the very *lack*

FIGURE 1.1. Summer visitors at the Maple Inn Farm in Newfane, Vermont, c. 1910. Visitors such as these represented a shift in how farm property was used and perceived in Vermont. Here visitors relax in chairs and hold a toy sailboat, while a woman who may very well have been the mother of the household takes a break on the porch. Photo by Will D. Chandler. Courtesy Vermont Historical Society.

of stability that characterized landscape and identity in rural Vermont at the end of the nineteenth century. Even as tourists embraced the state as a site for the nostalgic contemplation of rural life, those who served these tourists were forced to adjust to a host of new circumstances. Indeed, it is not hard to imagine the sense of disjunction with their immediate past that some farmers must have felt as they carried feed past a hammock and through a new croquet course to the animals now banished to the other end of the farmyard.

Nor is it hard to imagine that such changes generated social tension between hosts and guests, as each group struggled to define their relationships to one another and to the newly constituted farm landscape. Dona Brown offers a good example of these tensions in her discussion of "plain country fare." Many farm boarders came to Vermont expecting plenty of fresh, hearty farm foods, such as rich cream, butter, milk, eggs, and home-grown produce. What they often found on the family table was something very different. Traditional farm foods were often fried and bland, meals were often unvarying, and fresh vegetables were noticeably absent. After

enough visitors complained about farm food, Vermonters figured out that
they needed to make some changes in their menu. As always, that meant
extra work, not to mention the potential for resentment among residents
for whom a loss of control over diet may well have symbolized a loss of control over much more.[28]

Differences of opinion also emerged in Vermont about tourism's impacts on farming, and more specifically about the role that nostalgia should
play in defining the state's identity. On the one hand, Vermont's farm-
boarding economy offered hope to many rural families who were struggling
to get ahead within an economic climate troubled by rural out-migration
and by competition from other regions of the country (see chapter 2). To
alleviate their struggle, many agricultural reformers encouraged Vermonters to open their farms to tourists. As the VSBA's statistical secretary Victor Spear argued, "Now there is no crop more profitable than this crop
from the city, and it is one that comes directly to the farmer, and he should
encourage and promote this visiting from our city cousins."[29] Some farmers clearly agreed; for them, tourism offered a way to get ahead, a way to
buy another parcel of land or to replace antiquated farm machinery or domestic goods with the modern, labor-saving devices that so many craved.
That might mean making some changes around the home, but it did not
necessarily mean that farmers were willing to place themselves in an economic holding pattern based on a caricatured image of traditional farm
life. In fact, if sprucing up the farm and catering to visitors was what it took,
many must have seen the sale of nostalgia as the most progressive idea to
hit Vermont in decades.

But others wondered whether tourism really offered farmers a viable
economic strategy. A number of social critics and agricultural officials questioned the wisdom of tourist development in northern New England's rural
districts, citing what one called its "demoralizing effect" on the region's
agricultural economy and society. Tourism, opponents argued, made young
people question the very nature of rural life, tempting them to leave the
farm in search of the urban lifestyles modeled by their guests.[30] Seen in this
way, farm boarding might actually threaten Vermont's agricultural progress
by robbing the state of those who would carry on its development. As Brown
notes, many turn-of-the-century agricultural reformers in Vermont and elsewhere were busy promoting a more scientific and streamlined approach to
farming, particularly in the dairy industry. Tourism, they worried, would
siphon off the labor force necessary to make that transition, threatening to
keep the state in a perpetual condition of stasis, if not decline.[31] Opponents
of tourism therefore expressed their concerns according to a progressive

impulse that they believed was at odds with the spirit of nostalgia so central to farm boarding. Should Vermonters continue to deploy a tourist-based reputation for timelessness as a means for boosting economic development? Or were there other ways to define rural tourist discourse in Vermont — ways that moved the state ahead without looking to the past alone?

Tourism, Industry, and Progress in the Deerfield River Region

Although Vermont was often sentimentalized in the popular press as a time and place apart, Vermonters were by no means indifferent to the optimistic embrace of national expansion, industrialism, and economic progress so common among America's late-nineteenth-century middle and upper classes.[32] For some, the state's best expressions of economic progress lay not with farming but with industrial developments in logging, mining, and manufacturing. From the paper and lumber mills of St. Johnsbury, Newport, and Bellows Falls to the slate, marble, and granite quarries of Proctor, Dorset, and Barre, Vermont's late-nineteenth-century landscape was shaped by a wider range of economies than agriculture alone. Contrary to what we might think, promoters did not necessarily hesitate to deploy the state's industrial landscapes in broadsides and other advertisements, nor did visitors necessarily shun them as entirely offensive or contradictory to their expectations about rural life. In fact, Vermont's tourist and industrial landscapes often developed in tandem, sharing both a similar geography and a similar progressive language about modernity and economic growth. As much as agricultural spaces and nostalgia remained cornerstones of the state's emerging tourist trade, then, other types of tourism and other worlds of work informed landscape and identity in rural Vermont.

The Deerfield River Region in south-central Vermont offers a good opportunity to explore the confluence of tourism, industry, and progressive sentiments that characterized some areas of Vermont. The Deerfield River heads in the remote highlands of the southern Green Mountains before crossing into Massachusetts, where it eventually meets the Connecticut River. Falling over a thousand feet in elevation along its course, the river and its tributaries drain over three hundred square miles, much of which remained heavily timbered in the 1880s with virgin stands of hemlock, spruce, and beech. Despite a rough turnpike connection westward to the town of Bennington, and despite a handful of small-scale water-powered industries, settlement and resource development in the region remained hampered by isolation, rugged mountain terrain, and a cripplingly short growing season.[33]

FIGURE 1.2. The Deerfield and Connecticut River Valleys.

Some in the region bemoaned its lack of development, but to the industrialists Daniel, John, and Moses Newton of Holyoke, Massachusetts, that same lack of development represented a valuable opportunity to expand their lumber and papermaking operations. In the early 1880s, the brothers chose the village of Readsboro, Vermont, as the hub of a new inland empire that would extend north along the forested course of the Deerfield River. Because of its water power and proximity to large stands of timber, Readsboro was already home to a few modest sawmills, tanneries, and furniture factories. But with the arrival of the Newtons, the scale of industrial production in Readsboro increased dramatically. The brothers upgraded preexisting industries, purchased vast tracts of forests, raised the village's milldam to a height of over fifty feet, and constructed a new pulp mill, paper mill, and chair factory. The capstone of their efforts was a short, narrow-gauge railroad connecting Readsboro to the Hoosac Tunnel in Massachusetts. Although the new rail line went no further north than Readsboro, it was prophetically named the Hoosac Tunnel and Wilmington Railroad for another Vermont village located only a dozen miles to the north.[34]

The patterns of investment and natural resource development set in motion by the Newtons were being played out simultaneously in towns and

villages across Vermont. Annual production values for timber soared into
the millions of dollars during the 1880s and 1890s as investment from capitalists like the Newtons grew and as timber companies cut new sources of
spruce and hemlock from many of Vermont's higher slopes.[35] Readsboro
never achieved the same lasting prosperity as other mill towns in Vermont,
but its initial success did spawn a new degree of economic optimism
throughout the south-central part of the state. That optimism soared when
the Newtons announced tentative plans in 1889 to extend their rail line
north to Wilmington — a move that would place the rail line that much
closer to the remote forests on the upper Deerfield, reducing the length of
log drives on the temperamental river. The move promised to transform
the entire town of Wilmington, starting with its central village. With a population of just over one thousand people, Wilmington's village center was
more an agricultural hub than an industrial one. But Readsboro's growth
had not gone unnoticed by Wilmington's aspiring business leaders, nor
had the town's strategic location relative to the region's northerly forests.
Wilmington's geography presented an opportunity to expand their commercial and industrial economies, town leaders argued, keeping them "abreast
of our lively sister village of Readsboro."[36]

As in budding towns elsewhere in the United States at the time, Wilmington's optimistic leaders began an enthusiastic campaign to transform a
sleepy town described regrettably by one as being "fifty years behind the
times" into one of Vermont's premier industrial and commercial centers.[37]
Residents founded a newspaper, the *Deerfield Valley Times*, that quickly
became a voice for progress across south-central Vermont. Others produced illustrated monographs extolling the town's many opportunities for
investment. Local businessmen formed the Wilmington Board of Trade to
"secure the advantages which the position of the town offers to agriculture,
trade, and manufacturers." They installed a town clock so that residents
would not fall "behind hand in any business matters." And in an additional
sign of self-worth, they commissioned a handsome bird's-eye view for the
village, drawn in 1891 by a noted regional lithographer.[38] Their coup de
gras, however, was a four-day "town reunion" held in 1890 and commemo-
rated by a celebratory two-hundred-page book. Not unlike the "Old Home
Week" tradition that would spread across northern New England in the
coming decades, Wilmington's town reunion was designed to showcase its
economic promise to former residents who had since moved away, and to
encourage them and others not formerly familiar with the town to invest in
its future rather than merely celebrate its past.[39] All that was needed, everyone agreed as the 1890s got underway, was a railroad link to Readsboro.

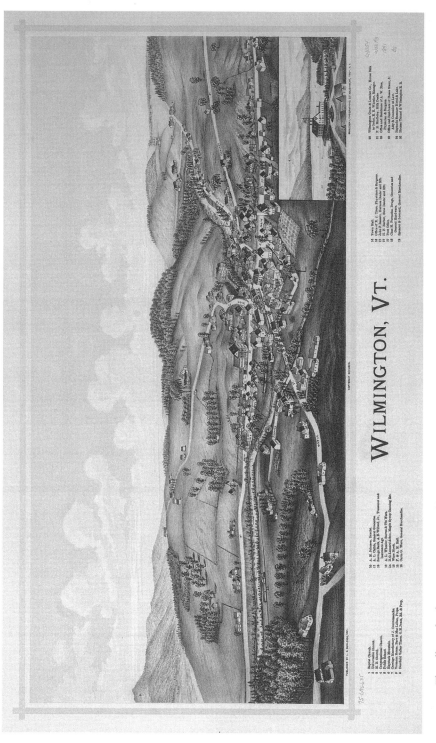

FIGURE 1.3: The village of Wilmington, 1891. Created by Lucien R. Burliegh of Troy, New York, this birds-eye view depicts a seemingly quiet country village. At the time this drawing was created, however, local leaders were working diligently to refashion Wilmington into a bustling regional center — a vision for the future based on a combination of industrial and tourist development. Library of Congress, Geography and Map Division.

Wilmington's dream came true in the fall of 1891, when town leaders negotiated successfully to have the Newtons extend their rail line north to the village. The move prompted a flood of progressive sentiments. "Surrounded by a large and fertile territory," the *Deerfield Valley Times* noted, "and located, as we are, in the midst of a lumber region, with good water-powers at command, it seems as if Wilmington must in time become a manufacturing town of some importance and a center of trade — in fact, the mecca [*sic*] of the surrounding country."[40] For many, that future seemed assured in 1893, when the Newtons announced their decision to move their regional headquarters from Readsboro to a new riverside complex just west of Wilmington's village center. Known as Mountain Mills, this sprawling industrial center soon became a village unto itself, with housing, a store, a hospital, and other services. At its center were a massive new dam and mill-pond as well as steam- and water-powered mills where timber driven down from the upper reaches of the Deerfield River was processed into lumber and pulp for reshipment by rail to Holyoke and other points south.[41] By most accounts, Wilmington no longer seemed "fifty years behind the times" — an impression not expressed in words alone, but one embodied for all to see in the town's developing industrial and commercial landscape.

That impression was embodied as well in Wilmington's emerging tourist landscape, the creation of which local entrepreneurs hoped would enhance the town's progressive reputation. Wilmington had its share of farm boarding enterprises in the 1890s, and it had its share of nostalgic visitors as well. But as in other communities across northern New England, Wilmington also had other, formal tourist destinations, including a lakeside resort hotel and an exclusive fish and game club. Resorts and clubs such as these were vital components of Vermont's turn-of-the-century tourist economy. An 1894 report on tourism completed by Victor Spear of the VSBA found that over fifty thousand people had visited Vermont in 1893, spending an estimated $500,000 on room, board, and livery services.[42] A decade later, another observer identified 150 resort destinations in Vermont, ranging from retreats on popular lakes such as Champlain, Memphremagog, Dunmore, and Bomoseen to mountain resorts in towns like Stowe, Manchester, Ripton, and Woodstock.[43] State and local leaders looked at numbers like these and saw an important avenue for economic growth. According to observers like Spear, for instance, the state's vacation numbers were promising indeed: "It is doubtful," he argued, "if any agricultural product, except the dairy product, is bringing as much money to the State at the present time as our summer visitors."[44]

A handful of resort hotels opened in the Deerfield River Region during

the 1890s and 1900s, the most elegant of which was Wilmington's Lake
Raponda Hotel. In 1890 the industrialists Moses and John Newton formed a
new group of investors whose goal was to develop resort hotels on a pair of
local lakes. One of these was Wilmington's Ray Pond, a small body of water
immediately east of the village that had earlier been enlarged by a mill dam
and that derived its unassuming name from the family who farmed its shores.
Ray Pond was popular with local fishermen, picnickers, and their guests —
who at times were forced to contend with the half-wild pigs that roamed its
shoreline — but there is little to suggest that it caught the attention of anyone
outside the immediate region.[45] But with the promise of a railroad link to
town, with the backing of powerful investors like the Newtons, and with a
calculated name change to the more intriguing-sounding "Lake Raponda,"
the lakeshore was ripe for development. By the summer of 1891 workers had
cleared land, built carriage roads, and constructed a modern, three-story, one-
hundred-person hotel, complete with verandas, billiard room, barbershop,
and elegant dining accommodations. Soon the hotel was boasting a diverse,
well-mannered, and discerning clientele who, as at other Vermont resorts,
came in search of scenery and sociability, recreation and relaxation. Although
the hotel's owners struggled during the national economic downturn of 1893,
and although the original hotel burned to the ground under mysterious circumstances in December 1896, a new, more elegant hotel followed soon
after and remained open for business into the 1910s. Other investors purchased the remaining available land around the lake and began selling lots
for private cottages. Like other lakeside resorts in Vermont, Raponda became
a nucleus around which a summer community of private homeowners
would grow — a community that would ultimately outlive the hotel itself.[46]

At the same time that developments at Raponda were getting underway,
plans were also taking shape to build a prestigious sportsmen's club on
hundreds of acres of land immediately to the west of the village. The plans
were set in motion by another group of brothers, Rollin, Asaph, and Frederick Childs, who had been born and raised in Wilmington and had since
become successful businessmen in Brattleboro, Vermont, and Springfield,
Massachusetts. Now, in the wake of Wilmington's 1890s town reunion, they
were turning their attention back home. They purchased land for the club
from a local farmer and a local lumber company, and in the shadow of
Wilmington's Haystack Mountain they established the Wilmington Forest
and Stream Club.[47]

Having been raised in small-town Vermont, it is likely that the Childs
brothers grew up with hunting and fishing, and their membership in the club
suggests that they retained an interest in these activities as well. But their

FIGURE 1.4. The Lake Raponda Hotel, c. 1893. Lakeside resort hotels of this stature and style were not unusual in turn-of-the-century Vermont. Here visitors gather at the hotel's main entrance — a place where seeing others and being seen oneself were always on the visitor's agenda. Courtesy of Pettee Memorial Library, Wilmington, Vermont (Margaret Greene Collection).

decision to start a sporting club was based on more than personal interests alone. The Forest and Stream Club was a serious business proposition that reflected the tastes and status of many American businessmen at the time. Since the 1870s groups of well-to-do American men had been organizing sporting clubs such as Wilmington's. The most famous of these was the Boone and Crockett Club, whose membership included some of the nation's leading politicians and whose political power gave impetus to the emerging conservation movement in the United States. Sportsmen's clubs gave men an opportunity to display their social status to others, express their support for fish and game conservation, and demonstrate their proud adherence to class-based codes of sportsmanship (such as shooting fowl only on the wing).[48] Wilmington's Forest and Stream Club was a product of this larger movement. The group's grounds and clubhouse were expressions of gentlemanly good taste, and its members considered themselves models of enlightened conservation. The club's exclusive membership list grew in time to include well-to-do professionals from Boston, New York City, Hartford, Providence, Chicago, and beyond — men described in the local press as being of "resolute will, sterling integrity, and of hospitable habit."[49] These members were screened by a strict referral process, by their ability to pay the stiff $1,000 stock-purchasing fee required for membership, and by their adherence to proper codes of sportsmanship. Their membership entitled them to fish on the club's property (which had been well-stocked), stay overnight in the clubhouse and cottages, use the bowling alley and sitting rooms, and drive, walk, and play golf on the club's manicured property.

The Forest and Stream Club and the Lake Raponda Hotel were not aberrations in a town whose leaders were otherwise busy courting industrial development. Nor did they owe their existence entirely to a nostalgic impulse among travelers seeking rural days gone by. Although signs of industry existed throughout the Deerfield River Region, visitors did not shun this or other similar regions in Vermont, nor did tourist promoters downplay the region's modernizing impulse in favor of a nostalgic image of rural life. Rather, Wilmington's new resorts and new industries grew in tandem with one another, connected by a shared regional geography, common investors, and a shared outpouring of progressive sentiment heaped upon each by a diversity of social groups. Tourism and industrial development therefore redefined the Deerfield River Region according to a shared vision of progressive modernity and economic expansion — a vision inextricable from the landscapes of rural work and rural leisure in which it was embedded.

This overlap between industrial and tourist development in Vermont towns like Wilmington was expressed in a number of ways. First, each was backed by common investors, such as the Newton brothers, who brought money and optimism to all their projects, whether industrial or tourist in nature. Investors spread their wealth into both tourism and industry, suggesting that each had an important role to play in the state's future. From the perspective of Wilmington's capitalists, it seemed only logical to invest in mills and resorts simultaneously; both were part of the broader progressive vision under which they operated. Second, state and local promoters often championed the potential of industry, tourism, and agriculture on the same pages, in the same paragraphs, and even in the same sentences. Wilmington's Board of Trade, for instance, produced circulars describing characteristics such as the region's "beautiful valley and all its advantages as a manufacturing and agricultural center; also as a summer resort."[50] To that, the town leader and capitalist Hosea Mann Jr. added, "Consider also the large territory of tributary productive farms, timber in any quantity, and valuable water power, with some of the most beautiful scenery in the world, and a summer resort noted for its delightful climate and healthfulness, and what more could be desired to start the nucleus of a thriving town?"[51] With such variety of apparent opportunities, it seemed only logical to assume that Vermont's future was bright with promise.

As attitudes like these suggested, there seemed to be nothing unusual about advocating tourism and industrial development in the same towns and in the same breath. Quite simply, the two were joined by a common goal: the creation of economic and material progress. Indeed, Wilmington's tourist developers turned time and again to the town's two major resorts as evidence of a progressive outlook taking root across the region. Each resort was a model of enterprise, initiative, and modernity, they proudly claimed, and in this sense, each was not that different from the town's industrial developments. Just as promoters celebrated the power and modernity of Mountain Mills, so too did they praise Raponda's modern comforts and amenities, including its up-to-date lighting and heating systems as well as its telephones, which allowed visitors to keep in touch with the business world back home.[52] And just as town promoters celebrated the aggressive business principles of its industrial leaders, so too did they praise resort proprietors for developing their property to its fullest economic potential. As the *Deerfield Valley Times* noted enthusiastically, Lake Raponda's "many elements of beauty and attractiveness" were being continually "developed and advertised" by its business-savvy developers. Transforming nature into profit, the lake's developers improved upon pre-existing "natural charms" and "natural advantages" by adding "all that money can procure for the comfort and convenience of guests."[53] These were places where one could still commune with nature, and from which one could still take a nostalgic ride into the countryside if one wished. But they were also places on the leading edge of modern resort design, places that town leaders were happy to use in the service of promoting a progressive identity for Wilmington as a whole.

Similarly, members of Wilmington's Forest and Stream Club were held up as models for aspiring local businessmen, in part because of their personal success, and in part because of their approach to managing their land. The club's primary constitutional objective was to "purchase, acquire, lease, hold and improve real estate" for the sake of providing "facilities for the rational enjoyment of out-door sports with rod and gun, and such other amusements as may naturally attach thereto to give relation from business cares."[54] But when members purchased the club's limited stock, they purchased into something that was more than a haven from business cares. Instead, club members bought into what was essentially a new business itself. The club's property included hundreds of acres, a small mill, a working farm, and thousands of maple trees, which club leaders planned to use for the production of maple sugar. To pay for maintenance and to provide a return on the investment of its shareholders, the club hired a local farmer to run their model farm — essentially putting their recreational landscape to work. For a time, the farm provided fresh produce for guests and generated revenue from the sale of dairy products and maple sugar. Optimistic members even displayed their maple products at the Columbian Exposition in Chicago in 1893, presumably in the hopes of expanding their market.[55] Through their creative efforts to blend work and leisure, then, club members became local models of initiative and business savvy. According to one observer, the Forest and Stream Club was a true "indication of push and progress," a foundational cornerstone for a broader program of regional development that combined tourist development with "the sound of the steam whistle, the din of industry."[56]

Town leaders often aimed celebratory rhetoric such as this at local residents in an ongoing effort to dispel critics, win support for tourist development, and boost civic pride. Yet it was also aimed at *potential visitors*, some of whom, promoters sensed, would come to Vermont with more than rural nostalgia in mind. Accordingly, promoters sought to reach a broader audience by painting a progressive picture — one that suggested that industrial and tourist development in Vermont towns like Wilmington were by no means incompatible with one another, and one that thereby implicated Vermont in the production of a larger cultural discourse in late-nineteenth-century America based on material and economic progress.

There were a number of reasons why promoters aimed progressive rhetoric at visitors, bundling tourism, industry, and town development together into a broad-based advertising campaign. First, one often gets the sense in the historical record that every visitor to Vermont in the 1890s and 1900s was considered a potential investor as well — one who might someday generate new working opportunities in the state's industrial and commercial sectors, not to mention its tourist sector. Booster organizations such as the Vermont Development Association, for instance, appealed both to Vermonters and to outsiders when they called for new investments in industry. But when they pitched the state to outsiders, they often did so by treating them *both* as tourists and as businessmen.[57] If visitors could be enticed to come to Vermont, if they saw, first hand, the promise of Vermont's forests, minerals, and progressive-minded citizenry, they would almost certainly want to invest in Vermont enterprises. Vermont had more than a timeless landscape to offer the businessman/tourist, promoters maintained; it had the potential for profit.

Second, promoters deployed modern motifs to reassure visitors that a trip to Vermont was not an act of self-imposed exile to an isolated wilderness inhabited by a backward people. This was not an unusual concern, either among visitors or among those who catered to them. "Significantly," David Richards writes in his history of Maine's Poland Spring resort, "patrons of the resort did not check their faith in progress at the front gate during their annual summer sojourn to Poland Spring."[58] Instead, visitors here and elsewhere sought places where they could insulate themselves from the more unpleasant aspects of progress in urban America — crowding, noise, filth — while not divorcing themselves entirely from the benefits of modern society. The reputations of Vermont towns like Wilmington were served well by direct rail connections, for instance, and resorts such as Raponda and the Forest and Stream Club were served well by their telephone service to the world back home. Communities throughout Vermont, one promoter noted, retained "sufficient nearness to the busier world outside to conveniently provide all the creature or social comforts that the habits of the generation demand. Her countryside, picturesquely rural and wonderfully diverse in scenic panorama, retains all the freshness of its virgin loveliness without the primitive privations or depressing isolation of the wilderness."[59] Visitors to towns like Barton, Vermont, for instance, found that it "combines the comforts of city life with the pleasures and health of life in the country." In Barton, one wrote, the visitor "may be at all times in touch with his own home and business in the city and yet at the same time enjoy the beauties of nature, the pure air and the purest of water."[60] Combining beauty, freshness, and a sense of difference from the city with all the best of modern amenities and easy connections beyond its borders, Vermont seemed to prove that modernity and picturesque scenery need not be mutually exclusive.

Finally, promoters celebrated progressive economic development in towns like Wilmington because that was what many middle-class visitors to the state would have wanted to see. A vacation to a town like Wilmington, promoters suggested, offered travelers more than a nostalgic escape to a place bound to the past. Rather, it offered an encounter with a thriving, modern place — a place that matched middle-class conceptions of a well-ordered, well-adjusted society. As the historian Cindy Aron has argued, many turn-of-the-century American tourists embraced places and experiences that reconfirmed their class-based commitment to social and economic progress, to diligence and hard work.[61] Perhaps it is not surprising, then, to find middle-class travelers the targets of promotions emphasizing Vermont's modernity and bright economic future. Wilmington, in the words of one Raponda promoter, won favor for being "one of the cleanest, most progressive and thrifty towns in the Green Mountain state." Its civic institutions, modern amenities, and commitment to commercial and industrial development seemed, time and again, to prove its citizenry's faith in progress.[62]

Indeed, some tourists took an interest in the industrial landscape itself, using their leisure time on occasion to contemplate scenes of industrial work. Travelers throughout the United States at this time often visited industrial landscapes in places like Niagara Falls and the coal-mining town of Mauch Chunk, Pennsylvania, where they embraced the sublimity of industrial power and the national economic progress it represented.[63] This was also true in Vermont towns such as Dorset, where visitors toured a well-known local marble quarry. The quarry's operators, one promoter noted, "are always pleased to show visitors about their works, where can be seen in operation the latest devices for quarrying marble."[64] Visitors to towns like Manchester and Rutland also toured quarrying operations where, as one impressed traveler wrote, "the vast accumulations of debris tell the tale of years of industrious burrowing in the earth."[65] And in Wilmington, tourists visited Mountain Mills to see the tremendous steam shovels, the impressive stonework, and the massive wheel-pit. They were encouraged, one promoter noted, to take scenic drives that offered views of "the smokestack of Mountain Mills [outlining] the glinting water of the bog pond."[66]

Of course, visitors to resorts like those in Wilmington would have spent the majority of their time on the resort's property, boating, fishing, resting, and socializing. And certainly some would have been troubled by the cut-over hillsides and smoking mill complexes that greeted them in places like the Deerfield River Region. But we should not assume that all visitors were unable or unwilling to reconcile such scenes with their conceptions of a properly ordered rural landscape. Nor should we assume that they considered such scenes an outright affront to their vacation experience. In fact, Wilmington's resorts were not as spatially segregated from industrial development as we might expect. The Forest and Stream Club's property was within smelling and perhaps even visual distance of the pulp processing complex at the nearby Mountain Mills, while the sounds of industry from the modest "Raponda Reclining Chair" factory (which opened in 1896 at the lake's outlet) would have surely echoed out across the water, greeting visitors as they boated and fished on the small lake.[67] The same could be said of other Vermont towns as well. Visitors to Newport — a mill town and tourist destination in northern Vermont — could have hardly failed to notice the town's noisy, sprawling, and smoking mills, or the commotion of the railroad yards located directly across the street from the popular Memphremagog House. Nor could cottage owners or pleasure boaters on Lake Memphremagog have missed the monstrous log booms floating past or resting in the village harbor.[68] Signs of industry were inescapable in many areas of Vermont at the turn of the century, including areas with flourishing tourist economies.

FIGURE 1.5. Industrial development at Mountain Mills, c. 1900. Lying just to the west of the village of Wilmington, the Deerfield River Company's Mountain Mills complex reflected a decidedly industrial vision for the region's economic future. Courtesy of Pettee Memorial Library, Wilmington, Vermont.

Part of the reason for this was the relative scale of industry in places like Wilmington and the tendency for that industry to exist within a landscape still recognizable as "rural." Visitors did not come to Vermont expecting to find nothing but smoking mills, urban development, and industrialized forests. But industrial development was not viewed as a problem and was even welcomed by visitors to Vermont as long as it did not completely undermine the sense of scenic difference and charm that brought them to towns like Wilmington in the first place. The majority of Vermont's turn-of-the-century commercial and industrial towns remained relatively small and surrounded by farm landscapes. Such places retained a sense of being rural while also seeming bustling and progressive. Not unlike industrial towns such as Lowell, Massachusetts, which had once aspired to blend industrial and rural life into a promising new type of community, late-nineteenth-century towns such as Wilmington suggested that a harmonious relationship might yet evolve between economic growth and scenic, rural charm.[69] One can see expressions of this in promotional photographs of manufacturing towns statewide, many of which used bird's-eye perspectives showing factories in the foreground with picturesque fields and forests rolling away to a distant horizon.[70] Economic development such as this seemed clean, under control, and less complicated than that associated with large urban centers in southern New England.

Representations of industrial development and economic growth found their way into tourist promotions because they spoke to a culture of travelers for whom landscapes of progress were themselves worthy of attention. Visitors continued to come to Vermont to escape into a nostalgic world of working farm landscapes, but many turn-of-the-century visitors were also willing to accept that scenic rurality was not defined by timelessness or agrarianism alone. Consequently, Vermont's tourist identity was informed by two worlds of work, moving beyond farming and nostalgia alone to overlap with industry and progress as well. But just as some debated the wisdom of using nostalgic tourism to define landscape and identity in rural Vermont, there were those who debated the meaning and value of progress emerging from its tourist development.

The Politics of Progress

Popular support for tourism in towns like Wilmington was rooted in a belief that visitors brought progressive social and economic benefits to a wide cross-section of Vermont society. To a certain extent this proved true, as the state's leisure economy generated new working opportunities, many of which were filled by women. In Wilmington, for instance, women sold fresh produce, berries, and meat to hotel kitchens, while others took seasonal jobs as cooks and launderers. Men participated in the tourist economy as well, finding work as builders, hotel managers, and livery personnel. Tourism also created new social opportunities for native Vermonters. Public parks, such as the new Shafter Grove in Wilmington, were often designed with visitors in mind but soon became places for local leisure and social interaction as well. Vermonters even started traveling more themselves. The Newton brothers offered Wilmington's residents special excursion rates on their rail line to places like Readsboro (where they toured the Newtons' mills), Boston, and Saratoga Springs, where one 1892 excursion drew three hundred people from the Wilmington area.[71]

Such benefits were certainly welcomed by many, but no strict consensus existed in communities like Wilmington about the meaning or value of progress, particularly when that progress was chalked up to tourist development. Not unlike concerns about nostalgia's impact on the state's agricultural economy, Vermonters struggled to make sense of tourist-based conceptions of progress as well. Reworking the rural landscape to accommodate tourist development meant reworking social relations too — social relations that increasingly revolved around discussions of work, leisure, and the relationship between them.

Despite repeated promises about the benefits of tourism, then, some residents resisted its growing presence in communities like Wilmington, at times even refusing to endorse or participate in the tourist economy at all. As in other Vermont towns, Wilmington had what the *Deerfield Valley Times* called its "chronic kickers" or "croakers," who remained opposed to any kind of change, whether done in the name of progress or not.[72] Although it is difficult to track their voices in the historical record, such "croakers" would have almost certainly been unwilling to adopt new roles as employees in the service of tourists or to welcome strangers into their homes and communities. If progress meant making concessions to leisured guests, if progress meant altering one's traditional way of life, some were not having it.

Others may have been less condemnatory but were still unable or unwilling to privilege the demands of a leisure economy over the demands of traditional rural work. In Wilmington, for instance, residents who earned a little extra money working at the Raponda Hotel at times placed the time- and weather-dependent needs of farming ahead of the resort. As one entrepreneur noted in 1892, his plans for improving the resort were being continually held up by the loss of local labor to haying.[73] Working at the resort to add cash to the household income was fine, but bringing in the hay was something that just could not wait.

Residents also raised more fundamental questions about whether the "progressive" interests of capitalists and resort proprietors necessarily matched those of the community at large. For instance, promoters felt compelled to take a defensive stance when they suggested a name change for Ray Pond, reminding readers in the *Deerfield Valley Times* that concessions such as these may seem unnecessary or be unpopular, but that they were, in fact, in everyone's best interest.[74] In a community with strong kinship ties and strong associations between families and the land they owned (such as the Ray family on Ray Pond), resistance to something as seemingly innocuous as a name change was indicative of deeper concerns. The needs of tourism were coming into conflict with local tradition and local culture, forcing many to wonder whose interests, whose idea of progress, was being served by the tourist economy.

More was changing in communities like Wilmington than names, however, and more was at stake than the symbolic meaning of local landscapes. Material changes were also afoot, as large and small tracts of land were now being transferred into the hands of tourist developers, many of whom lacked

FIGURE 1.6. Employees of the Bread Loaf Inn in Ripton, Vermont, 1899. Resort hotels offered new, seasonal employment opportunities, particularly for women such as these, most of whom would have been involved in cooking and cleaning. In some cases, young unmarried women took jobs in hotels as a temporary station in life, while in other cases, married women worked in hotels to supplement the family economy. Courtesy Vermont Historical Society.

an understanding or, worse, an interest in traditional patterns of land use. Starting in the late nineteenth century, some property-owning visitors to Vermont began posting their land, disenfranchising residents of the un-written codes of open access for hunting and fishing that, the historian Richard Judd explains, had been at the core of rural land use in northern New England for well over a century.[75] Changes such as this raised troubling and lasting questions about landscape control. When Wilmington's investors first unveiled their plans for a resort at Ray Pond, for instance, many locals raised their voices in alarm, fearing they would lose traditional fishing grounds to an exclusive group from outside the community.[76] Although their fears proved unfounded in the case of Raponda, residents were not so lucky with the Forest and Stream Club. Soon after the club was formed, its members posted hundreds of acres against public fishing and published a very clear notice about the property's status in the *Deerfield Valley Times*:

With malice toward none, but in a firm determination to uphold the law and maintain our rights, we will pay the sum of Fifty Dollars ($50) for information that will lead to the detection of any person or persons taking trout from the waters controlled by the Forest and Stream Club. We will also pay the sum of Twenty-five Dollars ($25) for the apprehension of any person or persons who may destroy or injure any property upon said lands.[77]

By posting their property, the Forest and Stream Club turned their leisure landscape into a symbol of power and exclusion, coding it with a legal language intended to control residents' behavior and send a signal about the club's position within local society. Language like this suggested suspicion and distrust—emotions that, when coupled with the posting of club property, would have almost certainly offended some local residents. But from the club's perspective, their suspicions and efforts to tighten control over their property were entirely justified. In an open letter to the town, one club member from New York City complained that locals had been poaching young fish from the club's property—fish that the club had paid to have stocked in their streams, fish that were meant to grow to full size for members' fishing rods only. If they failed to respect the club and their property, this member reminded readers, residents risked losing the money and prestige that the club brought to their town. In the interest of maintaining the club's good graces, then, he strongly encouraged all residents to see to it that poaching stopped.[78]

Some residents respected and even cooperated with the Forest and Stream Club's policies, particularly when it came to the cause of fish and game conservation.[79] And some would have undoubtedly welcomed the strict enforcement policies of the Forest and Stream Club, just as many would have been pleased with the fact that, on at least one occasion, the club stocked streams *off* its immediate property, where they were fair game for local fishermen as well.[80] At least one resident even entered into a legal agreement with the club designed to protect local streams. In 1891, for example, the club signed a ten-year lease granting them access to the stream on the property of Lucius Fox "for the purpose of stocking, propagating, fishing, and the protection of fish in said waters." For a small fee, Fox agreed to do what he could to make sure no one but club members used the stream, and he agreed to do nothing to "pollute the said waters, nor to the injury of the fish therein."[81] Such support for the club reflects at least a partial willingness to validate its presence within the community.

For other residents, however, posted property would have been an affront to traditional ways of life, a challenge to local control, and a call to resistance. Feelings like these were not unique to Vermont at the time but were instead expressed throughout the northeast wherever residents sensed

FIGURE 1.7. Southern Vermonters at a hunting camp, no date. Turn-of-the-century Vermonters like these often opposed hunting and access restrictions put into place by new landowners. Courtesy Special Collections, University of Vermont Libraries.

a loss of control to visitors. Across the border in New York's Adirondack Mountains, for instance, land use restrictions put in place by wealthy landowners prompted violent reactions from local residents who viewed outsiders as threats to their power of self-determination.[82] Wilmington's residents do not appear to have ever turned violent, but they did express their dissatisfaction and lack of respect for the club in other ways. At least one resident was caught stealing from the club's model farm, while others continued to fish openly on club property, despite repeated warnings.[83] Whether breaking the rules was an intentional sign of protest or not, it suggests that not everyone respected their new neighbors and the new, exclusionary landscape they created. If the loss of public access to streams was the price of tourist-based progress, then surely there were reasons for serious concern.

Finally, disagreements developed in Vermont about the relationship between the progress of industry and the progress of tourism, and more specifically about whether landscapes of industrial work and those of a leisure economy were as compatible as some would have liked to think. Although tourism and industry were frequently linked to one another by a shared discourse of progress, some Vermonters recognized very early on that progress in one sector could have consequences that challenged progress in another. And as often as Vermont towns such as Wilmington were depicted as having successfully blended the demands of tourism and industrial development into a harmonious scene, some expressed concerns that the long-term viability of that scene was, in fact, quite fragile. Even the pro-development editors of the *Deerfield Valley Times* published a commentary in which the writer argued:

We cannot cut the forests from the hills and expect these same hills to retain the beauty and attractiveness which the forests gave them. It should be remembered too that it is not altogether for our butter and milk and eggs that the people come to Vermont in the summer, for all these things can be bought in open market, but they are drawn hither by the attractiveness of these same mountains and lakes and streams. If for no other reason then let the forests remain and keep the streams pure because "it pays." The trees are worth more on the hills than in the mill yard, for they are among Vermont's chief attraction.[84]

Aesthetic concerns such as this applied to Vermont's popular village centers as well. From a distance, Wilmington appeared clean, in harmony with its natural surroundings, and uncorrupted by urban problems. Nearly all of Wilmington's buildings were painted white, one observer wrote, so that "As one sees the town from a distant hillside on a summer day the scenic effect is like that of a pearl in a great emerald bowl."[85] But closer inspection revealed problems on the ground, particularly where the North Branch of the Deerfield River cut through town. As the *Deerfield Valley Times* complained, industrial development and the apathy of residents had befouled the North Branch: "It is bad enough to have the sawdust from the mills running down the streams," the paper wrote, "but it is far worse to have the stream at the village bridge used as a dumping place for everything that people want to get rid of. It is both unhealthy and unsightly." Equally important, the paper added, were the potential effects of stream pollution on the town's tourist economy: "Some out of town people said a few days ago that this [pollution] was the worst thing they knew about Wilmington. Let us make an effort to abolish this nuisance commencing today."[86]

Communities across New England often approached problems like these by forming village "improvement societies." And although it is unclear if Wilmington ever officially created such a society, town leaders certainly raised the idea on more than one occasion. They did so, in part, as a way to make their town more appealing to visitors, as in the case of the annual, springtime calls made by the *Deerfield Valley Times* to spruce up the town before its summer tourist trade got under way. As the paper put it in 1893, "A little paint, new shingles and a little work here and there where needed will make a vast improvement in many places. We hope all will do their part towards making our town look more inviting than ever this year to summer visitors."[87]

Improvement societies throughout New England, Joseph Conforti has shown, were often filled with as many, if not more, summer visitors as local residents. And as such, they often focused on making villages appear timeless, sentimental, and decidedly non-industrial.[88] Massachusetts's *Springfield Republican* recommended a similar course of action for Wilmington when it argued in 1891, "The businessmen of Wilmington hope for an impetus in trade and manufacturing, although it will be to the town's advantage as a summer resort to retain its quaint repose and old-time individuality."[89] But this trend toward timelessness was not universally true of all of Vermont's village improvers, even those who were at work in its tourist towns. For a time during the 1890s and 1900s, village leaders hoped to modernize their towns without compromising their tourist appeal. Rather than lock their town in a caricatured image of stasis, rather than freeze their town in time, Wilmington's leaders, for instance, consistently pushed a modern agenda. They pressed hard for electricity, more sidewalks, new streetlights, and improved sewer systems, reasoning that such things were appealing both to village residents and to tourists as well. The same was true in other towns. Described by one magazine as a town "living in the present and for the future," Barton, Vermont, had its "wagon . . . hitched to a star rather than to the memories of days and deeds gone by." With scenic beauty, new civic buildings, sidewalks, telephones, and modern water, sewage, and electrical systems to work with, Barton's village improvement society worked hard to continue making the town a model of progressive growth — one that they hoped would attract investments in industry and tourism alike.[90]

Between the 1880s and 1900s, promoters, visitors, and rural residents used both nostalgic and progressive discourses to frame their conceptions of landscape and identity in Vermont. Their actions set in motion a process that would persist in the state for decades to come — a process defined by the redefinition and renegotiation of work-leisure relations. In some cases, social groups drew landscapes of agricultural work into the tourist trade, restructuring the form and meaning of those landscapes to advance an image of nostalgia. In other cases, they drew industrial and commercial towns into the tourist trade, using optimistic assessments about the state's future to reconcile diverse types of development under a unified banner of progress. The precise meanings of nostalgia and progress and their values relative to rural tourist development were never entirely agreed upon. Nor were questions and concerns about the growing power of leisure as a force for restructuring the state's everyday landscapes of work.

As Vermont's popularity grew among tourists, the spaces, meanings, and practices associated with rural work and rural leisure would become increasingly interconnected. A new middle landscape was emerging in Vermont — a landscape implicated in the social relations that developed between and among the state's visitors and residents, and one where the needs of working residents and the needs of leisured guests would begin colliding with growing frequency and intensity. New and persistent questions had emerged out of the state's early experiences with tourism. How, some now wondered, can Vermonters reconcile the needs of working landscapes with those of leisure landscapes? How can they continue to embrace material progress in places like the Deerfield River Region without allowing that progress to compromise the beauty of the state's landscape? More fundamentally, was tourism compatible with the rural way of life that made Vermont special in the eyes of tourists?

Such questions remained salient for decades in places like Wilmington, even after the town's turn-of-the-century rush of tourist development subsided. By the end of the first decade of the twentieth century, it was beginning to look like Wilmington's hope of becoming a booming industrial and resort center had reached its limit. The town's industrial development eventually declined as timber resources became more costly to harvest. In 1904 the Newtons sold the Hoosac Tunnel and Wilmington Railroad and Mountain Mills to a group of New York City investors, who pushed rail lines farther north into the wilds of the Deerfield River watershed before selling off most of their holdings to a regional power company. In the 1920s the power company constructed a new dam and hydroelectric station on the Deerfield River, flooding the former site of Mountain Mills.[91] The town's tourist development also reached an initial peak around 1900, and by the 1910s the Raponda Hotel and Forest and Stream Club were no longer in operation. Tastes were changing among tourists, and with changing tastes came new ways to define and interact with the state's rural landscape. In particular, growing numbers of visitors were now purchasing property for summer homes, choosing to stay in Vermont for more than a week or two at a time, and choosing to invest themselves more heavily in the future of the state. Theirs is the story that follows.

CHAPTER 2

ABANDONMENT AND RESETTLEMENT

The Promises and Threats of Summer Homes

In 1937 a journalist from the *Boston Herald* sent a questionnaire to the secretary of the Vermont State Chamber of Commerce, James P. Taylor, designed to elicit the cultural essence of Vermonters and their state. Exactly what, the questionnaire asked, did it mean to be a Vermonter? What made them seem unique among Americans? What were their "leading characteristics"? What was their future? Were Vermonters changing, and if so, were they going forward or backward? Taylor passed the questionnaire along to be answered by a handful of Vermont's civic leaders and intellectuals, all of whom responded by turning to similar kinds of cultural traits. Citing Vermont's predominantly Anglo population, democratic heritage, Puritan-inspired piety, and rough, character-building environment, they defined its residents as true embodiments of American ideals such as "self-reliance," "individualism," "independence," "thrift," "honesty," and "ruggedness." Vermonters may have begun to change slowly in recent years, respondents conceded, but characteristics such as these remained proud and persistent features of the state's cultural identity.[1]

Public discussions about the identity of Vermonters were not new by the late 1930s, nor were the characteristics to which the survey's respondents pointed in their answers. For decades, civic leaders, social reformers, farmers, tourist promoters, and visitors had all pondered, in their own ways, what it was that made Vermonters seem unique. All tended to assume that there was an identifiable cultural baseline in the state — something special that one could define as the "real," "true," or, to use a common contemporary term, "typical" Vermonter.[2] In the eyes of many, typical Vermonters literally embodied the state's rural identity; to know the typical Vermonter, they suggested, was to know the state as a whole.[3]

From the perspective of many social critics, the typical Vermonter was a decidedly virtuous character type — a true "Yankee" who lived a modest life on a farm or in a small village and acted as a caretaker for the nation's rural, Jeffersonian heritage. Commentators from inside and outside the state frequently defined the typical Vermonter as an American cultural ideal, an enduring example of all that made New England, in Joseph Conforti's words, an "American cultural homeland."[4] To be considered a typical Vermonter by such standards was a true compliment, indeed. Moreover, such qualities made the typical Vermonter a marketable tourist commodity. Favored notions of typicality in Vermont enjoyed a wide popularity in the tourist literature from the 1880s forward, informing the identity of rural people and the landscapes they created.

But there was another, simultaneous identity for the typical Vermonter — one that stood in sharp contrast to the rosy image so often directed toward tourists, and one that haunted Vermont from the Gilded Age through the Progressive Era and on into the 1930s. Just as some observers lauded their apparent rock-solid stability, a wide range of agricultural reformers, civic leaders, and social critics warned of their passing, citing rural out-migration and a seemingly faltering agricultural economy as leading factors in the decline of Vermont's otherwise privileged identity. Thousands of struggling farm families left the state in the five decades following the American Civil War, searching for better prospects elsewhere and leaving thousands of abandoned farms in their wake. For concerned Vermonters, the abandoned landscape created by out-migration represented more than a decline in agriculture; it represented a challenge to the best of typical Vermont. For all of their virtues, it seemed, the identity of typical Vermonters was never quite as stable as many would have liked.

State and local official turned to a number of reform programs to deal with this problem, often targeting out-migration and the state's abandoned farms as root causes in Vermont's crisis of identity. This chapter focuses on one of these reform efforts: the sale of farmhouses and farm property for use as summer vacation homes. Summer homes varied considerably in Vermont, from hunting and fishing shacks to lakeside cottages to manorial estates. But it was the former farmhouse that vacationers prized above all others, and it was former farmland on which these homes were typically located, regardless of their style or age. In this way, Vermont's summer homes became central to the reworking of rural Vermont, emerging as potent symbols and powerful agents in the transformation of Vermont property from work to leisure.

This chapter considers that transformation, arguing that, over the long term, it posed both challenges and threats to the future of popular, socially constructed conceptions of typical Vermont and its stylized, typical inhabitants. Organized efforts to market and sell farms as summer homes were part of a larger, optimistic reform effort to bolster society and economy in Vermont, thereby reviving and reproducing the state's better qualities as effectively as possible. Those who promoted the use of abandoned (and, in some cases, not yet abandoned) farms as summer homes redefined rural Vermont as a consumer landscape capable of yielding short-term profits for those who sold their land, and, ideally, capable of yielding long-term benefits for rural society as a whole. In this way Vermont's farm landscape became implicated in a broad reformist discourse designed to reproduce the kinds of cultural characteristics that gave Vermont its unique and cherished identity.

Summer home ownership was a very different kind of tourist trend than what we encountered in the last chapter, and summer home owners experienced and understood the rural landscape on new and different terms. Not only did summer home ownership depend on a transfer of ownership from farm families to individual visitors, it encouraged vacationers to stay in their homes for months on end—a trend that caused these "summer residents" to become invested in Vermont to a degree that other visitors were not. Thousands purchased land for summer homes between the 1890s and the 1930s, and as they did, they assumed control over hundreds of thousands of acres statewide. Consequently, summer home ownership's influence on landscape and identity in Vermont was more powerful and widespread than any vacation trends that had come before. Vacationers' class-based and gender-based decisions about land use on former farm property fueled a reworking of the rural landscape with lasting consequences for the nature of work-leisure relations statewide.

Such consequences made many Vermonters nervous, and by the 1920s some civic leaders were beginning to wonder if summer home ownership was not more of a threat than a promise. One way to get a handle on the situation was to shore up control over the kinds of visitors who purchased the state's property. State officials, tourist promoters, rural residents, and even summer home owners themselves recognized intuitively that landscapes have the power to both include and exclude social groups by virtue of their use and representation.[5] With that in mind, some tried to ensure that Vermont property fell into the "right" buyers' hands, thereby mitigating the social consequences of summer home sales and regrounding those sales in their original reformist mission. In the process, however, they grounded the state's rural tourist landscape in the reproduction of larger class and ethnic prejudices, both in Vermont and beyond. Rural Vermont had once again become a middle ground where differences of opinion about work and leisure had moved to the fore.

Typical Vermonters, Farm Abandonment, and Vermont's Crisis of Identity

Between the 1880s and the 1930s, journalists, historians, travel writers, and tourist promoters often described Vermonters in highly stylized and cele-bratory ways, coding Vermont (and New England as a whole) with a unique regional identity.[6] Vermonters were an honest, thrifty, hardwork-ing, community-oriented people, they maintained. They were a people who had managed, unselfconsciously, to preserve some of the nation's most cherished cultural traditions and ideals. Typical Vermonters were true Yan-kees, the most American of Americans. And Vermont was their stronghold.

Image-makers turned to a number of cultural characteristics to create a common — yet restrictive — idea about what it meant to be a typical Ver-monter. First, typical Vermonters were always represented as being of Eng-lish ancestry, making them stand out among some audiences as models of what it meant to be a "true" American. This felt particularly appropriate in a predominantly rural setting, like Vermont. As other geographers have noted, western cultures often code rural landscapes with ethnic and racially specific identities.[7] That pattern has often held true in northern New England, where representations of the region — from both within and without — have tended to highlight the dominance of residents who were white, Anglo-Saxon, and Protestant. Such representations were welcomed by generations of middle- and upper-class Anglo Americans, many of whom worried about the last-ing social, cultural, and economic effects brought about by immigration — particularly late-nineteenth-century immigration from eastern and southern Europe. Concerns like these felt especially acute in southern New England, where "immigrant cities" like Lawrence, Massachusetts, harbored what some saw as troubling if not threatening ethnic populations.[8]

By contrast, Vermont seemed a bastion of Americanism — an impres-sion that made the state and its people attractive to some tourists on ethnic grounds alone. In 1891, for instance, the VSBA assured potential visitors that rural Vermonters descended from settlers "almost wholly of the Puri-tan stock; almost all of English ancestry" — settlers whose blood still ran strong in the veins of Vermonters at the end of the nineteenth century.[9] Another writer echoed that sentiment twenty years later in the general-interest state magazine *The Vermonter,* arguing that "the racial characteristics of the Anglo-Saxon — purposeful and resourceful beyond compare — find best expression in the sturdy, vigorous and high-minded men of the Green Mountain State."[10] Generations of Vermonters had taken pride in claims such as these, and some had turned that pride into political action. The state had been a hotbed of nativism during the mid-nineteenth century, and anti-immigrant sentiments persisted in Vermont politics and society for decades.[11] But as the appeal of Vermont's alleged ethnic homogeneity took root beyond the state's borders, it moved into the realm of tourism. Vermonters had discovered that they had more than scenery to sell.

In addition to ethnicity, image-makers frequently depicted rural Vermonters as paragons of democracy and patriotism. A number of indicators were used to suggest this. For instance, historians, politicians, and tourist promoters frequently described Vermonters as modern-day "Green Mountain Boys" — a character type derived in name and sprit from Ethan Allen's heroic Revolutionary War militia, and one used to prove the Vermonter's apparently innate commitment to freedom, democracy, and social equality.[12] Others pointed to Vermont's annual Town Meeting Day in early March as an indication of the Vermonter's political merit. As in other New England states, Vermont's town meetings provided a forum for face-to-face debates and face-to-face voting on local issues, and in the opinion of some they represented the highest form of participatory democracy in the nation.[13] As argued in 1928 by President Calvin Coolidge (himself a celebrated native Vermonter): "If the spirit of liberty should vanish in other parts of the union, and support of our institutions should languish, it could all be replenished from the generous store held by the people of this brave little state of Vermont."[14]

Typical Vermonters were also celebrated for being independent-minded, self-sufficient farmers — bedrocks of rural stability in times of national uncertainty and change. This image of the steadfast Vermont farmer had been around for decades, but its national resonance deepened during the depression years of the 1930s. Many cash-strapped Vermont farmers continued to work their land in traditional ways (using horses instead of tractors, for example), and many continued to produce a majority of their family's food. Because of this, Vermonters seemed all the wiser for resisting the modernization, speculation, and economic excess that lay at the heart of America's financial crisis. The Vermonter's economic conservatism and Puritanical self-denial had apparently left them well prepared to deal with the trials of economic depression: they could grow and gather their own food; they were accustomed to hard work; and they were satisfied to make do with less.[15] Vermonters, it seemed, modeled the kind of stability, strength, and economic independence to which the rest of the nation should aspire.

Relying on a range of characteristics, then, image-makers constructed a highly stylized cultural identity for typical Vermonters. Yet although many wanted representations such as these to be true, a number of social and economic factors challenged and complicated the resident's privileged identity. As often as some Americans celebrated the typical Vermonter, others feared for their future, prompting a decades-long crisis of identity and a sense of insecurity about rural culture in Vermont.

For starters, Vermonters were far more ethnically diverse than their re- strictive, Anglo-based reputation suggested. Immigration totals in Vermont were lower by comparison to other, more industrialized New England states. Yet immigrants came to Vermont nonetheless, finding work on farms, in mill and manufacturing towns like Colchester and Springfield, and in quarrying towns like Barre and Danby. French Canadians, in particular, had been moving from Quebec to Vermont in significant numbers since the early nineteenth century, both to work in the state's forests and mills and to take up farming.[16] In fact, a greater percentage of Vermont's turn- of-the-century immigrant population lived not in cities but in rural areas. By 1930 first- and second-generation immigrants accounted for 30 percent (72,946) of Vermont's nearly 250,000 rural residents (farmers and non- farmers alike). A full 39 percent of these immigrants were French Canadi- ans, while 18 percent came from Great Britain and Ireland. Thousands more came from Eastern Europe, Italy, Greece, and Spain.[17]

Many Anglos resented the presence of immigrants in the state, denying them entrance into the ranks of typical Vermonters. The state's "'Pollacks,' Jews, French, 'Eyetalians,' etc.," the regional writer Clifton Johnson noted, lived "like pigs," and were "often the worse for liquor."[18] Others marginalized immigrants' contributions to Vermont society and Vermont culture, down- playing the extent of their effects on the state's Anglo majority in an at- tempt to secure the ethnic boundaries of typicality in Vermont. As one au- thor pointed out, "Foreign elements in Vermont have made no appreciable contribution to arts of manners and no changes in the ways of living — or thinking — of Vermonters."[19] Whether one condemned or merely down- played immigration's impacts on Vermont culture was ultimately less im- portant than the larger story at work here: Many observers persisted in their belief that immigrants and immigration threatened the privileged identity of rural Vermonters.

Even more threatening and troubling, however, was the persistent out-migration of rural Vermonters from many parts of the state. Following the American Civil War, populations in hundreds of northern New England towns decreased as thousands of farm and village residents abandoned the region in search of new opportunities in cities and on western American land. Vermont was hit particularly hard, in part because of the state's notoriously rugged, rocky farmland, and in part because of broader changes in the state's agricultural economy as the profitability of wool production declined and many farmers struggled to make a shift toward butter and cheese production. Out-migration reached a peak in the late nineteenth century, as populations fell in 81 percent of Vermont's towns between 1880 and 1890 alone. But despite this peak, out-migration persisted well into the twentieth century, particularly as young Vermonters chose futures for themselves beyond the state's borders.[20]

Initially, some optimists saw farm abandonment as a temporary problem, one likely to solve itself over time as people once again learned to appreciate the value of a New England farm. After all, one journalist wrote in 1889, these farms had "done more for the elevation of American thought and character and intelligence and happiness than all the railroads and millionaires and stocks and securities and corn and hogs that the great cities and the great West contain."[21] Available farmland in the western United States was sure to run out, another added, prompting a "reflux wave of population" that would wash back over New England's abandoned farms.[22] Most, however, saw farm abandonment as a grave and long-lasting problem. Out-migration from rural Vermont, many feared, was undermining the state's economy, social structure, and celebrated cultural identity. Indeed, New England's "abandoned-farm problem" lay at the root of what many Progressive Era reformers began referring to as a "degenerate" or "decadent" rural population. "The decadence of the rural districts," one wrote of northern New England, "the flow of population towards the great centers, and the consequent decline of rural industries and values, are disastrous features of our latest civilization."[23] This seemed especially true of Vermont, where out-migration appeared to have siphoned off the best and the brightest, leaving behind a populace that one unfavorable reviewer described as a mob of "shiftless, lazy, drinking, ragged paupers . . . a class that disgraces our nationality."[24] Although, as Hal Barron has shown, Vermont's late-nineteenth-century economy was actually more stable than "degenerate," the state's lack of clear growth lay in sharp contrast to other, expanding regions of the United States.[25] Consequently, the state's reputation as a homeland of "ragged paupers" took hold in some people's minds, chal-

FIGURE 2.1. Young boy with sheep-drawn cart, no date. For decades, agricultural officials and rural reformers feared for the future of young Vermonters like this. Would this boy remain in Vermont, or would he leave his family's hilltop farm for prospects elsewhere? In the opinion of many reformers, tourism had a complex role to play in shaping his future. Courtesy Vermont Historical Society.

lenging those who championed the typical Vermonter as a quintessential American.

State agricultural officials and Progressive reformers deployed a variety of programs and arguments to reconcile this imbalance. Some tried to convince departed residents to return to Vermont, while others tried to convince non-native farmers to buy Vermont farms. Inspired by the recent Oklahoma land rush, one newspaperman went so far as to propose a free-land giveaway in the hopes of attracting people to Vermont.[26] Others reformers tried to keep young Vermonters in place by generating new social and economic opportunities and by playing on their pride by urging them to prove the naysayers wrong. Advocates downplayed the state's reputation for soil infertility and low crop yields on the one hand and actively promoted modern, scientific farming techniques on the other. Quite simply, there seemed to be many good reasons to choose Vermont: land was inexpensive, soils were good, modern farming practices were taking root, and Vermonters were hardworking, honest neighbors.[27]

But despite their best efforts, reformers consistently failed to generate the kind of response they were after, and by 1890 Vermont's Commissioner of Agricultural and Manufacturing Interests, Alonzo B. Valentine, admitted that "but few of the sons of Vermont, in other states, will return for permanent residence."[28] That kind of conclusion prompted Valentine and others to cast about for other groups who might take their place. Following the lead of reformers in Maine, Valentine started a program to encourage Swedish immigrants to take up farming in Vermont. In 1889 he sent a recruiting agent to Sweden, and he began compiling information on more than 1,000 available farms and over 500,000 acres statewide.[29] To win support for his program, Valentine deployed a host of economic, social, and racial arguments: Swedes would bring welcome new markets for local goods and services; they would put fallow land back to work; they would revive the very spirit of communities hit by abandonment. And not least, they would pose no significant challenge to the state's Anglo-American majority. Swedes are "our cousins with like instincts of freedom, secular and religious," Valentine wrote. "They are sober, peaceful and religious in disposition, industrious and thrifty." Swedes were an "agricultural people" from a climate and latitude not unlike Vermont's, he added, and they loved their homes and communities with a devotion "surpassing that shown by other nationalities." Moreover, their bloodlines appeared entirely compatible with those of Vermonters. Not only were Swedes "physically tall and stalwart," he wrote, they had "blue eyes, light hair and cheerful, honest faces."[30] Roughly two dozen Swedish families relocated to Vermont's abandoned farms in 1890, most of whom proved to be perhaps a bit *too much* like their Vermont "cousins." In the end, nearly all who came abandoned their farms not long after they arrived, moving on in search of better opportunities elsewhere.

Efforts to reform rural society in Vermont continued through the turn-of-the-century Progressive Era and into the early decades of the twentieth century. By the 1920s, however, it was looking more and more like the challenges facing rural Vermont and the challenges posed to the cultural identity of its people ran deeper than immigration and farm abandonment alone. Indeed, once one peered behind the mask of the so-called typical Vermonter, one found a host of deeply rooted social problems, ranging from poverty to public health, from ignorance to extreme insularity. That opened up avenues for strong, if at times unfair, criticisms. "Some of the run-down Yankees" who remained in one hard-hit town, writer Clifton Johnson noted harshly in 1924, "are more disreputable than the foreigners — drinking, swearing, worthless decadents, strangely shiftless and irresponsible."[31] Dealing with deep-seated problems like this — no matter how real or imagined —

involved more than selling farms. It involved reforming rural Vermont on a truly broad scale.

Some of the state's educators took up the banner of rural reform, instilling civic pride in young Vermonters by reminding them that they were the "best representative[s] of the true American." As one educator argued in 1922: "We have the right sort of groundwork to begin with, hereditary and otherwise [to instill pride in Vermont's youth], so, who knows, it may be for us to eventually develop into a race of super-men and super-women. It would seem that the typical Vermonter's outstanding characteristics were conducive to something of that sort."[32]

Other reformers acted directly on what this educator called the hereditary "groundwork" of typical Vermonters by turning to eugenics, the scientific study of human heredity and the manipulation of bloodlines. Eugenics found support from American reformers and intellectuals during the 1920s, among whom was a University of Vermont zoology professor named Henry F. Perkins. In 1925, Perkins established a formal program he called the Vermont Eugenics Survey. In his optimistic view, eugenics offered a chance to address rural problems in Vermont by stopping them at the level of heredity. Over the next few years, field researchers working for Perkins fanned out across the state, collecting detailed sociological and genealogical data on so-called "notorious" families, many of whom were of French Canadian and Abenaki Indian heritage. Armed with this data, Perkins tried to "rehabilitate" troubled families through institutionalization, marriage restrictions, and even sterilization.[33]

Despite its insidious nature and relatively short life, Perkins's eugenics survey helped to initiate a larger program of rural reform carried out by a private civic organization called the Vermont Commission on Country Life (VCCL). Founded in 1928 and directed by Henry C. Taylor, a noted rural sociologist from Northwestern University, the VCCL consisted of two hundred reformers, intellectuals, and civic leaders. These participants were divided into seventeen committees, each of which was designed to make recommendations on a range of issues including health, education, agriculture, forestry, government, religion, and tourism. Among its members were well-known tourist promoters and reformers such as Walter Crockett and James P. Taylor, as well as nationally known writers such as Zephine Humphrey and Dorothy Canfield Fisher. VCCL members conducted research and compiled statistics on an astounding array of social and environmental factors affecting life in rural Vermont. In 1931 each of the committees published their findings and recommendations in a remarkable 385-page book entitled *Rural Vermont: A Program for the Future*. The book remained an important guide to public policy in the state for years to come and, as we shall see in greater detail later in the chapter, helped shape discussions about tourism's relationship to the state's abandoned farms.[34]

What all of these efforts reveal was the truly profound sense of anxiety that generations of Vermont reformers felt about their state's future. Despite the popularity of the typical Vermonter's imagined cultural characteristics, many felt that image and reality were disconcertingly out of synch. Social reformers tried for decades to reconcile that disconnect, in part by polishing a cultural identity tarnished by farm abandonment and by other, seemingly endemic rural problems. And although they could never succeed entirely in that effort, the image of typicality in Vermont remained promising in and of itself. When cast in the best light, the typical Vermonter remained an attractive, romantic cultural icon — one capable of attracting vacationers to Vermont, and one therefore capable of reproducing itself through the sale of abandoned farms as summer homes.

The Promises of Summer Homes

During the last decades of the nineteenth century, a growing number of professionals and well-to-do urban residents began purchasing New England's abandoned farms for use as summer homes. For rural farmers who had moved away or were looking for an out, vacationers became an unexpected and welcome new market. And for state officials and rural reformers looking to mitigate the social and economic consequences of rural outmigration, summer home ownership became a powerful new tool. Their goal in promoting summer home sales was not to transform the rural landscape entirely in the image of leisure, nor was it to supplant the state's agricultural economy entirely with tourism. Rather, they hoped to use leisure to supplement the agricultural economy, thereby improving social and economic conditions in Vermont's rural districts and stemming the tide of outmigration. Summer home owners would aid in this process by infusing new money and new ideas into those districts. Admittedly, some Vermont communities would not be the same in the wake of summer home sales, but for optimistic reformers they would be much better places. Summer home sales were intended to transform former spaces of work into new spaces of leisure, reworking the rural landscape according to a discourse of reform and reproducing favored notions of typicality in rural Vermont. In this sense, the abandoned landscape held the key to its own redemption, not through work but through leisure. And initially, at least, that felt like a promising thing indeed.

Vermont's turn-of-the-century tourist promoters used the term "summer home" to describe a wide variety of overnight accommodations, including hotels, farm boarding houses, and rented cottages. Increasingly, however, the term was used more restrictively to describe privately owned vacation properties. Such summer homes were typically of two types: remodeled abandoned farmhouses and new cottages (or even larger homes), many of which were built on lakeside lots. At times visitors purchased farmland and farmhouses that had been abandoned entirely, and at times they purchased land or houses from farmers looking for an excuse to move on. In almost all cases, however, they purchased land that was once used for farming and that would now be used for vacationing instead.

This link between the fate of farming and that of tourism was underscored by the fact that the Vermont State Board of Agriculture was Vermont's first de facto tourist agency. Originally created in 1870 to oversee Vermont's agricultural, manufacturing, and mining industries, the VSBA had established an abandoned farms commission during the 1890s whose job it was to market these farms to any and all potential buyers. To make that happen, the VSBA published a series of annual advertising books, each of which was distributed by the thousands to potential buyers inside and outside the state. The first of these, *The Resources and Attractions of Vermont*, was released in 1891. With an emphasis on Vermont's social and civic institutions, agricultural potential, and opportunities for investment in mining and manufacturing, the book catered almost exclusively to farmers and industrial capitalists. Notably, however, subsequent editions revealed a telling shift in focus. Increasingly these books devoted a greater amount of attention to the state's resort hotels, farm-boarding opportunities, and abandoned farms. And by the second half of the 1890s the term "summer home" had worked its way firmly into the titles of the board's annual promotions.[35] This growing emphasis on summer home sales would continue for decades to come, not only through the efforts of the VSBA and its successor agencies but through the efforts of local development associations and private real estate firms as well.[36]

Taken as a whole, summer home promotions redefined the value of Vermont's abandoned landscape according to a new language of leisure. At the same time that optimistic rural reformers were telling farmers that Vermont farmland was fertile and productive, tourist promoters were telling potential summer home owners that the same land was scenic, charming, and well-suited to the needs of the vacationer. "Cheap lands, admirably adopted for building summer homes are to be found in nearly every County," the VSBA wrote in 1892, "and many people of limited means are

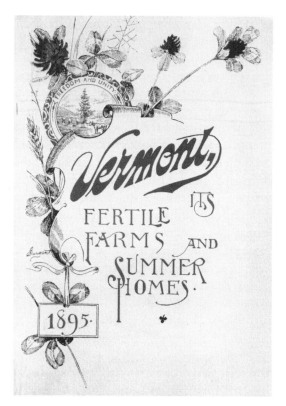

FIGURE 2.2. Selling Vermont farms as summer homes, 1895. This promotional booklet, like others published by the Vermont Department of Agriculture and later by the Vermont Publicity Bureau, hastened the transition of many farms from spaces of work to spaces of leisure. Courtesy Vermont Historical Society.

finding that in providing these homes, their families may enjoy the advantages of a summer in the country, and return [to the city] rich in renewed strength and health."[37] Sportsmen were also encouraged to take a closer look at abandoned farms in Vermont and across northern New England. As noted by the sporting magazine *Field and Stream*, landscape changes associated with farm abandonment were creating an ideal sporting environment. The "ruin and decay" left behind by abandonment was a blessing to the hunter, one writer argued. Not only were abandoned farms typically not posted against hunting or trespass, their overgrown pastures "made good hidings and furnished forage for quail and grouse."[38] In the absence of work, then, nature was transforming farmland back into forests, creating opportunities for recreation that might not otherwise exist.

It is difficult to know precisely how many Vermont farms were sold at the turn of the century for use as summer homes. It is safe to assume, however, that by the start of the twentieth century hundreds of properties and thousands of acres had already made the transition from working farm to leisured getaway.[39] Some town leaders were going to great lengths to attract buyers, particularly in districts where rates of abandonment were especially high. Voters in the town of Jamaica, for instance, agreed to waive the taxes for five years on properties whose owners — visitor or farmer — made improvements to abandoned farm buildings or erected new buildings on abandoned farm property.[40] Word of summer home sales and the benefits they promised was spreading among rural communities, just as word of farms-for-the-taking was spreading among potential buyers.

Summer home vacationing quickly took on a distinctive set of patterns, both in Vermont and in neighboring states. Each June, summer home owners returned to northern New England to take up residence, either for weekends and extended holidays or for the entire summer. The majority of the region's summer residents were from regional cities such as Hartford, Boston, or New York City, and most were the families of middle- and upper-class professionals — professors, writers, bankers, and businessmen. Some husbands were able to stay the entire summer with their families, while others left their families to return to work, visiting them on weekends and for occasional longer stays. Initially, summer residents arrived by train, hiring local farmers to cart their luggage and supplies from the nearest station to their summer homes. By the 1920s, however, most arrived by automobile. The automobile made it easier for visitors to travel back and forth to their summer homes on short notice, and it made it possible for visitors to purchase abandoned farms in communities well off main rail lines.

Among Vermont's optimistic reformers and civic leaders, summer home ownership promised a number of important returns. To begin, it promised to return abandoned property to depressed tax roles, bolstering town services in communities hit hard by abandonment. In addition, summer home ownership was also expected to jumpstart local economies, thereby making the state's abandoned hill towns, in one writer's 1908 opinion, "prosper and blossom as the rose."[41] Summer residents created new demands for goods and services, advocates argued, adding depth to local economies. They represented new markets for home repairs and farm produce, for instance, putting cash directly in the pockets of their farm neighbors. Farm wives could make extra money selling crafts like braided rugs to summer visitors with an eye for traditional wares, some argued. And if wooed properly, well-to-do summer residents might even funnel capital into Vermont's manufacturing, mineral, and agricultural development.[42] By whatever means, then, the sale of farms as summer homes was intended to improve economic prospects in rural communities, setting a course toward a more prosperous and stable future.

That process was more conservative than radical in its means and intended ends. No one intended for leisure to replace agricultural work as a primary land use in rural Vermont, nor did they intend for visitors to replace farmers as primary landowners in the state. By all accounts, supporters envisioned summer homes fitting unobtrusively into the state's rural fabric. Their ability to do so, of course, depended on the ability of reformers to create a balance between work and leisure. To make that happen, some promoters marketed Vermont's abandoned farms as both financial investments and vacation opportunities, urging well-to-do buyers to become absentee "gentlemen farmers" by hiring locals to manage their farms. The hope here was that absentee farmers would keep valuable land in agricultural production, creating job opportunities for rural residents and perhaps even modeling modern farming techniques to their neighbors.

It is difficult to know how many summer home owners continued to farm at least some of their land, but more than a few tried. If nothing else, hiring local farmers to cut hay from their pastures gave new owners a chance to protect distant views against encroaching forests and to maintain a pastoral feel on their property. That kind of arrangement seemed beneficial to all sides. As one observer wrote, "The millionaire — who can afford to do more, to build fine houses, breed high blooded horses and cattle, set out shade trees and farm for fun generally — he may find upon these farms the opportunity he wants to scatter his income, promote his health and happiness, and prolong his life."[43] For the wealthiest of summer visitors, farming "for fun" was a welcome diversion from their regular routines, and at the very least, a well-managed farm promised to pay for one's annual vacation expenses.[44] And, of course, whether visitors purchased an abandoned farm or one that was still inhabited, they were reminded of the importance of finding a reliable and knowledgeable labor force. According to one journalist writing in 1908, it was "useless to look to French Canadians or Italians as foremen. The best scheme is to engage as foreman the very man from whom you buy your farm."[45] With more than a touch of prejudice and bitter irony, such advice draws stark attention to the scope and scale of change in rural communities as former farmland was converted to new uses.

In addition to generating economic opportunities, the presence of summer homes in rural communities was expected to revive social institutions and bolster civic pride. Indeed, summer people often participated in social organizations and contributed to the improvement of schools, churches, and libraries.[46] Such changes, optimists felt, strengthened the likelihood that young Vermonters would remain at home, overturning that which had originally prompted "farmers' sons to forsake the scenes of their childhood

and to try fortune abroad." Whether real or imagined, visitors were expected to "bring culture into the rural districts . . . and an influence towards
appreciation of beauty and goodness is exerted over those who live permanently in the region."[47] By reviving social institutions, civic pride, and an
appreciation for "beauty and goodness," then, the sale of abandoned farms
as summer homes promised also to revive the very best in community
spirit — the very best of typical Vermont. The object was not to overrun traditional culture with new people and new ideas but to use the abandoned
landscape as a catalyst for change, to blend new and old in ways that made
year-round residents more likely to stay in their home communities.

Finally, summer residents were expected to improve the overall appearance of rural communities, anchoring the benefits of reform in the visible
landscape itself. This was true both in Vermont and in New England as a
whole. As one optimistic observer noted of summer home ownership in
northern Massachusetts, "To begin with, the lazy, shiftless, and ignorant
farmer, whose premises are an eye-sore to every passer-by, and whose dull
effort returns but the most meager existence, is being crowded out, and
forced to seek a place better disposed to tolerate him."[48] By driving out
those who did not appreciate the attractive qualities of their surroundings,
summer home sales promised to scrub away any visible signs of decay, bringing out the latent charm and beauty that many saw in the region's built
landscape. This certainly happened in many summer communities, as visitors cleaned up old abandoned places, giving them a renewed, tidy, and inhabited look. Thanks to summer people, "public and private grounds had
been made unusually attractive in simple but systematic ways," argued an
observer in the *New England Magazine*. "This was noticeable not only in
the village itself, but along the roadsides in the outlying parts of the town,
conveying the impression of general thriftiness and growing prosperity, far
indeed from anything like that of degeneracy."[49]

And that reversal of degeneracy was just the point. To survive, it seemed,
turn-of-the-century typical Vermont needed an infusion of new money,
new people, and new ideas. Perhaps a limited conversion of abandoned
land from work to leisure was just what was needed to make that happen.
Perhaps the abandoned landscape offered more promise that it seemed. At
least that was the hope. As we shall see in the final section of this chapter,
however, not everyone was convinced that the sale of farms as summer
homes was a benign path to progress. As the 1890s passed into the early
decades of the 1900s, a growing number of Vermonters would begin ques-
tioning the wisdom of summer home sales and the effects they were having on landscape and identity in Vermont. Before we can explore why they

did so, we first need to consider what the summer home landscape looked like, and what effect its development had on social relations between visitors and residents in Vermont.

Gender, Class, and the Reworking of the Abandoned Landscape

Summer home sales transformed rural Vermont quite literally into a consumer landscape, where potential land owners shopped around for "ideal" abandoned farms and farm property. By converting those farms into a new leisure landscape, summer visitors used them to perform the gendered and class-based identities that guided their approach to rural land and life. Reworking the rural landscape to accommodate the demands of middle-class leisure required considerable effort, much of which was provided by women. The aesthetic they created reflects upon class- and gender-based interpretations of rural landscapes as well as upon the growing power of summer visitors to direct the future of work-leisure relations in rural Vermont.

When visitors thought about "ideal" abandoned farms and "ideal" summer homes, certain aesthetic qualities typically came to mind. Potential buyers combed the New England hills from the late nineteenth century forward, searching for vacation properties such as that described by author and part-time Vermont resident Sinclair Lewis:

For fifteen hundred to three thousand dollars, part down, you would, if you poked about long enough, find a hundred-acre farm with a solid old farmhouse of eight or ten rooms. It might have running water; it would not have a bathroom, electric lights, or a telephone. There would be one or two magnificently timbered old barns; one of them would make such a studio or minstrels' hall as to draw tears from Christopher Wren. It would be two thirds of the way up a mountainside, protected from too shrewd a wind but looking ten or twenty miles down a valley between hills checkered with pastures among small forests of pine, maple, and poplar. With luck, it would have a trout stream. There would be one or two or three miles of dirt road, narrow, crooked, very decent in summer, foul in early spring and late autumn. Five or ten miles away would be a [gracious] village . . . You can spend thirty dollars on modern improvements—i.e., fifteen for kerosene lamps and fifteen for a fine tin tank to bathe in—and have a stoutly comfortable place, and authentic home, for the rest of your life.[50]

An ideal abandoned farm of the sort described by Lewis required qualities such as expansive views, rich historical associations, good craftsmanship, and strong bonds between its built and natural forms. Expectant buyers searched statewide for places such as this, always hoping "that around the next curve in the hitherto untried road we shall come upon the place of our dreams—the ideal combination of brook and view and trees and pasture—and posted upon tree or fence will be the sign, 'For Sale. Inquire of _____.'"[51]

Once visitors found property that suited them (and ideally, property with a useable farmhouse still on it), their next step was to transform that property according to popular tastes in leisure.[52] Gender was a central context through which that transformation occurred, making the rural tourist landscape both a product of and a force for shaping gender identities among women vacationers in Vermont.[53] At the turn of the century, many middle-class Americans equated rural landscapes with femininity and urban landscapes with masculinity. As the geographer Francine Watkins has argued, rural places in Western culture are often "coded as static, traditional and feminine. As such the rural is one site where women exert an unusually large representational influence on the identity of a place."[54] That influence was often quite pronounced in Vermont, where patterns of summer home ownership reinforced traditional links between women and the rural landscape. As noted earlier, fathers of summer families often remained at home in the city to work, visiting their families in Vermont by car or by train on weekends and longer holidays. For many turn-of-the-century visitors, that would have felt like an entirely appropriate arrangement—one that reflected middle-class conceptions of domesticity in which men were expected to be in charge of work and public life and women were expected to be in charge of leisure and family life. When applied to the summer home, this arrangement felt particularly important for the welfare of middle-class children. According to one journalist, "A big city is no place for children in summer." Far preferable was the rural summer home, where children could "live outdoors and become healthy, strong, useful, and good natured." Mothers should foster an interest in outdoor activities, exercise, and nature study, he advised, using the summer property to rear well-adjusted and healthy children.[55]

Many were undoubtedly pleased with this kind of arrangement, but not all women were happy with the idea of spending a comparatively isolated summer in rural Vermont. Although some well-to-do summer residents brought domestic servants to Vermont to help run the summer home, rural communities still offered fewer amenities and fewer social opportunities than urban New York or Boston. Many women undoubtedly missed their husbands and their close friends and confidants back home; one woman in southern Vermont missed New York society so much that she convinced

FIGURE 2.3. An abandoned-farm-turned-summer-home, c. 1915. This summer home, called "Cold-brook Cottage" by its owners, is representative of many others found throughout northern New England at the start of the twentieth century. Although it was once a working farm, Coldbrook Cottage's tidy grounds no longer suggest any connection to the daily routines of farm life. From the collection of Hildegarde Hunt von Laue.

her husband to sell their abandoned farm to another buyer.[56] But other women saw the summer home as a liberating opportunity to foster an independent space and an independent identity for themselves. As one New England historian has found, summer homes offered women unique opportunities to enhance their creativity and self-esteem, particularly in comparison to their lives back home. Vacationing mothers in Vermont formed casual social networks with other summer residents nearby. They planned new kinds of outdoor activities for their children, and they relaxed their patterns of dress. In the absence of their husbands and in the absence of urban social conventions about things such as clothing or outdoor physical labor, vacationing women often enjoyed new freedoms and a new degree of power and self-confidence.[57]

Women also assumed a measure of control over the physical transformation of Vermont's abandoned farms, moving those farms away from their associations with agricultural work and more toward the recreational needs of their vacationing families. Meeting the needs of their families always demanded time and energy from middle-class women, and the summer home was no exception. Women took an active role in building and remodeling their vacation homes, in addition to running them once they were

in use. Women worked hard to insure that summer homes reflected well
on themselves and their families, whether by supervising local workers, co-
ordinating the efforts of full-time domestic workers, or adding their own
labor to the house and grounds. Naturally, this required a great deal of effort. Converting former farmhouses into summer homes often involved
tearing down decrepit sections of the house, adding porches and windows,
installing new water and sewage systems, and in some extreme cases burning homes to the ground and starting all over again. And whether one was
remodeling an old farmhouse or building a new cottage on former farm
property, the grounds always needed attention as well. Where would flowerbeds be located? How much of the property should remain open, and how
much should be allowed to grow back to forest? For help, turn-of-the-century
vacation home owners could turn to magazines such as *Ladies' Home Journal, House Beautiful,* and *Country Life in America.* Scores of articles in
publications like these offered women a steady stream of advice on how to
create a summer home that was not only comfortable but reflected the
class status and good taste of its owners.

Vermont's abandoned landscape became a palimpsest on which vacationers inscribed a new rural aesthetic based on leisure and consumption
rather than on productive agricultural work. As the scale of their efforts grew,
as summer homes spread into communities statewide, vacationers exerted
an increasing degree of power over landscape and identity in rural Vermont. They expressed that power, in part, by transforming older vestiges of
their property's identity in favor of new ones. The process of naming the
summer home offers a good example. Rural Vermonters typically referred
to homes informally by the names of their owners — the "Lowe place" or
the "old Abel farm." Naming a farm for its inhabitants connected Vermonters with place, community, and a sense of generational continuity. Summer visitors approached naming differently. As national women's magazines reminded readers, choosing a name for the summer home was not to
be taken lightly, for that choice would define the identity of a place and its
inhabitants. One journalist cautioned owners to avoid "cheap" and "hideous
combinations" that would "show poor taste." Another suggested looking to
the surrounding countryside for creative names like "Nearthebay," "Uplands," "Meadowbank," and "The Downs."[58] By assigning formal names to
summer homes, and in some cases by displaying that name on a sign out
front, vacationers symbolized their ability to redefine the identity of rural
property in Vermont.

Summer residents also redefined their property by converting old spaces
to new uses. At times they did so by reversing practical changes made by

former owners in favor of their own ideas about what constituted a proper farmhouse. One journalist reported how his family reopened bricked-up fireplaces that former owners had forsaken, presumably in favor of more heat-efficient woodstoves. Opening the old fireplaces promised to "restore the interior to nearly its original form," mapping a new aesthetic onto the interior of the house.[59] Visitors also adapted farm outbuildings to serve new recreational purposes. A maple sugarhouse might become a painting studio, a chicken coop a writer's getaway. The same journalist reported how he converted a corncrib into a "play-house" for his children and in so doing made the corncrib "a never ending source of fun and pleasure."[60] On another abandoned farm, "the low rambling shed connecting the [house and barn] was turned into a kitchen, lavatory and large closet for tennis racquets, golf clubs, sport clothes, etc."[61] Changes such as these transformed former spaces of work into new spaces of leisure, reworking the identity of farmhouses and farmyards in telling ways.

Never, though, did the vacationer wish to erase the past entirely from their farms. The abandoned landscape was not a blank canvas but was instead coded with an agricultural identity often defined according to traditionalized conceptions of typical Vermont. That identity was partly responsible for drawing visitors to Vermont in the first place, and it was therefore something that summer residents hoped to absorb into the overall feel of their summer home. That desire was particularly strong among visitors who purchased property with a farmhouse still on it. For some, the fact that typical Vermonters had once lived in their summer homes gave those homes an attractive cultural essence and a rich set of historical associations. Abandoned farmhouses offered visitors an opportunity to glimpse, first hand, the impressive work of Vermont's Yankee homebuilders, whose skill, craftsmanship, and determination seemed to be etched into the very stability and simplicity of hand-hewed beams and straight rooflines. Indeed, one argued, every feature of New England's farmhouses "showed stability and integrity and was in unimpeachable taste."[62] Many visitors felt that summer property should be developed in such a way as to preserve that taste, and at times summer people expressed a perceived need to act as caretakers for the state's cultural heritage. "The spirit of the builders was there," one noted. "It was our job to bring it forth."[63]

Women had a special role to play by making farmhouse décor a model of simplicity and tactful thrift, thereby reproducing the popularized heritage of the Yankee farmers who built them.[64] This applied to the grounds as well, where women were reminded to resist the mistaken tendency to impose the formal style of city gardens on the "naturalness of the great outdoors." Instead, they were told, "Let us have our country homes surrounded by the plant life of the country!"[65] "An old farmhouse usually has a more lovable atmosphere about it than a new one," wrote another woman for the *Ladies' Home Journal*. "In repairing ours we tried to find out what gave it this homelike quality, and we studied to keep it. I believe it lies not only in the proportions of the house itself, but also in its relations to the land about it."[66]

Whether remodeling an abandoned farmhouse or building a new summer home on former farm property, middle-class women were often at the forefront of an emerging rural aesthetic. As overseers and as active laborers, women mapped new leisure identities onto former working spaces, redefining those spaces according to broadly conceived class- and gender-based conceptions about the rural landscape. Their efforts produced more than a new look in rural Vermont. As we shall see, they also produced new kinds of social relations between summer residents and their rural neighbors.

Coming Together and Keeping Apart

Summer home sales were a key component in Vermont's turn-of-the-century tourist trade. Although no official records were kept as to the total numbers of farms or the total farm acreage sold for use as summer homes, conservative estimates from the late 1930s revealed that between 150 and 200 properties were being sold to non-residents for permanent or summer homes each year.[67] Those kinds of figures were also likely to have applied to prior decades, as Vermonters had been consistent in their efforts to attract new property owners to the state. In 1911 the state legislature relieved the VSBA of its tourist duties by creating the Vermont Bureau of Publicity to handle all manner of tourist promotion. Over the next two decades, the bureau produced an elaborate line of promotional material, including substantial book-length guides with titles like *Vermont: Designed by the Creator for the Playground of the Continent* and *Vermont: The Land of the Green Mountains*. Local and private promoters joined the bureau's efforts as well, raising the availability of tourist literature with each passing year.[68]

In time, summer home sales generated distinct "summer colonies" in lakeside and mountain towns across the state. Summer colonies were localized clusters of summer homes — abandoned farmhouses and new cottages alike — that were often owned by close-knit groups of friends and families and often aligned along a particular road, valley, ridge, or lake. Such

places typically formed from a chain-migration of visitors. A pioneer summer family would purchase property and then invite friends and family to
visit. Taken with what they found, these guests would then begin their own
hunt for nearby abandoned farmhouses or abandoned property on which
they could establish a vacation home of their own. Some colonies, like the
well-known summer community in Dorset, became known as havens for
intellectuals and artists.[69] Others were identified by their associations with
lakes such as Fairlee, Dunmore, Bomoseen, and Champlain. These lakeside communities typically consisted of a mix of rental and private properties and were denser than summer colonies that crystallized out of proxi-
mate abandoned farms. But whether a summer colony was located on
lakesides or along a back country road, they all created new social dynamics in their host towns. Summer residents and rural Vermonters would now
have to figure out what it meant to be neighbors.

   The well-known summer colony along Caspian Lake in the northern
Vermont town of Greensboro offers a good example of how a colony formed
and how its members fit into local society. During the 1890s a chain-
migration of intellectuals from Princeton University and Harvard University began purchasing lakeside property and constructing cottages along
Caspian Lake. Within a generation, Greensboro's summer colony had become known as a home to writers, artists, and members of the educated
elite from some of the nation's most prestigious institutions. Clustered com-
pounds of families and close friends sprouted along the lake's shores, where
summer visitors wrote, painted, picnicked, hiked, swam, and sailed, while
other friends and relatives purchased homes in the village center and farmhouses in the surrounding area.[70]

   This influx of visitors, Vermont author Lewis Hill recalls, transformed
his home community forever. Hill's recollections of his Greensboro boyhood in the 1930s describe a rural society defined by hard work, tradition,
and insularity on the one hand and change and uncertainty on the other
as residents adjusted to cars, radios, electricity, and, not least, summer
people. Greensboro's "campers" — as they were known locally — elicited a
number of responses from the town's year-round residents. Some liked having summer people around, partly for economic reasons, and partly because
they were a novelty. Others resented the lack of parking in town during
summer and complained about finding the store and post office crowded
with strangers. Worse, many were dismayed by behavior that seemed to test
the bounds of propriety: that fact that some female visitors drove, wore shorts,
and smoked, Hill recalls, filled certain locals with shock. As a simultaneous source of interest and distrust, then, summer visitors became a wel-

come addition to the town as well as a nuisance that some were happy to see leave again each year.[71]

Responses like these were fairly typical in towns throughout Vermont and elsewhere in northern New England as visitors and year-round residents learned to interact across what must have often felt like a wide cultural divide. At times, those interactions occurred through the context of leisure. Lewis Hill recalls playing with visiting children, for instance, and he recalls how summer people mixed with residents at community dances and other social events.[72] Likewise, summer visitors recall residents allowing them access to their land for picnics, accepting visitors, in one's words, "with a degree of patient kindly tolerance."[73] At other times, those interactions occurred through the contexts of work, employment, and buyer-seller exchanges. Most summer residents were dependent on local labor and local knowledge to remodel and repair their homes, do the heavy work of landscaping, and stock woodsheds and ice houses in the off season.[74] In addition, summer residents were dependent on local shopkeepers for groceries and on their country neighbors for eggs, milk, and vegetables. Quite simply, summer people could not have survived without the help of local farmers.

For this reason alone, summer people and rural residents could not have avoided interacting with one another, even if they wanted to. Work-related encounters between visitors and residents offered opportunities for the two groups to get to know one another, although at times those encounters were not entirely smooth. Employee/employer interactions at the summer home could be quite complex, some journalists warned summer visitors, and social relations between the two could turn sour fast if they did not learn to treat their hired hands properly. In a 1938 *Harper's* article, "How to Live Among the Vermonters," social critic and summer home owner Bernard DeVoto warned visitors to be careful not to treat the natives as servants but as something more akin to neighbors who were helping around the place more out of generosity than economic need. Typical Vermonters will not bow to their employers or change in any way to satisfy summer residents, he advised. Instead, summer people must change their expectations of employees if they wish to live successfully among their rural neighbors.[75] Those who treated their hired help with respect, another commentator advised, could expect rewarding relationships and rewarding encounters with rural culture. As *House and Garden* magazine explained, urban residents who treated their rural employees as equals and who adopted rural standards of work would find themselves surrounded with "native folk who understand you and whom you understand."[76] As a testament to that and related claims, one summer resident explained how she had carefully nurtured her buyer/seller relationship with a local berry peddler, until she finally earned approval and acceptance by the local woman, who dryly noted, "I do like you, you're so common."[77]

But although visitors and residents often came together — whether through the context of work or of leisure — they often kept apart from one another as well, maintaining different outlooks on landscape and identity in rural Vermont. Their separation was enhanced, in part, by the clustered (and sometimes cloistered) nature of Vermont's summer colonies. Tight-knit and often contiguous colonies allowed visitors to withdraw from their local Vermont communities if they wished, becoming socially (if not economically) self-sufficient entities. As one summer resident recalled of his colony years later, "We were very, very provincial. We were a very tight community within ourselves."[78] Indeed, it is easy to imagine tight-knit groups of vacationers who were quite content to socialize among themselves and encourage the growth of their summer communities along established social lines. Like all social groups, many summer people would have preferred to remain in the company of people they knew and liked.

It is also easy to imagine that some rural residents looked on summer colonists as being elitist and exclusionary to the point of producing antagonism and resentment. That was the message of journalist Bruce Bliven's sharply worded 1923 article "Rock-ribbed," published in *New Republic* magazine. Although Bliven was not a native Vermonter, he assumed the voice of a native in an attempt to demonstrate the opinions some Vermonters held about social relations between themselves and vacationers. "Long ago, the village consisted of the general store and the livery stable," he wrote. "Today it consists of the general store and the garage, plus, from the day of the first dandelion until the day the snow flies again, the tea-room up on the hill in the old Oliphant place. The tea-room might as well be in Albuquerque, for all the use we natives make of it."[79] Of course, the fault for not visiting the new tea room at the "old Oliphant place" may well have rested as much with rural provincialism as with the provincialism of summer visitors. Yet the hesitation and impression of exclusion to which this example points, and the suspicion and concerns that some residents in places like Greensboro felt toward their town's summer colony, should draw attention to the degree of disconnect that was developing between visitors and residents. Each side, it seemed to some critics, was staking out their own social and physical terrain, and as they did, their actions raised new and troubling questions about the relationship between tourism and typical Vermonters.

FIGURE 2.4. Vermont farm family, no date. For many turn-of-the-century farm families, summer home owners brought new economic opportunities in the sale of goods and services. They also brought a new social order and competing visions about the meaning and value of the ground on which they stood. Courtesy Vermont Historical Society.

The Threats of Summer Homes

Turn-of-the-century summer home owners spruced up farms and villages in Vermont, and in some cases they strengthened local churches, schools, libraries, and civic organizations through generous investments of time and money. In this regard, summer home sales produced the kinds of results that many advocates had in mind. But by the 1920s Vermonters had a few decades of experience with summer homes to look back on, and what many of them saw was not always good news. To begin with, many abandoned farms remained unfilled (by one 1929 estimate, Vermont had one thousand), and young people were still leaving Vermont in numbers greater than state officials would have liked.[80] Equally important was the fact that summer communities and year-round communities did not always feel quite compatible with one another. In fact, some were now arguing that summer home sales threatened the entire fabric of "typical" rural life in Vermont. Some social critics tried to address that concern by re-grounding summer home sales in their original reformist mission. But that, it now seemed, required a more careful approach — one informed by both subtle and overt attempts to prevent the "wrong" kinds of vacationers from buying property in Vermont. As a consequence of that approach, Vermont's abandoned landscape became implicated in the reproduction of broader prejudices in American society rather than the celebrated ideals of democracy and harmony so often defined as typical of rural Vermont.[81]

Vermonters of all backgrounds and professions tended to arrange vacationers hierarchically. For many, property-owning visitors ranked at the top of the list, as compared to resort-goers or transient automobile tourists (about whom I say more in chapter 3). On the whole, some Vermonters reasoned, visitors who purchased property seemed to come from a wealthier, more educated, and more desirable class. They took a greater interest in their adopted summer communities, they contributed to the local tax base, and they spruced up places that had once been eyesores. Many summer home owners agreed, and some formed home owner associations to protect their interests and their particular brand of leisure. For example, Greensboro's summer residents (joined by a handful of year-round residents) founded the Greensboro Association in 1933—a group, The Vermonter reported, that "seeks to preserve the inexpensive simplicity and charm of the Caspian Lake colony." Its goals were to "foster the desirable features of the community life and combat any disruptive or unwelcome innovations. It will use its influence, one may rest assured, to prevent the erection of any cheap dance halls or noisy amusement-park devices and will wage a fight against any unsightly displays of advertising."[82] From this writer's perspective, then, non-property owning visitors whose vacationing tastes did not match that of association members would now be less welcome in Greensboro.

Although they rose to the top of the vacationers' hierarchy, not all potential property owners were equally desirable to all Vermonters, for not all were deemed to be a good match for the typical Vermonter. First, there were issues of ethnicity to consider. In an echo of Vermont's ethnically charged Swedish immigration experiment during the 1890s, some early-twentieth-century Vermonters targeted homebuyers based on their ethnic background. The efforts some Vermonters made to prevent Jews from vacationing in Vermont express this clearly. Jews were historically a distinct minority in Vermont, although they settled in the state as far back as the 1850s, finding work as traveling merchants and shop owners and creating small but reputable Jewish communities in towns like Poultney, Rutland, and Burlington.[83] By the early decades of the twentieth century, Jews (drawn primarily from regional cities) were turning to Vermont for vacations as well. Although Vermont never developed the same popularity among vacationing Jews as New York's Catskill Mountains, they came to Vermont nonetheless, both as individuals and in groups, and they ultimately succeeded in making the state a summer home for Jews as well as Gentiles.

Vacationing in Vermont was not always easy for Jews, many of whom were plagued by the dogged anti-Semitism that marked so many aspects of early-twentieth-century American culture.[84] Prejudice was a fact of life in Vermont's tourist trade, and the state was a well-known bastion of anti-Semitism during the early decades of the century. Vermont had no state-level anti-discriminatory laws in place for hotel owners until the late 1950s, at which time one survey found that 45 percent of the state's hotels still actively discriminated against Jews.[85] To minimize the number of Jewish visitors to Vermont and to control their effects on the state's tourist identity, anti-Semitic Vermonters knew they had to maintain control over the rural tourist landscape itself. During the early decades of the century, state-sponsored advertisements for private cottage rentals sent both subtle and overt signals to prospective Jewish visitors. Advertisements published by the Vermont Bureau of Publicity, for instance, specified cottage owners' preferences for a "Christian clientele" or for "gentiles only."[86] Hotel owners in Vermont also asked for "references" as a means for subtly weeding out unwelcome guests. "We must be sure," the manager of the prestigious Lake Champlain Club explained, "that the applicant will fit into the scheme of things."[87] Of course, fitting into the "scheme of things" could have many different interpretations — ethnic and otherwise — but to Jewish travelers in the 1920s and 1930s, the deeper message of exclusion embedded in statements such as this would have been clear.

Anti-Semitic entrepreneurs who limited access to rental and resort properties also exerted some degree of control over the sale of vacation property in Vermont. A casual visit to the state, after all, was known to be the first step toward purchasing property.[88] Some Jews did make that step from casual visitor to property owner during the 1920s and 1930s. In Wilmington, for example, Jewish visitors purchased the town's former Forest and Stream Club — once a haven for Gentile businessmen — and developed it as their own private summer resort, known among some locals as "Jew Town." Examples like that, and knowledge of Jewish summer enclaves in New Hampshire and the Catskills, would have been a call to arms for residents who were eager to make sure that property fell into the "right" hands. Among at least one group of Vermonters, for instance, the threat of having a Jewish presence in their town prompted coercive action. During the 1920s a development company owned by residents in Greensboro and Hardwick wrote into all of its property deeds the following: "No part of the herein conveyed property shall be leased or sold to any member of the Hebrew race."[89] Obviously, one example does not prove a larger rule. But the 1920s were a time when Vermonters earned regional notoriety for the strength of the Ku Klux Klan in their state. And as author Lewis Hill recalls, some young Vermonters during the 1930s openly supported the German-American Bund, a pro-Nazi group formed on American soil.[90] Overt and subtle signs of prejudice would have made it clear to outsiders that some kinds of visitors were more welcome than others and that Jews, in particular, were not considered a good match for the kind of typical cultural identity that many hoped to perpetuate in the state.

In light of such anti-Semitism, it is not hard to imagine that some Jews failed to find virtue in the idealized conceptions of typical Vermonters that were so often used to promote the state. In a tongue-in-cheek article that nonetheless carried serious undertones, one 1930s Jewish visitor recounted the skepticism of his parents when he decided to purchase property in Vermont:

My father stood by, repeating to himself. "No telephone, no lights, no running water."

"People have lived there before, Papa."

"Sure! But not Jewish people!"

"What's the difference?" I demanded.

My father looked at me. "My son," he said sorrowfully, "if a young man of thirty five don't know yet what is the difference between Jewish people and the *goyim* [Gentiles], then it's sad."[91]

Although the author of this article never condemned or praised his rural neighbors directly, the concerns he recounted in this passage reflected a clear sense among some Jews that Vermonters could be less welcoming, noble, or democratic than their ideal image suggested.

In addition to ethnicity, discussions about the desirability of summer residents were also shaped by class and occupation, and, by extension, by ongoing changes to work-leisure relations inspired by the rural tourist economy. Although Vermont had well-to-do property owners who built mansions and estates in towns like Manchester and along the shores of Lake Champlain, the state never developed the same kind of upper-class status as Newport, Rhode Island, or the lower Hudson River valley. Most of Vermont's summer residents were middle- and upper-middle-class families of professionals for whom modest and tasteful displays of wealth on rustic-looking farm properties were preferred. Nonetheless, these were people of a different class — people who were able to afford a second home and enjoy months of leisure time in Vermont each year. That alone was enough for some rural residents to wonder with suspicion whether the visitor's use of their property for leisure was not undermining traditional patterns of farm work and the traditional work-based identity that so many rural residents embraced. Again, work-leisure relations were moving to the forefront of discussions about rural tourism, in this case as related to questions of class, summer home ownership, and the typical Vermonter.

Summer residents often took a genuine interest in rural work as part of their summer vacation, visiting local farms and even, as Lewis Hill recalls, sending their teenaged sons to help their farm neighbors—an experience some felt would be good for the boys.[92] Another summer visitor recalls helping local farmers with tasks such as removing potato bugs from plants but added that she and her friends soon grew tired of the work and left it to them.[93] Interactions with local farmers aside, summer residents' encounters with rural work frequently occurred on their own summer properties and in this way merged conceptually with their idea of rural leisure. Puttering around the house and grounds, chopping a bit of wood, clearing some brush at the edge of a former pasture, and maybe even joining in to help local employees with a job here and there could feel like a personal, if imagined, engagement with rural farm culture.[94] For urban professionals who spent much of their time indoors, work like this—at least in moderation— could actually be a fun way to escape their normal working routines. And in this sense, work performed on the summer home was very different from that of their farm neighbors. Summer home owners worked on their properties as part of a larger engagement with leisure, not because they needed to. Theirs was work done to improve and maintain the appearance of a *second* home, not work done to pay for or keep their only home. In this sense all summer residents, regardless of their status back home, were in a different social and occupational class than their rural neighbors.

No matter how strongly and genuinely visitors and residents may have wanted to bridge the cultural divides between them, this fundamental difference in how they experienced the rural landscape remained a chasm that was difficult to cross. Turn-of-the-century reformers and state officials had always hoped that the sale of farms as summer homes would reenergize rural communities, thereby reinforcing and reproducing the very best in Vermont's cultural identity without radically transforming the nature of rural society. By the first decades of the twentieth century, however, farmers and reformers alike were beginning to express concerns about the lasting consequences associated with converting farms to summer homes. Indeed, if you asked rural Vermonters to describe their definition of an ideal, "typical" rural community, they would not likely have described a world where farms were sold to non-Vermonters who let most of their land grow back to trees or who "played" at work on their property. Moreover, the ability and willingness of Vermont's visitors to put their own work aside for extended periods each summer would have struck some as threatening to the kind of work ethic they wished to instill in their children. Lewis Hill recalls neighbor boys from the summer colony suggesting to him that they sleep in the hayloft — something that had never even occurred to Hill before — and then convincing him to jump in the hay. Although Hill recalls having a good time, he felt compelled to keep their "horseplay" secret from his parents, who would not have approved of such things in the barn.[95] Even officials from the United States Department of Agriculture worried about the "examples of extravagance" that summer people brought into rural communities, and the tendency for those examples "to create unrest among farmers' boys and girls."[96] Farm children, such concerns suggested, were expected to embrace a world defined by hard work and modest leisure, not a world where one went to the country to rest.

Some pointed to summer people to prove their case. Just look at the consequences that a leisure lifestyle had on the vacationer, journalist Bruce Bliven argued on behalf of the native Vermonter. These were soft-handed, lazy people, he wrote — people who were unforgivably ignorant of country ways:

How is it possible for anyone to be so preternaturally stupid and yet possess so much money? By any standard of the North these folk are weaklings. They cannot pitch hay; they are grossly ignorant of every process of farm life; they are not shrewd, for they pay without a word whatever price is demanded of them — yet they have the limousines and we have not, and from the pockets of their men folk come inexhaustible wads and wads of green or yellow bills, a magic supply.[97]

What truly mattered from this perspective, then, was not one's financial wealth but one's intimate knowledge of rural land and life — a knowledge that could only be gleaned through a work-based encounter with the rural landscape. No amount of money in the world could reproduce that encounter.

Criticisms of summer home ownership and its effects on the farm economy were also expressed by those who argued to limit the sale of farms in certain areas of the state. Some Vermonters conceded that summer homes made sense on upland farms, where soils were thin and farming was hard. But they were not willing to advocate the spread of summer homes onto more productive farmland or into towns that still had a relatively strong farm economy. As one member of the Vermont Farm Bureau argued, "Let's get summer people on the hill farms, like Ripton, Goshen and places not adapted to farming. But do not let's try [sic] and encourage them to live in good agricultural towns. There is a morale that you are like [sic] to break down when you get country gentlemen farmers in among you. I would rather see Vermont progress along agricultural lines than become the playground for New England."[98] Advocating a controlled sale of farms, then, this critic reminded his audience to treat summer homes as a supplement to a working farm economy, and he urged caution about selling farmland too quickly to visitors. Of course, no one could keep Vermont farmers from selling their land to summer people if they wanted to, especially if that land had been abandoned and especially if the selling price was higher than the land was worth for farming.

Arguments like these suggested the presence of deep concerns among Vermont farmers about the profound effects that summer home ownership was having on Vermont. Sharing that concern were sympathetic intellectuals and reformers such as the author and Vermont transplant Zephine Humphrey. In 1923, Humphrey published an article about summer home owners in *Outlook* magazine entitled "The New Crop." In her article, Humphrey outlined the larger transition occurring in Vermont away from farming and toward tourism, and she raised questions about the wisdom of Vermont's "wholesale giving over to the service of the tourist." The handfuls of well-meaning outsiders who purchased abandoned farms a decade or two before (like herself) were one thing, she argued. But in recent years an uncomfortable number of houses had been bought up and an uncomfortable amount of farm acreage had been divided up to build new summer homes. There seemed to be too many people buying and building in Vermont, she worried, causing irreparable changes to the landscapes and social structures of communities statewide. Much of her concern stemmed from what she saw as the summer resident's all too common lack of concern for rural traditions — a pattern she found reflected in the summer home itself:

The summer tourist takes little heed of the real nature of the social life in the midst of which he spends his vacations. Oh, he thinks he takes plenty of heed when he buys or builds him a house! The more intelligent he is, the more carefully he conforms to old fashions . . . He knows how to achieve the most thoroughly charming, harmonious effects. But they are not always quite real, these effects; paradoxically, his harmonies do not always ring true.[99]

More troubling still was the lifestyle that many summer visitors now pursued — a lifestyle of leisure rather than work, of self-indulgence rather than modest sobriety:

Moreover, when he has finished and furnished his house, what kind of life does the summer resident proceed to live in it? A plain, old-fashioned, laborious life, full of manual work, bounded by early hours, instinct with simple neighborliness? That is the kind of life the native New Englanders, when left to themselves, still lead, thus proving themselves true to the spirit of the old days, even though they have cast aside many of the old trappings. But the summer resident stops after making his environment what seems to him suitable, and his clothes and habits conform, not to the old days, but to the very new.[100]

For Humphrey, then, there was reason to think critically about the kinds of relationships that were developing in Vermont, and about the negative consequences that summer home ownership seemed be having on landscape and identity in the state. But, she argued, there was still hope. If this "new crop" were to grow in a positive way — if they were to reinforce rather than undermine Vermont's popularized cultural identity — they would have to leave behind their city ways, "become simpler in life and in dress," and "refuse to give elaborate entertainments or to wear sophisticated clothes in their cottages."[101] Said another way, the new crop would have to make a more concerted effort to conform to an idealized vision of the typical Vermonter.

And that was where class again intersected with discussions about work-leisure relations and the desirability of summer home owners. Outside of changing offending vacationers' behavior, the next best way to mitigate the effects of summer home ownership in Vermont was to insure that summer homes were sold to the "right" kinds of people. Just as certain ethnic groups seemed more and less appropriate in Vermont, certain social and occupational classes seemed more and less appropriate as well, more and less likely to contribute to a social climate in which the typical Vermonter's cultural identity was preserved.

Recommendations made by the VCCL's Committee on Vermont Traditions and Ideals on summer home ownership provide a good example. The committee consisted of eleven Vermont residents — at least six of whom were writers or professors — and its task was to "find ways and means of preserving the distinctive values . . . which have to do largely with immaterial rather than material possessions of any kind, and which are easily recognized in the history, literature, and the general life of the people."[102] Said another way, their task was to protect and perpetuate what they saw as the very best of Vermont culture, the very best of typical Vermont. That task involved protecting Vermont's cherished traditions and values *from* tourism as well as *using* tourism as a means for reproducing them. From the committee's perspective, some vacationers were better for this than others: "It is apparent," committee members noted, "that the state offers a pleasant environment for authors, artists, college teachers, and others in the same general classification." They added: "It would be fortunate for the state and its people if more and more men and women of this desirable type sought Vermont for summer or permanent homes. They are far more valuable to Vermont as summer residents or as habitual dwellers in the state than other classes that might be mentioned."[103] Through recommendations like this, then, the committee members positioned themselves as self-appointed representatives of Vermont opinion and as watchdogs of the state's cultural heritage. By making a determination as to whom the best summer home owners were, they placed themselves in an intermediary position between the visitors they sought to attract and the rural people they sought both to valorize and to protect.

Among the committee's members was one of Vermont's most famous writers, Dorothy Canfield Fisher. Fisher was born in 1879 into a Midwestern academic family with ancestral roots in Vermont. After earning a Ph.D. in French from Columbia University, she and her husband moved to her family's old homestead in Arlington, Vermont. For decades, her novels and articles earned her a national reputation, and her promotional pieces made her a powerful and earnest voice in the creation of Vermont's public image.[104] Fisher's interpretations of Vermont culture were often colored by a reverence for the past and in this way were not unlike those produced by others who were writing about Vermont in the early decades of the twentieth century. In one statement on the character of Vermonters, Fisher recounted with pride her grandfather's suggestion that Vermont be turned into a national park "so the rest of the country could come in to see how their grandparents lived."[105] A strong reverence for tradition and stylized conceptions of rural Vermonters informed Fisher's writings, but that, of course, does not mean that she was prejudiced. In fact, Fisher was a well-known crusader against racism and anti-Semitism, both inside and outside Vermont.[106]

But when it came to summer home sales, even Dorothy Canfield Fisher felt it necessary to target the "right" class of people. That was the message, for instance, of a 1932 booklet on summer homes she wrote for the Vermont Bureau of Publicity. This thirty-page promotion, called simply *Vermont Summer Homes*, echoed many of the sentiments expressed by the Committee on Vermont Traditions and Ideals. In the booklet's first lines, Fisher explained that she was addressing a very specific audience — one that included "those men and women teaching in schools, colleges and universities; those who are doctors, lawyers, musicians, writers, artists — in a word those who earn their living by a professionally trained use of their brains."

In a non-exclusionary gesture that with hindsight falls short of the mark, she also welcomed an audience of "those others not technically of that class but who enjoy the kind of life usually created by professional people. If your tastes, your outlook on life are generally in common with the classes I have named, please consider yourself one of my audience." These were the people, Fisher explained, who would find Vermont most appealing and most welcoming. But that, of course, raises a question: If one's tastes were not of that class, should one consider oneself unwelcome by Fisher, by the Vermont Bureau of Publicity who published her work, and by nearly everyone else in the state of Vermont? By the tone of the letter it might have seemed so to some readers; after all, Fisher noted, her pen was merely that "which writes what is felt all over the State."[107]

Why were people who made a living through the "professionally trained use of their brains" so attractive to promoters like Fisher? For starters, some professionals (particularly college professors) had the kinds of work schedules that would allow them to spend all or part of their summers in Vermont. And as modest as their income might have been in many cases, professionals were often better off financially than other travelers for whom purchasing property was not an option. In addition, it would have made sense to intellectuals like Fisher and her like-minded neighbors in the Dorset area (including Zephine Humphrey) to attract more people like themselves to their communities. From the perspective of the professionals who tended to produce Vermont's promotional materials, other professionals would have been a welcome addition to Vermont society, even if only for the summer season.

Fisher's appeal to families of professionals made sense in another way as well. In her view and that of others like her, such families posed less of a threat to the future of typical Vermonters than others. In fact, Fisher argued, the professional and the typical rural Vermonter had more in common than they might expect: both had strong "mental and moral qualities," both valued family and friends, both were hardworking, both were modest, and both were beyond the shallow worship of money. As she told prospective buyers:

In many ways Vermonters will seem like country cousins of yours, sprung from the same stock . . . We simply love the fact that your women folks do not feel it necessary always to wear silk stockings, and that your men folks like to wear old clothes. You value leisurely philosophic talk and so do we. We like the way you bring up your children, and we like to have our children associated with them. You like to be let alone a good deal, and we like to let people alone. In other words *we like you*. And when a Vermonter admits that he likes somebody, it means a good deal.[108]

An old Vermont Cape Cod cottage-type of farmhouse recently acquired by PROF. WALTER HENDRICKS of Chicago. Prof. and Mrs. Hendricks, both writers, plan to spend their summers here, and perhaps eventually make it their permanent home. The old house has several fireplaces with old Dutch ovens.

FIGURE 2.5. Summer home promotion with original caption, 1932. This photograph and caption appeared in Dorothy Canfield Fisher's *Vermont Summer Homes*, an extended essay advertising summer homes primarily to an educated class of potential buyers. Buyers such as writers and professors shared a great deal in common with local residents, she argued, and were therefore preferable to others. Vermont Department of Tourism and Marketing, successor to the Vermont Publicity Bureau.

One can imagine that suggestions of compatibility between summer residents and their rural neighbors appealed to potential buyers, who obviously would have wanted to feel comfortable in Vermont. But they also appealed to those for whom the sale of summer homes was also an exercise in rural reform. Two generations of reformers had hoped that summer residents would help stabilize Vermont society, creating a climate where the best qualities of the typical Vermonter could thrive. In this sense, summer home owners and the leisure landscapes they produced on the state's abandoned farms *had important work to do* on behalf of Vermont's cultural identity. As Fisher explained, "Aware as all cultivated people are of the immense value of regional color in a standardized world, you will help us keep what we have, rather than making our young people ashamed of it, as do some of the ignorant among the well-to-do . . . You will value what is worthwhile in our inheritance, and so will help us perceive what is best in our traditions, and help us hold to it."[109]

Best of all, the benefits from that arrangement would apparently flow both ways. Summer residents would not only help Vermonters "hold to" their traditions, they would feel those traditions rubbing off on them as well. Through their intimate encounter with the rural landscape and the people therein, summer home owners would gain access to a world where typicality was still defined in terms of stable, time-honored relations between people and place: "There is something else of value we feel Vermont has to offer you and your children," Fisher told potential buyers, "something which the modern world seems determined nobody shall have. This is stability." Amidst a world of mobility and change, the Vermont summer home offered vacationing families a rooted sense of place. The summer home may be children's "only chance to learn how much richness and depth is added to life by *belonging* somewhere," Fisher told potential buyers. "Vermont towns and villages, you see, are above all static, [they] provide that experience of unchanging stability that is such a rest to nerves assaulted by the modern haste to change for change's sake."[110]

With hindsight we can see the irony embedded in statements like these. Fisher's claim that society and culture in Vermont were stable was challenged by the very task outlined for *Vermont Summer Homes*—that is, the sale of property to people from outside the state. Moreover, by maintaining the position that summer home sales would stabilize Vermont society and reproduce the best of typical Vermont, Fisher ignored the complex reality of decades of summer home ownership in her state. As growing numbers of properties made the transition from work to leisure, the reformer's dream of stability would become ever more elusive, and their best intentions for the future of Vermont would continue to be complicated by the social tensions that marked the state's tourist landscape.

Typical Vermonters and the rural ideal they embodied were decades-long sources of both celebration and concern. As we have seen in this chapter, promoters, reformers, and civic leaders often marketed the state's abandoned landscape to vacationers in an effort to create new social and economic opportunities, thereby bolstering the best of what they defined as a typical identity in the state. In the process, however, they set in motion a series of changes that posed new threats to the future of that identity. New kinds of vacationers and the Vermonters who encouraged them to come, it seemed, were reworking the rural landscape in ways that reproduced more than a celebrated image of a wholesome rural culture.

What emerged from the transition of abandoned farms into summer homes was a new kind of middle landscape — one where leisure's power to redefine spaces of rural work had grown, and one where little consensus existed about the wisdom and direction of that process. Whether one feared or embraced the profound and lasting changes associated with that middle landscape, few could help but notice the changes it had brought to early-twentieth-century Vermont. In particular, few could help but notice the growing economic potential of Vermont's *entire* rural landscape. Rather than leave that landscape as it was — trusting in its organic, scenic appeal — Vermonters and their visitors would begin transforming its spaces and meanings at dramatic new scales. Their efforts made the state accessible to millions of new visitors and set in motion intensifying debates about tourism and the reworking of rural Vermont.

CHAPTER 3

Accessing the Rural Landscape

Driving, Hiking, and the Making of Unspoiled Vermont

Vermont's tourist economy grew markedly—both in scope and scale—between the decades of the 1910s and the 1930s. Profits in the state's tourist sector increased steadily into the 1920s, and although the Great Depression of the 1930s cut into the total amount of money spent by tourists, the number of visitors who traveled to Vermont annually hovered consistently between 1,200,000 and 1,500,000 throughout the decade. Vermont's popularity was strong enough, it seemed, to weather tough economic times, and by 1941 industry observers were reporting that Vermont's tourist economy was bringing roughly forty million dollars to the state each year.[1] In addition to the ongoing trade in summer homes, growing numbers of tourists patronized new motels, cabin camps, and tent camps, pursuing a range of recreational activities such as automobile touring, hiking, horseback riding, boating, and bicycling. Activities such as these focused more attention on movement and on action than vacations spent on farms or socializing on the front porch of resort hotels. Vermont's early-twentieth-century tourist economy was growing and changing, as were the ways in which people defined and encountered the state's rural landscape.

As we shall see, automobile travel had a great deal to do with such changes. The percentage of visitors who drove to Vermont rose steadily during the 1920s and 1930s, opening more of the state's out-of-the-way places to more people than ever before. But Vermont's expanding tourist economy was also a product of the concerted, on-the-ground efforts made by Vermonters (as well as their guests) to create new recreational opportunities for automobile tourists, hikers, campers, boaters, and others. The spaces and identities they created and the social tensions inherent in their efforts are the subjects of this chapter.

Two keywords come up repeatedly in Vermont's tourist literature between the 1910s and the 1930s, and it is around these keywords that this chapter is framed. The first of these is "unspoiled." Not unlike their use of the adjective "typical," promoters at this time commonly deployed the adjective "unspoiled" to describe rural Vermont. As they had for decades, promoters continued to call attention to Vermont's lack of excessive urbanization, industrial development, and immigration — a condition many now described as "unspoiled." But additionally, promoters added, Vermont was unspoiled by excessive tourist growth. Vermont seemed to have escaped the worst expressions of tourist development, particularly by comparison to other, competing vacation destinations such as Maine, New Hampshire, Cape Cod, and Florida — places, commentators liked to point out, that were overrun by crowds, cheap hotels, billboards, amusement parks, and the often berated hotdog stand.[2] Somehow Vermont's rural landscape seemed to have escaped these symbols of poor taste and mass appeal. Somehow it still seemed as timeless, organic, and scenic as ever.

In 1931 the Vermont Bureau of Publicity officially adopted the adjective "unspoiled" into what would become a decade-long advertising campaign called, simply, "Unspoiled Vermont." In the opinion of Charles Edward Crane, author of the popular book *Let Me Show You Vermont* (1937), the campaign neatly captured the state's "chastity," its "rural simplicity and thrift." Visitors to Vermont, Crane argued, felt as if they had entered a timeless place where they could escape from the "sophistication, ambition, speed, crowd, and all the complexes of urban civilization." Of course, Crane admitted, Vermont had its problems and blemishes, but "the slogan [unspoiled Vermont] is still supremely good as a watchword against wantonness. It keeps uppermost in the mind of native and visitor alike the necessity of keeping Vermont as sweet and pure, as unadulterated, as her maple sugar."[3]

The second keyword found in tourist literature at the time is "accessibility." Even as tourist promoters warned of the threats of excessive and tasteless tourist development, they made the increased accessibility of Vermont's unspoiled mountains and valleys a primary focus of their tourist campaigns. Typically, the word "accessibility" referred to two different things. First, accessibility was defined quite literally by the ease with which visitors moved around in Vermont and by the numbers of tourists that the state

ONLY A STEP TO
UNSPOILED
VERMONT

AN UNCROWDED SCENIC FRONTIER

FOR HEALTHFUL COUNTRY LIFE AND FUN

FIGURE 3.1. Unspoiled Vermont as both accessible and distinct, 1939. The success of Vermont's "unspoiled" reputation hinged as much on its accessibility to travelers from the urban northeast as on its rural qualities. Vermont's accessibility to the city as well as its inherent difference from the city is captured nicely by this state-sponsored brochure. Courtesy Vermont Historical Society.

could accommodate. Although automobile travel and the state's proximity to regional cities put Vermont within easy reach of millions of potential travelers, those travelers also had to be able to get around *inside* the state once they arrived. That meant improving road networks for drivers as well as recreational networks for hikers, boaters, or horseback riders. But accessibility was also defined by the ease with which visitors encountered unspoiled rural beauty along these networks of roads and trails. Unfortunately, not everywhere in Vermont was as scenic as its unspoiled reputation suggested; correcting that fact was a central goal in the campaign to make unspoiled Vermont accessible to all its visitors.

Such a goal demanded that landscape and identity in rural Vermont be reworked on a truly statewide scale, thereby expanding the reach of tourism into every rural hamlet and every farmyard, along every main highway and every back road. Making Vermont accessible yet keeping its appearance unspoiled required willingness on the part of Vermonters and visitors alike to manipulate anywhere and everywhere in service of the state's leisure economy. It encouraged people to think in terms of traditional rural tourist discourses such as timelessness and organicism, but it also demanded that they transform the rural landscape to make it more accessible while simultaneously maintaining its unspoiled qualities. Their efforts added complexity to rural Vermont's reputation for organicism by implicating the state's landscape in a simultaneous and often quite frank embrace of development and change. In that embrace lay opportunities to expand tourism and to achieve one's goals relative to rural tourist development. It never paid to leave the state's rural landscape alone, yet it never paid to let it become "unspoiled" either. "Keeping unspoiled Vermont unspoiled," as one journalist called it, was an ongoing and interventionist activity.[4]

The efforts of those who engaged in that activity were complicated by two fundamental problems. First, unspoiled Vermont was a socially constructed concept, the definition of which was by no means agreed upon by all those involved. Second, few could agree on the degree or kinds of accessibility they should embrace in order to make the state more appealing to tourists. What, exactly, did it mean to be unspoiled? What kinds of recreational activities and landscapes should be made more accessible in Vermont? Who should be included in these efforts, and who should not? Just as a widespread willingness to manipulate the state's rural landscape complicated its uncontrived reputation, the debates that emerged from questions like these complicated its reputation for harmoniousness and stability. Rural Vermont remained a middle ground where social groups struggled to embed their perspectives on leisure and work in both the form and meaning of the state's rural landscape. By the 1930s they understood very clearly that their ability to do so was dependent on their willingness to get involved directly in the reworking of rural Vermont. As this chapter suggests, driving and hiking became central contexts through which this played out.

Accessing Unspoiled Vermont by Automobile

A great deal of Vermont's emerging success as a tourist destination depended on the quality of its road surfaces and the quality of the rural scenery along its roadsides. Tourist promoters and entrepreneurs responded to this fact by emphasizing the modernity of the state's highways as well as the scenic beauty and romance of the roadside landscape. They blended the old with the new, suggesting the accessibility and compatibility of modern highways and timeless roadside scenery statewide.

The primary catalyst for this, of course, was the automobile. As automobiles became more affordable to American consumers during the 1920s, the popularity of driving as a leisure activity grew nationwide. Automobiles granted American travelers a new degree of spontaneity in their travel decisions and a new degree of freedom to explore areas farther from the confines of railroad lines.[5] But by putting more people on the road, automobiles also drew attention to the substandard condition of the nation's roadways, adding renewed strength to longstanding "good roads" campaigns designed to improve road surfaces, grades, and signage.[6] Even before the advent of automobile travel, rural reformers nationwide had backed road-improvement programs in the hopes of widening markets for farmers and tying rural communities more closely together. Their efforts intensified as automobile travel became more popular. Organizations in Vermont like the Grange and the Vermont Good Roads Association, for instance, helped shepherd new highway laws through the Vermont legislature, consolidating the state's control over the funding and maintenance of roadways, first for horse-drawn travel and later for automobiles.[7]

Road-improvement programs were critical to making rural Vermont more accessible to vacationers. Automobile tourists began arriving in Vermont as early as the first decade of the twentieth century, using their automobiles to reach their summer homes more easily or just to drive around the countryside for a few days, taking in rural scenery as it passed by their windshields. Their numbers grew dramatically in the 1920s, until by the mid-1930s 90 percent of those who visited the state each year arrived by car.[8] As the popularity of automobile tourism increased, so too did calls for road improvements, particularly among entrepreneurs who equated inaccessibility with a loss of tourist dollars.[9] Their worries were not unfounded. Despite modest improvements to its road system in the 1920s, Vermont was infamous for having some of the worst roads in New England. To the dismay of Vermonters in the tourist business, that impression only deepened after the disastrous autumn flood of 1927 wreaked havoc with roads and bridges statewide.[10] Ironically, however, the flood was a blessing in disguise. It forced Vermonters to improve their road system faster than they might have otherwise, and by one highway commissioner's assessment it put the Vermont State Highway Department "thirteen years ahead of its schedule."[11] Within a year of the flood Vermont's roads were well on their way to recovery, and within three years the state was boasting over five thousand miles of roads surfaced with gravel, macadam, or concrete.[12]

A variety of state and local organizations advertised Vermont's road improvements to vacationers. The State Highway Department, the Bureau of Publicity, and the Vermont Department of Conservation and Development (which assumed the bureau's duties for a time in the mid-1930s) produced annual travel guides like *Vermont Motor Tours* and a short-lived but ambitious magazine entitled *Vermont Highways*, all of which were backed by sophisticated surveys and statistical analyses of visitors' preferences and demographics.[13] By the second half of the 1930s the Vermont State Legislature was allocating between $35,000 and $45,000 annually for film, radio, and printed publicity, much of which was designed specifically for automobile tourists.[14] Private promotions and full-length books such as Crane's *Let Me Show You Vermont* and Walter and Margaret Hard's best-selling *This Is Vermont* (1936) added to the media campaign, as did the Federal Writers' Project guide *Vermont: A Guide to the Green Mountain State* (1937).

Promotions like these spread the gospel of accessibility by reiterating Vermont's commitment to modern roadways and assuring automobile tourists that they could drive around the state with ease. But when they spoke of accessibility, tourist promoters often referred to more than actual mobility and more than modern roadways. Accessibility also referred to the ease with which drivers could encounter the kinds of roadside scenery they were coming to expect from rural Vermont. By comparison to the celebrated modernity of the state's road system, its roadside scenery was typically defined in romantic, nostalgic terms. This became particularly true during the 1920s and 1930s, as Vermont's reputation as a quintessentially New England state grew, making it a stand-in for an emerging regional mystique.[15] This was a time nationally when writers and artists who subscribed to "re-

gionalism" celebrated geographic distinctiveness as a hallmark of national
identity, drawing attention to differences between New England, the South,
the West, or the Midwest and suggesting that each had its own readily definable problems and appeals.[16] Regionalism was more than an academic
or stylistic exercise, however; regional identities informed travel decisions
as well as the visitors' understanding of what they encountered on the road.[17]

Vermont's early-twentieth-century tourist identity commonly drew from
a number of familiar regional icons, including farms, fields, and distinct
village centers. As the Vermont Commission on Country Life put it in 1932,
"Our visitors from other states do not expect or seek a standardized common-
wealth. They expect something distinctly beautiful in our colonial houses,
in the steeples of our churches, in our village greens, in the neat and thrifty
appearance of our farms and our homes."[18] Others, such as the popular
American tastemaker Wallace Nutting, agreed with statements like these,
arguing that pastoral farm scenery gave Vermont its beauty and sense of distinction. Nutting's 1922 book *Vermont Beautiful* emerged as a standard text
on that position. "Of course," Nutting argued, "we all enthusiastically
admit that a fertile, well cared for, well-watered, well-wooded countryside,
with grazing herds, neat farm buildings and fair roads, is the most delightful vision that bursts on weary mankind. It calls us back to paradise, or what
is better, to making a paradise of our own."[19] Nutting's book includes hundreds of photographs featuring streams, grazing livestock, working farmers,
and tree-lined rural lanes stretching out ahead toward enticing bends in
the road, and in this way it helped teach travelers to embrace auto tourism
as a means for viewing Vermont's pastoral scenery. Notably absent from all
of the book's pictures, however, is the car itself. Instead, Nutting chose to
privilege a decidedly anti-modern view, carefully constructing his photographs to create an imagined rural world where residents lived as they
might have a century before.[20]

Others represented Vermont's villages in similar ways. The state's rural
hamlets, one pamphlet reminded visitors, were "usually true to the New
England tradition of white houses set under majestic maples and elms."[21]
Being true to that tradition meant having iconographic church steeples,
village greens, and country stores. It meant offering passing motorists an
opportunity to feel as if they had entered a place about which only a few
travelers knew, a place still visibly connected to the nation's rural past.
"Real Vermont" villages had been "screened from the tide of travel," one
wrote, and thus had retained "their old-time features."[22]

In addition to pastoral views and quaint village centers, roadside representations often included a relatively new rural icon — the covered bridge.

FIGURE 3.2. Covered bridge with signs, no date. By the middle of the twentieth century, the covered bridge had become an icon of Vermont's rural roadways. Because of the signs tacked to its entryway, however, this bridge and others like it would have been a source of concern for Vermont's roadside beautifiers. Courtesy Special Collections, University of Vermont Libraries.

Generations of bridge builders in Vermont and neighboring states had used roofs as a practical measure to protect their bridges' expensive timber frames, and in time their practicality became a source of charm to tourists. Scores of Vermont's covered bridges were lost in Vermont to the flood of 1927, most of which were replaced by bridges with modern steel frames. The two hundred or so covered bridges that remained, however, evolved into proud icons of traditional craftsmanship and the wisdom of local, non-professional knowledge. Moreover, covered bridges were public spaces located directly on the road itself; as such they offered drivers an intimate, easily accessible encounter with rural tradition. As the architectural historian Herbert Wheaton Congdon noted in his 1941 book *The Covered Bridge* (tellingly subtitled *An Old American Landmark Whose Romance, Stability and Craftsmanship Are Typified by Structures Remaining in Vermont*), Vermont's covered bridges were "an essential part of a romantic landscape." They expressed "links that bind us to an earlier, simpler way of life," he argued, and they carry one's imagination "back to the land of half-forgotten dreams."[23]

Representations of pastoral beauty, village centers, and covered bridges all reproduced traditional, tourist-based conceptions of rural landscapes, suggesting that nostalgic and romanticized rural scenes were easily accessible to motorists. But importantly, Vermont's sense of regional distinctiveness and its ability to compete on the tourist stage were not dependent on impressions of timelessness alone. In fact, although some image-makers, such as Wallace Nutting, chose to cut the automobile out of their pictures in favor of an anti-modern image, others did not. As often as the automobile was absent in promotions, it was also there in the picture, coasting along new, well-maintained highways. Those highways and the automobile itself were what made it possible for visitors to connect with the pre-modern rural identity they envisioned along the state's roads. Visitors demanded modern roads and modern travel as well as a timeless sense of tradition. Unspoiled Vermont may have been equated with "oldness" and "tradition," but it was not necessarily undermined (or "spoiled") by all expressions of modernity. In fact, just the opposite was true. As they had done a generation before, visitors continued to expect modern development in rural Vermont as much as they expected romantic scenery. And with modern development came the expectation that rural landscapes were something other than stable or timeless spaces.

Roadside Beautification and the Construction of Unspoiled Vermont

Automobiles and modern roadways made the *entire* state of Vermont more accessible to travelers than ever before, turning every back road, every village, every view, and every farmyard into a potential tourist attraction. Unfortunately, however, not everywhere in Vermont was as unspoiled as some would have liked, and by opening the entire state up for viewing (and reviewing), Vermonters also had to ensure that the entire state looked as good as its unspoiled reputation suggested. To make that happen demanded more than attractive brochures and celebratory rhetoric about unspoiled rural space. It demanded an investment of time, money, and energy. It demanded a commitment to reworking the rural roadside according to the lofty dictates of rural leisure, as compared to the often messy realities of rural work. If roadside scenery succeeded in appearing "unspoiled," it did so not necessarily by some innate quality of the rural but by virtue of the efforts that Vermonters and their visitors put into formal and informal programs of roadside beautification.

Being willing to support roadside beautification was one thing. Defining the precise meaning and implications of the operative word "unspoiled" was something else entirely. Considerable variation existed as to how different social groups identified and set about creating unspoiled roadsides in Vermont. Here we explore this variation across three different contexts of roadside beautification: attempts by professionals to clean up and manage the public right of way; attempts to convince private landowners to maintain their property; and attempts to restrict the size and placement of billboards along the state's roadways. Taken together, these contexts reveal the new kinds of work that were being used to create new kinds of leisure landscapes, and they reveal an intensification of the social politics that accompanied the reworking of rural Vermont. The state's roadsides, we shall see, became important sites for the negotiation of landscape and identity in early-twentieth-century Vermont.

State highway officials often took the lead when it came to formal roadside beautification efforts, in particular by redesigning and repairing the public right-of-way in the wake of road improvement projects. The State Highway Department added a landscape architect to its staff during the early 1930s, for instance, whose main task was to insure that roadsides "have all the finish and the delightful features and quality of a work of art, heightening the appeal of landscapes by contributing harmonious attractive foregrounds."[24] Although such goals sounded lofty and artistic, the processes involved were fairly straightforward. Highway officials had their crews plant grass and trees to beautify slopes and to control erosion. They redesigned roads and added pullouts to offer better views. And they worked with volunteers, civic organizations, and local governments to tidy up roadsides near village centers in the hopes of encouraging travelers to stop and spend a little time and money before passing on.

Formal efforts like these were directed by new kinds of professionals operating under new conceptions of land use and development — professionals whose jobs included enhancing rural tourism by reproducing unspoiled Vermont in the material landscape itself. Broadly defined, Vermont's professional roadside beautifiers included politicians, planners, and highway officials, all of whom used political power together with formal planning and engineering to create attractive roadside scenery along the public right-of-way. Theirs was a new approach to land use in Vermont — a state whose residents had a reputation for opposing state-led planning initiatives, land use controls such as local zoning ordinances, and federal programs such as those put in place during the 1930s as part of Franklin D. Roosevelt's New Deal. But the early twentieth century was also a time when popular and professional support for such measures grew nationwide, and within this climate of acceptance many of Vermont's roadside beautifiers turned to professional planning and policy. They got help from the state legislature, which passed an enabling law in 1931 giving towns the right to pass local zoning ordinances if they chose to do so. Although most towns waived that right for at least three more decades, some saw this as an important symbolic step capable of empowering roadside beautification. In addition, the state legislature established the Vermont State Planning Board in 1933 to serve as an advisory group and as a liaison to the National Planning Board, the federal agency responsible for implementing New Deal projects. The board would continue to operate until 1945 as a planning and promotional agency, although its power and its achievements remained relatively limited in scope.[25]

One of Vermont's most powerful advocates of land use planning and formal roadside beautification was the reform-minded secretary of the Vermont State Chamber of Commerce (VSCC), James P. Taylor. Born in New York in 1872 and educated at Colgate University and Harvard University, Taylor joined the VSCC in the 1920s after an initial career in education. While at the VSCC, Taylor became an advocate for a wide range of programs in state governance, rural reform, good roads, and tourism. Although the VSCC was not state sponsored, it had strong links to state-level politics, and under Taylor it operated essentially as a state-level tourist agency alongside the Bureau of Publicity and, later, the Department of Conservation and Development.[26]

Taylor used his position in the VSCC to extend the good roads movement beyond road *surfaces* to the *sides* of roads as well. All of the states in New England, he argued, were engaged in a "new beauty contest" — one waged "by roads and roses," and one in which victory was measured in tourist dollars.[27] Roadsides, Taylor advised, should offer a "perfect foreground for the moment to moment succession of varied scenic pictures which reveal the character of a state to all, citizens and travelers alike, who travel the open road."[28] For two decades Taylor worked hard to create such "scenic pictures," lobbying state officials for funding to support roadside beautification, urging town officials and local chambers of commerce to adopt beautification programs, and driving thousands of miles throughout Vermont documenting the best and worst of Vermont's roadside scenery.[29]

His efforts, along with those of the landscape architects and state highway officials with whom he worked, were carefully coordinated, professional, and broader in scale than anything up to that point. They were also quite public. We might be tempted to think that Taylor and others hid their work behind a polished, tourist-directed veneer designed to appear entirely uncontrived to the visitor's eyes. But that was not always the case. Professional roadside beautifiers often publicized their efforts, displaying their plans and achievements proudly in outlets like *Vermont Highways* and *The Vermonter*. Rather than depicting rural Vermont as a place unspoiled by an innate organicism — by virtue of just being "rural" — they advertised their work as essential to the future of unspoiled Vermont and as a reflection of the state's commitment to the beauty of its rural landscape. In this sense, the active construction of roadside scenery in Vermont became another selling point for promoters; people like Taylor, visitors were told, were doing all they could to protect the distinctiveness and beauty of unspoiled Vermont, making it accessible to all who drove the state's roads.

At times that meant supplementing the labor of paid professionals and highway workers with that of local volunteers. Roadside beautifiers routinely turned for help to local women's organizations and rotary clubs, for instance, encouraging groups such as these to spread the message of roadside beauty among their neighbors and encouraging them to get out there and do the work of keeping unspoiled Vermont unspoiled.[30] Residents were encouraged to remove dead trees, prevent the clandestine dumping of trash, and organize local work bees to pick up debris along the public right-of-way.[31] Taylor even managed to organize a "Community Roadside Improvement Day" in 1931, during which volunteers across the state removed dead trees, planted over a thousand new trees, and hauled away truckload after truckload of trash. The goal of the project's organizers, Taylor explained to readers in *Vermont Highways*, was not to make Vermont look antiquated. Instead, it was intended to demonstrate in a very public way "that the communities of Vermont are progressive, and that they intend to make and keep our Vermont beautiful and attractive both for ourselves and the traveler within our borders."[32] Rather than depict Vermont as inherently unspoiled, then, Taylor and others hoped that efforts like the Community Roadside Improvement Day would institutionalize roadside beautification, making it a civic duty. By such efforts, he and others hoped to build an attractive reputation for Vermont as a place committed to the creation and maintenance of unspoiled roadside scenery.

Cleaning up Vermont's immediate roadsides was an accomplishment in itself, but it meant little if private property beyond the right-of-way appeared less than unspoiled. Because automobile tourism made the entire state of Vermont accessible to tourists, every farmyard, every village, every turn in the road was a potential tourist attraction. Many of these places, however, were far less appealing than those typically captured by the promoter's camera or pen. Poverty was more widespread during the 1920s and 1930s than stylized images of tidy farmhouses and attractive village greens often suggested, and many Vermonters simply could not afford to spend any

time or money on the appearance of their property. Depression-era photography as well as comments from concerned Vermonters and concerned
visitors all suggest that rural Vermont had its share of junked cars, rusty
farm machinery, rundown buildings, and trash heaps.[33] As one Vermonter
put it, "It is a pity that our people are not more united in preserving the
beauty of their surroundings. Sometimes one sees deplorable dwellings
and barns, an old wagon under trees, little spots of plowed ground. Probably
no money but plenty of dogs — one of the practically abandoned farms."[34]
Thankfully, another commentator hastened to note, such "instances of ugliness" in Vermont were rare and were typically confined to the state's few
urban pockets. But where they did exist, she added, they came as "something of a shock; they seem so un-Vermontish."[35] Whether "instances of
ugliness" were urban or rural, common or not, anyone traveling the state's
roads with a lofty vision of unspoiled "Vermontish-ness" in mind could not
have helped but be struck by the cluttered and unkempt appearance of
some rural homes. Indeed, the more visitors were told that Vermont was
unspoiled, the more such places were likely to stand out in their minds.

Many roadside beautifiers defined private property — and the private farmyard, in particular — as a key ingredient in the creation of Vermont's unspoiled identity. "In the ultimo," Taylor wrote in a series of exchanges with
one Vermont politician, "the owners of property who have as apparent pets
the sylvan dead or wrecks of buildings, really ought to do something about
it themselves from their initiative." He added, "Government has been
building roads and taking care of the immediate roadsides. Now people
need to be good sports and do their share of the work, lest the effect be
spoiled."[36] Unfortunately, at least from the perspective of Taylor and others, not all rural residents were ready to make their private farmyards the
public domain of the tourist's gaze. That is not to say that Vermonters were
necessarily opposed to the idea of tidy farmyards but rather that many did
not embrace the coupling of accessibility and unspoiled Vermont that
guided roadside beautification efforts.

Quite simply, many Vermonters were less than enthusiastic about having visitors prowling their back-road hamlets, peering into their daily lives
with a discerning, outsider's critique. There was nothing special to see along
many of the state's roads, some argued, so why bother with all this talk
about making them so accessible? Etta Wilson, the town clerk of Lunen-
berg, Vermont, put it succinctly, agreeing with Taylor that road signs were
important along the town's main roads but disagreeing that the same logic
applied to its back roads: "Our back roads, which lead nowhere in particular, have no signs," she argued, "but that hardly seems necessary we think."[37]

FIGURE 3.3. Abandoned farmhouse near Newport, Vermont, 1936. This photograph is indicative of the kind of "un-Vermontish" scene that roadside beautifiers deplored. During the 1920s and 1930s, many state and local officials worked hard to persuade Vermonters to clean up and maintain property within view of passing cars. Photograph by Carl Mydans, Library of Congress, Prints & Photographs Division, FSA-OWI Collection.

Why was it not necessary? Because from Wilson's perspective, only those who knew their way around were likely or inclined to be driving on her town's back roads. Why open up those roads — with all of their beauty and all of their blemishes — to anyone and everyone? That kind of reasoning, however, was exactly what state officials did *not* want to hear. After all, their goal was to make Wilson's "nowhere-in-particular" more accessible and more attractive to visitors who might want to explore its depths.

Short of coercive legal controls over private property, how could advocates of roadside beauty convince Vermonters to bring their property in line with the standards of tidiness and unassuming timelessness set by unspoiled Vermont? One way to do so was to promise residents an economic reward for their efforts. Tourist officials had been telling Vermonters for decades that tourism was good for the state's economy as a whole, and that *all* residents stood to gain, in one way or another, from its continued expansion. Yet that logic was difficult to connect to one's personal life, particularly if one was not directly involved in the tourist economy. Partly for this reason, it made sense to state officials to encourage more Vermonters to get directly involved in the state's tourist economy. The easiest way for many Vermonters to do so was to take in passing motorists, converting their private homes into "tourist homes." Growing numbers of automobile tourists meant that growing numbers of residents in villages and on farms could put their private property to work in a tourist capacity, serving the cause of roadside beautification as they did. After all, an attractive farmyard could mean the difference between success and failure for host families, making private property and the efforts of individual property owners central to the mission of roadside beautification and the reworking of rural Vermont.

Automobile tourists patronized a number of different types of accommodations beyond private tourist homes, including cabin camps, motels, and campgrounds. Tourist homes, however, offered travelers an inexpensive and comfortable alternative to these options, where visitors expected to eat a good meal, sleep in a comfortable bed, and enjoy all of the benefits of modern plumbing and electric lights. Indeed, an overnight stay with a Vermont family was more of a financial necessity to many tourists than a deliberate nostalgic retreat, although farms, in particular, undoubtedly held many charms for urban travelers. When these homes failed to meet the modern expectations of visitors, they ran the risk of being critiqued, in one travel writer's words, as "antiquated and uncomfortable."[38] For Vermonters, accommodating automobile tourists offered a chance to make a little extra money for things such as household improvements, medical and education costs, or new farm machinery. Fully one third of those traveling in Vermont during the 1930s stayed in private homes, making any average household with an extra room or two a potential source of income and prompting hundreds of Vermont families living along major roadways, adjacent to popular lakes, and even in remote corners of the state to open farm and village homes to tourists (see figure 3.4).[39]

One of the best glimpses we have into Vermont's tourist homes is Dorothy Canfield Fisher's fictional play *Tourists Accommodated* (1934). Inspired by real-life experiences among Fisher's neighbors in Arlington, Vermont, the play captures a variety of frustrating, humorous, and pleasant encounters between host families and their auto-touring guests — encounters that resonated with readers and live audiences statewide. *Tourists Accommodated* tells the story of the Lymans, a hardworking, tight-knit Vermont family whose members decide to take in travelers to raise money for the daughter's education. "There's a stream of gold running right past the door all summer long," one family member stated of automobile travelers. "All you've got to do is to have gimp enough to dip your spoons in and take out your share." Throughout the play, a variety of tourists with stage names like

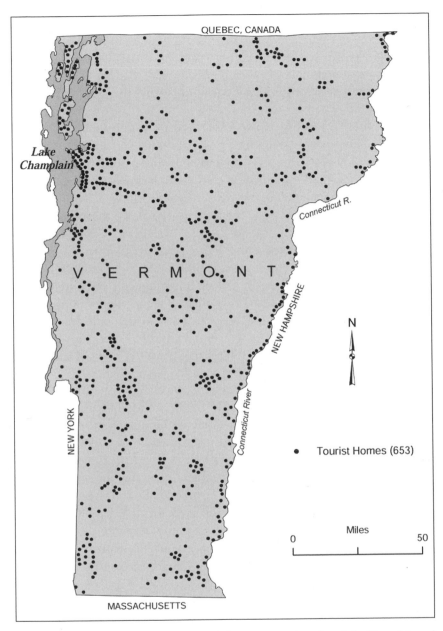

FIGURE 3.4. Vermont tourist homes, 1930. Aside from concentrations along Lake Champlain and the roadways along the Connecticut River/New Hampshire border, tourist homes were dispersed fairly evenly throughout the state. Tourism's reach by 1930 had spread into all corners of Vermont. Excluded from the category "tourist homes" are cabin camps, hotels, campgrounds, and children's camps. Adapted from Vermont Commission on Country Life, *Rural Vermont: A Program for the Future* (1931).

"Silly Tourist," "Pretentious Tourist," and "Artist" stop at the Lymans' house. Each exemplifies character traits that would have been familiar to many Vermonters: some are condescending; some are thick-witted; others are kind and considerate. All, however, were designed to evoke situations and emotions familiar to Vermont audiences, capturing the spirit of life for a host family.[40]

Scenes from *Tourists Accommodated,* coupled with insights from other sources, suggest the degree to which host families reworked landscape and identity on the family farm according to new, leisure-based demands. In this sense tourist homes in the 1930s were similar to farm boarding enterprises in the 1890s and 1900s. But unlike earlier farm boarding, accommodating tourists in the 1930s depended to a greater degree on the transformation of the farm's appearance *from the road.* One's success in this business demanded a willingness to make one's property appealing to visitors who were just passing by. Some visitors made plans ahead of time and stayed for a week or two, but other passersby arrived unannounced, lured in by roadside signs reading "tourists accommodated." Success therefore depended on first impressions, on bringing one's property in line with the popular images of unspoiled Vermont that travelers carried in their heads. The private farmyard, in this sense, was not only central to the success of the host family; it had work to do on behalf of roadside beautification.

State promoters and tourist officials considered that an important charge — one that rural families would likely need help with to perform properly. After all, if automobile travel made the state's entire rural landscape a potential tourist attraction, then every action by every resident reflected on the state as a whole. The Vermont Agricultural Experiment Station offered advice designed to help host families make the transition to a tourist economy, as did the Vermont Chamber of Commerce and the Vermont Bureau of Publicity, both of which hosted a widely publicized conference in 1930 devoted to the theme of roadside hospitality. Local papers and the *Journal of Home Economics* reported on the conference, noting the confluence of roadside beauty, private property, and tourism in passages such as this:

Whether or not the tourist stops at your house depends a great deal on whether the place impresses him. If the yard is littered with farm machinery or children's toys and the hens are strolling about at their will, or if the yard does not show that someone has any thought and care to the planting and arrangement, the tourist will very quickly decide that within the house he would find no cleanliness and comfort. The best advertising that you could do is time and money spent in making your

FIGURE 3.5. Sunnyside Farm, c. 1930. The success of tourist establishments like the modest Sunnyside Farm depended in part on their ability to appear clean and inviting to passing motorists. In this sense, they helped reinforce the larger agenda of roadside beautification in unspoiled Vermont. The sign out front reads "Meals at all Hours, Dinner, Rooms, Bathroom, Inspection Invited." Courtesy Vermont Historical Society.

yard neat and attractive and creating a carefully-cared-for look through the planting and care of trees, flowers, and shrubs.[41]

Advice on how to create this "carefully-cared-for look" was typically targeted at women, who, as in generations past, were the main labor force behind the farm boarding economy. According to one study of twenty-three "boarder and tourist establishments," women performed an average of 91 percent of the labor associated with accommodating tourists, adding nearly $400 on average to the annual household income.[42] In *Tourists Accommodated*, for example, Mrs. Lyman feeds and cleans for both her family and her guests, using the evening hours to balance the family's account books while her husband reads the newspaper. Women were also reminded to be patient with eccentric visitors and to learn about travel opportunities in Vermont in order to offer travel suggestions to their guests. Women, many argued, should learn to think of themselves as Vermont's ambassadors of hospitality.[43]

Women also were expected to patrol the front lines of moral authority in the state's tourist trade, maintaining an unspoiled social reputation for Vermont, in part, by maintaining an unspoiled roadside landscape. Local officials in Vermont, as elsewhere in the United States at the time, often treated automobile tourists with suspicion. Such tourists often stayed in makeshift and unregulated tourist camps on the edge of town, for instance. They were anonymous, mobile, and, some feared, of questionable morals. They were here one day and gone the next.[44] Vermonters like Mrs. Lyman who accommodated tourists worried about such issues as well: single women traveling alone, couples whose marriage affiliation was suspect, ethnic travelers, and any number of other undesirable visitors might stop by one's farmhouse unannounced. "Why," Mrs. Lyman confides to her aunt, "how can you tell whether they're even *married* or not, nowadays, with no morals to speak of."[45] One of the best measures of defense against such concerns, tourist officials and roadside beautifiers would have reminded her, was the appearance of the farm from the road. Attractive farmyards set standards for proper behavior, and clean yards meant a better chance of attracting clean people.

Although it is difficult to measure in the historical record, some Vermonters would have resisted the idea of redesigning their lives to fit the expectations of tourists, and, of course, many rejected the option of accommodating tourists in their own homes. Even those who did take in passing motorists were forced to deal with many inconveniences that, at times, must have made them question the wisdom of their choice. Aside from the additional work involved in maintaining roadside flowerbeds or keeping up appearances in a working farmyard, host families were at times obliged to give up their own beds to tourists. As one writer recounted of her summer stays with a family in Vermont, the entire host family and its hired hands (nine people in all) moved each year into a large storage room to create space for their guests. One can only imagine, as the author does, the heat of the room, the lack of privacy, and the "closeness and density natural to a room that has been overcrowded with people whose clothes and bodies reek of hard labor in fields and barns."[46] Such arrangements could be particularly hard on children, who, as Fisher's work suggests, often resented having to be on their best behavior around guests. Moreover, visitors could make self-conscious adolescents painfully aware of their family's economic status and isolation relative to the outside world. Considering what farm boarding demanded of them, it is no wonder that the Lyman family from *Tourists Accommodated* sighs in collective relief at the end of the summer tourist season. Only then did they regain control over the use and meaning of their private property.

Questions of control over landscape and identity in rural Vermont were at the heart of roadside beautification efforts during the 1920s and 1930s — a fact perhaps best exemplified by Vermont's 1930s billboard debates. The

use of roadside signage in the United States grew tremendously during the
early decades of the twentieth century, particularly as growing numbers of
Americans took to the roads in automobiles. Private businesses and advertising companies adopted a commercial view of the nation's roadsides,
erecting modest signs as well as massive billboards advertising everything
from cars to chewing gum. As larger and more numerous signs crowded
American roads, local reformers, civic groups, and national organizations
stepped up their calls for legislative controls on roadside advertising, citing
both aesthetic and safety concerns. In response, many American towns
passed modest ordinances designed to control the placement and size of
roadside advertising in the public right-of-way — a move to which advertising companies often responded by building larger signs on private property
set back from the immediate roadside.[47]

Similar patterns unfolded in Vermont, where the growing popularity of
automobile tourism made the state a fertile market for roadside advertising.
Local businesses displayed messages on barns and covered bridges, while
regional and national advertising companies constructed larger billboards
that they then rented to advertisers. Despite modest, state-level regulations
enacted in the 1920s, thousands of signs peppered Vermont's roads, including 744 billboards by one 1936 count.[48] Individual residents and civic
organizations such as the Grange, the Daughters of the American Revolution, and the Rotary Club responded to Vermont's rising numbers of signs,
voicing their opposition to the commercialization of Vermont's roadsides,
both on public and private property alike. Roadside scenery belonged to all
Vermonters, they argued, and the public's view from the road should not
to be marred for the sake of private financial gain.

Opponents of outdoor advertising organized themselves into a number
of groups, the most influential of which was the Vermont Association for
Billboard Restriction (VABR), formed in 1937 and headquartered in Springfield, Vermont. The VABR was overseen by a committee of five prominent
men from around Vermont who lobbied the state legislature for new billboard restrictions and helped coordinate the activities of local committees
in towns statewide. Springfield's anti-billboard committee took some of the
VABR's most significant early steps, using handbills, petitions, boycotts,
and condemnatory letters to fight for the removal of a number of billboards
in their town. If they could create a climate that was uncomfortable to advertisers, the committee reasoned, local and national businesses would refuse to rent space from billboard owners, forcing them to pull up stakes.
Their plan worked, and within a year and a half Springfield's offending
signs had been removed. Others drew confidence from Springfield's suc-

FIGURE 3.6. Anti-billboard postcard, c. 1937. For many Vermonters and visitors, billboards were an invasive stain on an otherwise beautiful and harmonious rural landscape. Courtesy Vermont Historical Society.

cess, fueling the power of the VABR and fanning the flames of conflict between outdoor advertisers and roadside beautifiers.[49]

Vermont's anti-billboard campaigns of the 1930s shared similarities with those in other parts of the nation, including a focus on safety and aesthetics as well as a general opposition to the unregulated commercialization of the nation's roadsides. In addition, Vermont's experience reveals another story line — one rooted specifically in the context of rural tourism and defined by a struggle between residents and non-residents to control the form and meaning of unspoiled Vermont. Both sides of the billboard debate framed their arguments according to an insider/outsider binary, and as it turned out, one's social status as a resident or a non-resident mattered deeply to one's credibility, and more specifically to one's right to shape the future of work-leisure relations along the state's roadsides.

Many billboard opponents defined their crusade as an effort to protect the roadside landscape from outsiders whose loyalty was to their own commercial success and not to the welfare of Vermont's residents. Opponents argued that national and regional billboard companies were "parasites," or power-hungry, opportunistic outsiders with little regard for local communities and little concern for the protection of rural scenery. By their logic, even Vermont-owned businesses who rented space from billboard companies were guilty of undermining the welfare of the state as whole.[50]

To make their case, opponents crafted a two-part argument linking the protection of roadside scenery to the welfare of Vermont. First, they maintained that billboards were incompatible with the entire concept of unspoiled Vermont. It was true that billboards offered automobile tourists information on services such as food and lodging, but the general sentiment among those who opposed billboards was that a few confused visitors here and there was not as much of a problem as the clutter and safety hazards that billboards created. Participants in a 1938 VSCC conference on "keeping 'unspoiled Vermont' unspoiled" placed billboards at the forefront of "highway scenery problems," for example, while the Vermont State Planning Board argued in 1939 that "'unspoiled Vermont' will be spoiled with a degree of rapidity similar to the growth of the tourist business, if the increase in outdoor advertising is permitted to pursue a normal and unchecked course."[51] The very idea of unspoiled Vermont, it seemed, left little room for the commercialization of the roadside landscape, particularly when that commercialization originated with businesses from outside the state. Second, billboard opponents asked Vermonters to consider the widespread economic benefits of unspoiled rural scenery. Roadside beauty was good for the state's tourist economy, billboard opponents stressed, and what was good for tourism was good for all of Vermont.

With hindsight, arguments like these contain an intriguing contradiction. The welfare of all Vermonters, they suggest, did not depend on the commercial interests of non-residents, but it did depend on the leisure-based interests of another group of outsiders—vacationers—some of whom were now joining the state's anti-billboard campaigns. Not all of Vermont's visitors were passive consumers of scenery, nor were they content to assume that scenic beauty in Vermont came without effort or cost. Although year-round residents led the charge against billboards in towns like Springfield, and although non-residents were unable to vote on land use proposals of any kind, some vacationers became actively involved in public debates about the future of landscape and identity in unspoiled Vermont. This was especially true among those who owned property in Vermont, many of whom were eager to insure that nothing be done to mar the views from their land or reduce its resale value, and none of whom would have needed billboards to find local services as passing automobile tourists would have. Dorothy Thompson Lewis, author and wife of Sinclair Lewis, articulated their view by reiterating links between the protection of roadside scenery and the welfare of the Vermont economy. In a 1937 letter to the *Rutland Daily Herald*, she argued that Vermont "has beauty to sell," and she urged residents to remember that their state's economic interests depended on protecting that beauty in the name of tourism. "Many tourists who come to see [Vermont]," she wrote, "remain and live — if not all year, then part of it, as we do . . . They come to live with its fields and villages and mountains and people, not with advertisements. They can get those better on Broadway."[52]

Billboard opponents welcomed support from non-residents. But for billboard supporters, the activism of outsiders like Thompson proved that billboard opponents were themselves beholden to outside interests. If billboard owners and outdoor advertisers were labeled as opportunistic outsiders, then what, they asked, were tourists? People like Thompson arrogantly assumed the voice of an adopted native, they complained, when in fact their stylized, leisure-based conceptions of rural life were running roughshod over the needs of legitimate business interests, many from right there in Vermont. For evidence, billboard supporters could point to comments like those made by Sinclair Lewis in an address to the Rutland Rotary Club. Lewis argued that his decision to purchase property in Vermont was based on the belief that the state was "precisely the opposite" of the all-too-common American "desire for terrific speed and the desire to make things grow." Vermont would do well, he suggested, to protect its heritage of beauty and its slower pace of life, for in these things lay the state's attraction to cultivated and educated non-residents — people who brought lasting financial rewards to the state.[53]

The "desire to make things grow" criticized by Lewis, however, was precisely the kind of attitude that Vermont's business leaders embraced. It was also the kind of attitude that the Vermont State Chamber of Commerce was created to support — a fact that put James Taylor and the VSCC in a difficult position. As part of their support for tourism and roadside beauty, the VSCC came out in opposition to billboards — a move, some business owners maintained, that undermined the organization's entire raison d'être. The VSCC, they argued, should not be turning its back on the business interests of working Vermonters, especially not in favor of vacationers. As one wholesale manager complained, billboards "are a part of America. They have contributed their share to the up-building of the mercantile and industrial life of America . . . The [Vermont] Chamber of Commerce could do much better to devote their time to other purposes than the destruction of a legitimate mercantile venture, which is exactly what the billboard industry is."[54]

Yet for many in Vermont's business community, tourism was also a legitimate business venture — one that promised to create new jobs for rural Vermonters, and one that just happened to be dependent on the opinions of people from outside the state for its success. Indeed, few by the late 1930s could have failed to recognize tourism's growth statewide. If tourism's success depended on restricting billboards, roadside beautifiers argued, then restricting billboards would work to the benefit of working Vermonters and leisured guests alike. That logic reflected a core principle now guiding the reworking of rural Vermont: the state's rural landscape was not timeless, organic, or fixed in stone; rather it was a place that required reshaping and redefinition in order to reach its full potential as a tourist destination. More than ever the challenge now was how to rework the rural landscape in ways that met the demands of visitors while also meeting the economic demands of working Vermonters.

The degree to which those demands overlapped would remain an open and increasingly contentious issue, and it would be put to the test along the state's roadsides repeatedly in the coming decades, before flaring up a generation later over the billboard issue once again. For now, however, Vermonters tried to strike a balance on the issue of billboards. In 1939 the state passed legislation tightening restrictions on the size and placement of roadside signs and mandating the removal of many billboards statewide. By one estimate the number of Vermont's large, commercial billboards was reduced by more than half between 1938 and 1944.[55] By catering to both sides of the debate, architects of the state's new billboard legislation hoped it would meet the demands of the many different social groups now claiming some right to dictate the form, meaning, and future of unspoiled Vermont.

Accessing Unspoiled Vermont on Foot

Hiking emerged as part of Vermont's expanding tourist economy at roughly the same time as automobile tourism. Vermont's best opportunities for hiking were among the higher elevations of the state's Green Mountains, where visitors sought connection to a different kind of unspoiled landscape — one that was wilder, more forested, and more remote than the settled roadsides in the valleys below. Hiking's popularity grew tremendously in the northeastern United States between the 1890s and the 1910s, thanks to an emerging cultural embrace of wilderness and outdoor recreation, and thanks to the promotional and trail-building efforts of hiking organizations such as the Appalachian Mountain Club. Compared to neighboring states, though, hiking in Vermont was slower to take root. The state's turn-of-the-century hikers were less organized, its trail networks were less developed, and its terrain was widely considered less inspiring and less wild than New Hampshire's White Mountains or New York's Adirondacks.[56]

Trends like these were changing by the 1910s, however, as Vermont's Green Mountains grew more popular and more accessible to hikers than ever before. There were a number of reasons for this. First, the ecology of the Green Mountains was changing at the turn of the century, as forests crept back over large areas that had once been open, working landscapes. Scholars estimate that loggers and farmers had cleared about two-thirds of Vermont's total acreage by 1880. But subsequent reductions in logging, coupled with farm abandonment and farm restructuring, allowed forests to begin to return to many parts of Vermont, particularly its upland pastures and more rugged mountain terrain. That process was gradual, of course, and even by the 1930s over 60 percent of Vermont's acreage remained in farmland.[57] Nevertheless, the return of forests had a profound impact on the look and feel of the state's rural landscape. As patterns of rural work changed in Vermont — as the acreage of land touched by farming and logging declined — new kinds of landscapes and new kinds of recreational activities, such as hiking, were becoming more common than they once had been.

New efforts in forest conservation also contributed to hiking's increasing feasibility and popularity. Early-twentieth-century Vermonters made a number of concerted, government-sponsored efforts to protect the state's forests for timber, watershed protection, and recreation. Between the 1900s and the 1930s the state and federal governments purchased hundred of thousands of acres in Vermont's Green Mountains, establishing state forest reserves on Mount Mansfield and Camel's Hump Mountain as well as the much larger Green Mountain National Forest, which was established in 1932 across hundreds of thousands of acres. State and federal reserves became important sites for recreational activities including hiking, camping, hunting, and fishing. The Green Mountain National Forest, in particular, was home to many of Vermont's best-known hiking trails, and by the mid-1930s it had become an important selling point for tourist promoters.[58]

Hiking's popularity was also enhanced by its enthusiasts, many of whom worked hard to raise its profile by increasing access to Vermont's forests and peaks. That process began formally in 1910, when James Taylor established an organization called the Green Mountain Club (GMC). Before becoming secretary of the VSCC, Taylor had been headmaster of a Vermont academy. Taylor was committed to making outdoor recreation an important part of his students' education, but he was also continually frustrated by the relative lack of hiking trails in Vermont. That was where the GMC came in. "The object of the Club," the GMC's constitution stated, "shall be to make trails and roads, to erect camps and shelter houses, to publish maps and guide books, and in other ways to make the Vermont Mountains

play a larger part in the life of the people."[59] Although the GMC's constitution did not specify who "the people" were, Taylor hoped that Vermont
residents as well as visitors would lace up their boots and take to woods.

The GMC's primary objective was to construct and manage what they
called the Vermont Long Trail — a 265-mile hiking trail slated to run north
to south along the crest of the Green Mountains from Massachusetts to
Canada. That was obviously an ambitious and complex task. A trail of this
size required money, surveyors, trail builders, and agreements with hundreds of landowners to secure a right-of-way along its route. It required organization, vision, and a willingness to act on a scale unprecedented
among American hikers at that time. Proposed roughly a decade before
Benton MacKaye outlined plans for an "Appalachian Trail" to run from
Maine to Georgia, the Long Trail was unlike anything that had come before. Construction got underway first in the popular Mount Mansfield/
Camel's Hump region, and by 1913 the GMC had patched together a series of logging roads and footpaths, extending the trail 150 miles through
central and northern Vermont. But it was not until the late 1920s that the
trail was essentially complete. The GMC maintained primitive shelters all
along its route so that hikers could walk the entire trail at once, if they
wished. Or, they could park their car at one of the trail's road crossings and
hike for a day or two until they reached the next road. At that point, they
would have to turn around or arrange with a local farmer or a friend to
drive them back to their car.[60]

The Long Trail quickly became popular among vacationers, for whom it
offered access to unspoiled Vermont's less-domesticated landscape. Promoters adopted the nickname "footpath in the wilderness" to describe the Long
Trail, and they billed its scenery as wilder and more remote than that found
in Vermont's valleys below. Just as roads and roadsides offered access to Vermont's unspoiled cultural landscape, then, hiking trails and forested moun-
tainsides offered access to its unspoiled natural environment. Here in Vermont's forested uplands, hikers, trail builders, and promoters were adding
depth to the idea of what constituted unspoiled scenery, expanding its meaning to accommodate new kinds of spaces and new kinds of tourist activities.

But how did Vermont's unspoiled "wilderness" compare to the state's
unspoiled landscapes of farms, fields, and villages? What was the trail's relationship to the road, and what was the mountain's relationship to the valley below? Three trends define the relationship between Vermont's Green
Mountain wilderness and the state's broader rural landscape. First, for all
of its claims to wilderness status, the Long Trail (as well as other hiking
trails) often passed through a landscape profoundly shaped by work — a land-

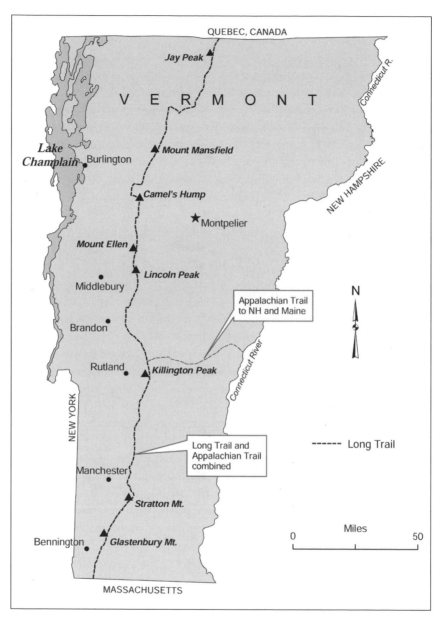

FIGURE 3.7. The Vermont Long Trail. The 265-mile Long Trail runs the entire length of Vermont from Massachusetts to the Canadian border, cresting some of the state's highest and best-known mountains along the way.

scape that was often still being farmed or logged or had only recently been abandoned. Hikers' reports from the 1920s, for instance, recount the many people, roads, pastures, farmhouses, cut-over timberland, and logging camps that marked their journeys. This was particularly true during the early decades of the trail's history, and it was particularly true of the more settled areas of southern Vermont, where as one hiker reported, "For mile after mile, up hill and down, largely through pastureland under the hot sun, the Trail wound on."[61] The prospect of soda or ice cream at country stores just off the trail was a welcome one for hikers, as were the prospects of fresh produce, milk, a warm bed, or a ride into town offered at the farmhouses located along its route. None of this is meant to suggest that the Long Trail was not also rugged, remote, and even dangerous, or that forests were not dense and expansive throughout a great deal of the Green Mountains. What it does suggest, though, is that expressions of rural work and an established rural society were inescapable and accepted parts of unspoiled scenery along Vermont's footpath in the wilderness. Promoters and hiking enthusiasts may have been crafting a new identity for the state as a place of wild natural beauty, but they were never able to escape entirely its identity as a place shaped by rural work.

Second, those who represented Vermont's Green Mountain wilderness in promotional literature often did so in ways that echoed representations of its settled landscape. Both, in this sense, were integral and related parts of the state's larger scenic package. Like Vermont's pastoral valleys, for instance, its wild places were often described as being pure, timeless, peaceful, and uncontrived. The Green Mountains, Vermont's Secretary of State Guy Bailey noted in 1913, were symbols of "serenity and dignity and nobility and the things that are unchangeable and eternal." Moreover, he added in a passage that was quoted often in the decades to come, the Green Mountains were just as accessible to the visitor as the state's rural valleys:

In many respects the Green mountains [sic] satisfy the mental conception of what a lofty mountain should be quite as well as do some peaks much higher, because here the surrounding country is correspondingly lower. Most of the great mountain heights are isolated from settled regions, but the Green mountains are surrounded by cultivated fields and pleasant villages and are not far removed from the habitations of men. In other words the Green mountains are high enough to satisfy the desire for height and vision, but are not too high to be accessible.[62]

The Green Mountains, then, were large yet not too large, inspiring yet non-threatening. They were wild yet still linked to the settled countryside below. In a word, they were accessible.

Finally, Vermont's mountains and its settled valleys were connected by a common goal — economic growth through tourist development. According to its architects, the Long Trail was meant to be a catalyst for economic growth, not necessarily along its wilderness corridor but in the valley towns below.[63] Originally, the trail was envisioned as a spine from which a series of looping trail systems would drop down from the highlands to connect with nearby villages. Here in "mountain centers" like Brandon, for instance, visitors were expected to spend money for supplies, lodging, and transportation. "The coming of the Long Trail is working a change," one optimistic observer wrote, "and creates interest wherever it passes. Towns are now beginning to take into account whatever trail may exist in its limit and work out routes by which this trail may be brought in touch with the Long Trail."[64] By making the high-elevation wilds of Vermont more accessible to hikers, then, the trail's advocates hoped that hikers' pocketbooks would be made more accessible to local merchants as well. It did no one any good if hikers stayed up there in their forested heights without taking some time out to spend some money in town.

Hiking's emerging popularity and the completion of the Long Trail made it possible for Vermonters and their visitors to draw new kinds of spaces into the state's tourist trade, adding depth and variety to the meaning of unspoiled Vermont. Hiking trails offered visitors access to places that felt wilder and more remote than the state's roadside landscape, expanding unspoiled Vermont up from the valleys below and redefining the meaning of mountainous terrain statewide. But even as new kinds of spaces were added to Vermont's expanding tourist trade, they retained connections to the state's more settled and more obviously "rural" landscape. That landscape was hard to escape entirely in Vermont, even in the state's unspoiled wilderness environment.

The Politics of Landscape and Identity on the Long Trail

Not everyone with an interest in the Long Trail agreed about how to manage or use it, nor did they agree about how to manage or define the unspoiled landscapes to which it offered access. Many GMC members looked on with pride as the Long Trail's popularity grew during the 1920s and 1930s. Others worried openly about the ability of Vermonters to maintain control of their state's trail, about the *kinds* of visitors the trail was attracting, and about the ways in which those visitors were using it. There was no way to control access to the trail — no gates, fees, or fences — meaning that it was difficult to control its use and ultimately the public identity it assumed. Nonetheless, some hiking enthusiasts sought ways to include and exclude certain groups and certain kinds of hiking. The politics of landscape and identity sparked by their efforts reveal the intensification of social conflicts surrounding concepts like accessibility and unspoiled Vermont.

Building Vermont's 265-mile Long Trail required a great deal of time, money, and labor on the part of the GMC. To make construction and management more efficient, the club divided its members into geographic chapters (or "sections"), each of which constructed and maintained sections of the trail in a given part of the state. But progress was slower than expected during the club's first years, and money was always tight. So it was welcome news when the Vermont Forestry Department approached the GMC in 1912 with a plan to help. The Forestry Department was planning to build a fire patrol trail along the central spine of the Green Mountains, they explained, and if the club was willing to help pay for its construction, they could adopt it as part of the Long Trail. The GMC took them up on the offer, and within a year the Forestry Department had opened a new fifty-mile work trail for its foresters and the GMC had added fifty miles to the Long Trail's length.[65]

Despite this bit of good news, two issues soon came up that challenged the management structure of the GMC and complicated the status of the foresters' work trail. Together, these issues raised new questions about the power of non-residents to shape land use decisions in Vermont and prompted new debates about work-leisure relations in a landscape touched by tourism.

The first of these issues emerged from a 1916 proposal to allow non-resident GMC members from New York and New Jersey to form their own section within the club. Initially the GMC's sections were based solely in Vermont and were comprised of resident members: the Burlington Section, for instance, consisted of residents from the Burlington area who maintained the trail in the relative vicinity of Burlington. Non-resident members were classified as "guests" of the GMC and were not allowed to organize their own sections. But now some of these members wanted to take a more active role in the trail's management. After all, the GMC's treasurer Theron Dean pointed out, many had "taken more long tramps on our trails than have our Vermont members."[66] The GMC's constitution made no room for such a request, however, meaning the issue would require debate and a vote by the group at their annual meeting. Within a few years of its creation, then, the GMC was forced to confront the growing influence of non-residents over the future of the state's leisure landscapes. As in the case of roadside beautification, non-residents were not simply buying into the idea of Vermont as an uncontrived and timeless place. Rather, they were eager to reshape the state's landscape according to the specifics of their own leisure-based demands.

A second problem developed in 1916 as well when a non-resident GMC member named Will S. Monroe completed a new hiking trail of his own design parallel to the Forestry Department's work trail. His trail was far superior, he argued, and it should become the official route of the Long Trail through this part of the mountains. From the start hikers had criticized the forestry trail for being too much of a work road and not enough of a recreational path. It followed logging roads with little in the way of views, they complained, and it was oriented away from challenging summit pitches. In short, it was not what hikers had in mind when they took to the woods. Monroe—a New Jersey professor who taught summer courses at the University of Vermont—was among those who complained outspokenly about the trail. And now, after spending his own time and money on the project, he was proposing that the GMC adopt his new forty-mile stretch of trail running south from Camel's Hump. He called the new route the "Skyline Trail."[67]

Of the two trails through this area, Monroe's was clearly better for hikers, and many in the GMC praised him and the non-resident hikers who had helped build his trail.[68] But now that those same non-residents were asking to form their own section within the GMC, and now that Monroe was suggesting a new route for the Long Trail, others voiced concern. None of this may have ever been an issue had Monroe not stepped on the toes of forestry officials as well as a powerful minority in the GMC with personal ties to the Forestry Department. Among others, that minority included the club's president, Mortimer Proctor, and its secretary, Roderic Olzendam, both of whom interpreted Monroe's outspoken criticism of the forestry trail as an affront to the good will of the Forestry Department and an unwelcome attempt by an outsider to dictate the trail's future.[69] Questions about Monroe's trail and about the possible creation of a "New York Section" within the GMC therefore forced club members to confront this issue of outsider influence at the same time that they confronted the status of the foresters' work trail. And just as Vermonters would do fifteen years later in the context of 1930s billboard debates, some in the GMC maintained that outsiders had no right to dictate the future of landscape and identity in Vermont.

On another level, debates about the management of the Long Trail raised fundamental questions about work-leisure relations and about leisure's growing power as a force for shaping land use in rural Vermont. Supporters of the Forestry Department's trail argued that it served a welcome *dual purpose,* meeting the needs of working Vermonters and hikers at the same time.[70] But many Long Trail enthusiasts were not satisfied with this dual role, regardless of whether they were from Vermont or not. The Forestry Department's trail missed the mark, they argued, by favoring the work-based needs of foresters over the leisure-based needs of hikers. And despite the fact that the Long Trail would ultimately intersect with landscapes of work throughout so much of its length, it seemed in this case that work and leisure were served best when kept apart.

The status of Monroe's Skyline Trail and the proposed New York section were decided at the GMC's annual meeting, held in January 1917 at the club's headquarters in Rutland. After a few tense exchanges and a predictable round of opposition from some members, the club approved both proposals, adopting Monroe's trail and adopting the proposal to create a non-resident section.[71] The GMC's decision validated the growing influence that non-residents were having on recreational land use throughout Vermont and reinforced their desire to keep work and leisure at a distance, at least along this section of the trail. Of course, that segregation was hard to achieve elsewhere. The landscape of early-twentieth-century Vermont was shaped so profoundly by human activity that hikers were not entirely able to escape expressions of rural work, whether along the trail's immediate sides or from its vantage points along the way. Promoters may have touted the Green Mountains as being unspoiled, but that did not necessarily mean that they were untouched by human hands.

The Long Trail's popularity grew tremendously in the years following the GMC's contentious meeting in 1917, and many new hikers were welcomed into the ranks of the trail's enthusiasts. Women hikers were among them. By the 1920s many GMC members viewed women as compatible with the kind of lasting identity they hoped to create for the Long Trail. That was a relatively new status for women at the time. The American outdoors was a decidedly masculine place at the turn of the century, and, in the minds of many, participation in activities such as hiking or hunting remained a proving ground for middle- and upper-class conceptions of manhood.[72] By the 1910s and 1920s, however, barriers to outdoor recreation among women were breaking down, as women challenged restrictive conventions about domestic femininity, dress, and physical activity. Hiking offered a specific opportunity to transcend traditional gender boundaries, and among young women its popularity grew accordingly.[73]

On a local level, the Long Trail became a space through which women hikers expressed their identity as outdoor enthusiasts. Women took an ac-

tive role in the GMC from its earliest years, and they took to the trail in
growing numbers during the 1920s — so much so that the GMC's official
publication, *The Long Trail News*, offered specialized advice to women
hikers on issues such as clothing and equipment.[74] In the summer of 1927
women hikers burst into the public eye when three young women from
Vermont and New York, Kathleen Norris, Hilda Kurth, and Catherine
Robbins, became the first women to hike the trail's entire length. Dubbed
the "Three Musketeers" by an adoring press, the young women posed for
photos and interviews as they worked their way up the length of the state.
Upon the completion of the women's month-long trip, representatives
from the Vermont State Chamber of Commerce met the women on the
trail, showering them with praise and touting their achievement as a great
moment in Long Trail history.[75]

Promoters deployed gender through examples like this to raise the Long
Trail's profile and simultaneously to enhance Vermont's identity as an accessible tourist destination. Above all else, the publicist Walter Crockett
wrote to James Taylor, Vermonters should not "convey the idea that the
Long Trail can be made such an easy and pretty place that it can be used
at any locality chosen for a garden party, for society ladies." But, he added,
they should also not make the trail seem overly challenging for fear of discouraging women or inexperienced hikers.[76] Women were becoming an
important new market for the trail. It was therefore in the best interest of
the state to portray the trail as a safe, fun, and accessible vacation opportunity. "The Long Trail Is Safe for Women Hikers," *The Vermonter* magazine
proudly announced in a headline from 1929.[77] Indeed, *The Long Trail News*
added, the Three Musketeers "walked without male escort and without
arms, and experienced no trouble whatever."[78]

Nonetheless, commentators also expected women hikers to behave in
ways that did not challenge the trail's identity as a "footpath in the wilderness." As the author Charles Edward Crane wrote, "The Long Trail scarcely
has the salt of danger, and its mild adventures are often sought by women —
though of course they must realize it is no garden path or place for high
heels."[79] In other words, women were welcome on the trail as long as they
did not try to make the trail something less challenging, less masculine. The
trail's "rugged character needs to be maintained," Walter Crocket added,
and it should not be understood as a place suitable "for persons who are
weak or ill or lazy or unwilling to abandon temporarily some of the con-
veniences of city life."[80] Obviously the very nature of a hiking trail in the
mountains limited the extent to which the trail's identity or its physical
landscape could ever change, but the larger point behind comments like

FIGURE 3.8. The "Three Musketeers," 1927. With the sign above them pointing to Jay Peak, the "Three Musketeers" pose near the end of their famous journey on the Long Trail. Imagery like this would have helped to persuade audiences that the Long Trail was a safe space for young women, both physically and morally. Courtesy Vermont Historical Society.

these remained clear: the Long Trail and Vermont's unspoiled wilderness were open to all hikers just as long as those hikers abided by certain conceptions of what a hiking trail should be.

The good-natured exploits of women like the Three Musketeers seemed to do just that, but the same could not be said of all classes of potential hikers. In fact, the growing popularity and accessibility of hiking in Vermont were mixed blessings at best. Attracting greater numbers of hikers, after all, meant a greater possibility that the "wrong" types of people might find their way into Vermont's unspoiled forests. As the Vermont historian Hal Goldman has shown, some GMC members actively discouraged Jews from joining the club, again reflecting an anti-Semitic undercurrent in Vermont's tourist history.[81]

Other GMC members worried less about the religion or ethnicity of hikers and more about the ways in which hikers used the trail, and for a time, the exploits of Irving Appleby of Roxbury, Massachusetts, became the centerpiece of their concern. In 1926, Appleby hiked the entire 265-mile length of the Long Trail in fourteen days and five hours—an incredible and highly publicized feat that set off waves of enthusiasm among other young hikers, who vowed to beat his record. Brash, temperamental, and self-aggrandizing, Appleby became a powerful new spokesperson for hiking in Vermont: the *New York Times, Boston Globe,* and *Christian Science Monitor* all covered his exploits, and hiking outfitters in Boston hired him for promotions.[82]

Appleby is easily one of the most colorful figures in Vermont's tourist history, but for the GMC he was a thorn in their side. Some GMC members welcomed the free publicity Appleby generated, but others were not so sure he fit in with the kind of image they had in mind for the Long Trail. Far better, it seemed, was the wholesome publicity the club received from the Three Musketeers, who, Taylor noted, "do not care to go in for the record stuff."[83] Part of the problem with Appleby was his uncouth and self-promoting behavior. His advice to the Vermont Bureau of Publicity on how to promote tourism more effectively, and his presumptions to speak on behalf of the GMC, rubbed many people the wrong way.[84] As 1926 passed into 1927, many GMC members were beginning to wonder if Appleby was too much of a liability for the club to associate with him publicly. In November 1927 the club formally soured their relationship with Appleby by asking him to furnish a written itinerary documenting the daily times and distances of his almost unbelievable record-setting hike. There were too many troubling inconsistencies in the stories that he was bantering about in the press, the club said; they just wanted to set the record straight.[85]

Appleby was incredulous. In a vicious letter to the GMC that the club sarcastically reprinted in the April 1928 edition of *The Long Trail News,* Appleby included an itinerary for his 1926 hike as well as an itinerary for an even faster hike he claimed to have made in 1927. In a burst of anger he threatened to "ram . . . down the throat" of any club member the lies they were spreading about him. He reminded the GMC of just how famous he was, and of how gracious its members should be to him for the free publicity he brought to the trail. And he compared his style of hiking and his kind of publicity directly to others, asking why the GMC chose to target him as a threat: "Why do you not question or criticise [sic] the 'Three Musketeers' as to whether they ever covered the Trail or not?" he asked. "You didn't question or criticise the girls for just one good reason," he sputtered, "*because they were girls,* young, pretty, popular and unusual."[86]

The GMC's criticism of Appleby hinged not only on his outlandish personality but on the threats that his style of hiking posed to the trail's use and identity. So-called "marathon" hiking was not the kind of thing the club had in mind when they built the Long Trail, nor did it match their conception of an unspoiled wilderness experience in Vermont. They envisioned the trail as a place of leisured communion with nature, not a place for hurry and competition reminiscent of city life. By the spring of 1928 many in the GMC felt that it was time to reassert their control over the trail's identity by marginalizing Appleby once and for all. In April the club passed a resolution of non-support for Appleby's style of hiking and for any use of the trail for "self-advertising and commercial purposes, for freak stunts and speed records." In a telling passage, the editors of *The Long Trail News* described the GMC's opinion as such:

Does [speed hiking] not tend to bring the Trail into disrepute with real mountain lovers, for whom it was designed? Unless this tendency were checked, there is no knowing to what lengths it might go . . . We might, in the course of a year or two, meet on the Trail stunt artists, with their sweaters emblazoned with the names of firms who made their boots and other articles of equipment. Record smashers would dash by, accompanied by camera men who would take motion pictures of them in various attitude and performing various functions. Marathon runners would rush along in both directions . . . Hot dog stands would spring up at short intervals to accommodate the crowds . . . And the genuine nature lover, for whom the Trail was intended, would be crowded out of the cabins and off the Trail entirely.[87]

A cascade of disasters like this left little doubt about who, in the GMC's mind, *should* have access to the trail (a nebulous category of "real mountain lovers") and who *should not* ("stunt artists" like Appleby). Moreover it leaves little doubt about the intensifying politics of identity surrounding the reworking of rural Vermont. A familiar cluster of questions was cropping up time and again in Vermont. Who would control landscape and identity in a state so profoundly influenced by tourism? What roles would work and leisure play in deciding the future of Vermont? What kinds of leisure activities were preferable and what kinds were not? And perhaps most importantly: Could Vermonters reconcile the concepts of accessibility and unspoiled Vermont? Was it possible to attract more and more visitors without compromising the beauty of the rural landscape? Questions like these took center stage in the late 1930s, as competing opinions about accessibility collided on the road to unspoiled Vermont.

Debating Accessibility in Unspoiled Vermont

In 1933 a New York State native and retired civil engineer living in Vermont named William Wilgus proposed the construction of a scenic parkway in the state. Similar in style to Virginia's Skyline Drive, then under construction in the Blue Ridge Mountains, Wilgus's "Green Mountain Parkway" would run north to south along the crest of the state's Green Mountains. Many Vermonters liked the idea. The parkway would create construction jobs during tough economic times and it would increase access to Vermont's unspoiled mountains, adding depth to the state's expanding tourist trade. Not every one agreed, however, and as soon as Wilgus's proposal went public, newspapers, civic groups, residents, and visitors all began taking sides. For the first time in Vermont history debates about tourism assumed a truly statewide scale. Opponents set the terms of that debate, framing it according to the language of accessibility and unspoiled Vermont. Some focused specifically on the Long Trail, arguing that increased automobile access to the Green Mountains threatened the future of hiking in Vermont. Others saw the proposal in larger terms. Perhaps, they suggested, Vermonters should consider the threats that accessibility posed to the entire concept of an unspoiled rural landscape.

Wilgus's parkway proposal was not entirely unprecedented in Vermont, but the timing of his proposal made it stand out by comparison to others that had come before. President Franklin D. Roosevelt's administration had recently earmarked ten million dollars in relief funding for each state to use on a large development project of their choice. To secure their ten million, Vermonters had to approve a project like Wilgus's parkway and then come up with whatever additional funding was needed to construct it. Those who supported the parkway plan, however, had to convince Vermont's fiscally conservative and staunchly Republican voters both to embrace Roosevelt's New Deal programs and to put up an estimated eight million additional dollars to build the mountain road. Wilgus tried to win over skeptics during the summer of 1933, writing articles and giving lectures throughout the state. He fielded complaints that his proposal would take "tainted money" from the federal government, that it was impractical or extravagant, that it would bring "undesirables" to Vermont, and that it would harm wildlife and forests. He assured critics that the road would pass mostly through state and national forest reserves, limiting the need to purchase property or rights-of-way along its length. Its construction would employ thousands, he explained, and once completed it would bring a steady stream of new tourist dollars to Vermont. Besides, he added, state gas taxes could easily finance maintenance costs, and the National Park Service (NPS) might someday take over the road's management, eliminating any financial burden for Vermonters.[88]

The NPS actually acted more quickly than expected. With the state's backing, Wilgus pitched his plan to the federal agency in the spring of

1933. NPS officials picked up on the idea quickly, envisioning the Green Mountain Parkway as part of a larger system of federally funded parkways they hoped to build along the entire length of the Appalachian Mountains. Landscape architects from the NPS and engineers from the Vermont Bureau of Public Roads teamed up in 1934 to survey a route for the road, and within a year they released a full report detailing their findings and their plans. The parkway would have a minimum buffer of five hundred feet on either side, and its right-of-way would be expanded at times to protect lakes, valleys, hillsides, and other scenic areas. Picnic grounds, shelters, and bridle trails would line its sides, but no gas stations, restaurants, or motels would be constructed along the road itself. Instead, the parkway would intersect with roughly twenty existing highways, giving motorists access to services in nearby towns. The road would terminate in a 20,000-acre "wilderness area" surrounding Jay Peak in northern Vermont, where travelers could stop at a full-service visitor center or connect to another proposed parkway running up into New Hampshire and Maine.[89]

In the opinion of the nationally renowned Boston planner and NPS consultant John Nolen, the Green Mountain Parkway was a wise and cost-effective plan capable of rendering the entire northeast "more quickly and more agreeably accessible to the nation."[90] To many, that made perfectly good sense. After all, Vermonters had been making scenery more accessible to visitors for decades. Whether visitors traveled by foot or by car, their ability to access Vermont's unspoiled landscapes was central to the success of tourism in the state.

The Green Mountain Parkway seemed entirely compatible with that logic, but that did not mean it was universally popular. Vermonters came up with a host of reasons to oppose the parkway — reasons, other Vermont scholars have shown, that ranged from opposition to the New Deal to persistent concerns about attracting the "wrong" kinds of people to Vermont.[91] One of the most compelling reasons to oppose the plan, however, was its potential effects on hiking. In his recent book, *Driven Wild*, the environmental historian Paul Sutter has argued that America's cultural embrace of wilderness evolved, in part, from concerns about road building and the spread of automobile tourism into wild, natural environments. Sutter views the idea of wilderness less as a "recreational ideal" than as a "recreational critique" — an alternative offered as a response to increasing pressures from automobile tourism.[92] Opponents of Vermont's Green Mountain Parkway voiced their concerns in similar ways, arguing that Vermont's unspoiled mountains and their value for hiking were under threat from automobile tourism. In fact, it took hiking enthusiasts no time to figure out that the parkway would run parallel to the Long Trail throughout the entire length of the state, opening up terrain that had formerly been accessible only to hikers.

In light of this fact, the GMC's initial response to the parkway was predictable enough. In the summer of 1933 the club published a statement of opposition, arguing that the parkway "would mean the abandonment of the Long Trail of the Green Mountain Club and would commercialize a section of the State that has so far been unspoiled but has been opened up by the Green Mountain Club's Trails to lovers of the outdoors in its natural state."[93] A road and a hiking trail situated in the same corridor were entirely incompatible, the club maintained; any attempt to construct a road adjacent to the Long Trail would be considered a threat to the interests of the GMC. NPS planners did what they could to reassure the GMC by arguing that the road would intersect with the Long Trail only six or seven times, and that it would typically run no less than a quarter mile from the trail, thereby staying out of earshot of hikers. GMC members were not convinced, however, nor were they likely pleased when NPS planners proposed a new state-long hiking trail of their own design. The Long Trail had too many problems, the NPS argued: use of the trail was down; it was poorly constructed; its shelters were often vandalized, and their sanitary conditions were unsafe. It seemed that only the federal government could provide hikers *and* motorists a safer, more agreeable means for accessing the wilds of unspoiled Vermont.[94]

The parkway clearly threatened the interests of hikers in Vermont, prompting the GMC to reiterate its opposition in 1934 by a vote of 272 to 196.[95] Nonetheless, opposition from the GMC was not a foregone conclusion. Prominent members of the GMC supported the project, arguing that it was in the best interest of the state as a whole, and that it reinforced rather than undermined the club's constitutionally mandated mission to serve the public good by making "trails and roads."[96] As the *Burlington Free Press* asked, did this mandate not dictate "that the mountains should be made accessible [by the GMC] in more ways than one?"[97] Even James Taylor — progressive reformer, architect of the Long Trail, and advocate of wilderness adventure — supported the parkway, arguing that it fit the larger mission of accessibility advocated by the club: "That was the way some of us felt about the [Long Trail] in 1910," he wrote, "and the Parkway seems just a natural and inevitable second chapter in the development."[98]

The fact that so many GMC members supported the plan suggests that there was something broader at work here than a self-interested attempt to protect the club. Debates about the Green Mountain Parkway went be-

FIGURE 3.9. The proposed Green Mountain Parkway, 1934. The Green Mountain Parkway's route—marked on this National Park Service map by the heavy north-south line in west-central Vermont—followed essentially the same route as the Long Trail, sparking debates about how best to define and access recreational space in unspoiled Vermont. Courtesy Special Collections, University of Vermont Libraries.

yond the GMC's efforts to protect hiking, and they went beyond issues of wilderness protection. Rather, these were debates about the wisdom of accessibility and about its compatibility with popular conceptions of unspoiled Vermont. Would Vermonters make their state more accessible to growing numbers of tourists, or would they try to limit the tourists' numbers and the kinds of activities they pursued? Was the goal of unspoiled Vermont to attract more people to the state, or to set restrictive standards on development? With no blueprint for accessibility and no clear definition of "unspoiled" to work with, the parkway became a symbolic referendum on the future of tourism in Vermont.

Parkway supporters and parkway opponents both used fears about spoiling unspoiled Vermont to their advantage. For instance, some opponents defined unspoiled Vermont as a place free from the kinds of commercial development they feared the parkway would bring. "Vermont is now virtually the only unspoiled state, and only the mountains continue to be unspoiled," the *Rutland Daily Herald* proclaimed. "Attempts to commercialize them, especially by extravagant and impracticable schemes, should not be encouraged."[99] Supporters responded by reminding Vermonters that commercial development would be prohibited along the parkway's route. The parkway, John Nolen wrote, would have "no hot dog stands, no tourist camps, in fact, no unsightly structures of any kind. Nor would it have any billboards." Contrary to what some might think, he argued, the parkway's entire objective was to *maintain* the unspoiled nature of Vermont's wild uplands. Many places along the parkway's proposed route were likely to be opened by roads someday, supporters pointed out. Without parkway status, no one could guarantee that these future roads would be free from commercial development.[100]

Many opponents refused to buy into arguments like these, maintaining instead that the parkway would bring an unsustainable number of visitors to Vermont. No matter how well-controlled the parkway itself might be, towns in the valleys below the parkway would be overrun by the "hot-dog stands, filling stations, cheap amusement parks, [and] unsightly lodging shacks" that seemed sure to follow.[101] Just look at other tourist destinations around the country, they insisted. Wherever there were large numbers of tourists, there was a greater potential for the kind of sprawling, cheap, "hot-dog stand" development that so many feared. Whether it was in the mountains or in the valleys, that kind of development did not match spirit of unspoiled Vermont.

Aside from numbers, opponents also feared the *kinds* of visitors that they felt a parkway would bring to the state. As argued by the GMC's Roderic

Olzendam in a letter to David Howe, the influential editor of the *Burlington Free Press*, "Why do you want to make it so easy for all sorts and conditions of men and women from the eastern states to flood your 9,000 square miles? You can't put a sieve at the Massachusetts end of the highway and let in only those people whom you think would be good residents."[102] Property-owning summer residents expressed similar concerns. The influential Greensboro Association (90 percent of whose 140 members were summer people) announced its official opposition to the parkway, while non-resident members of the GMC (many of whom were also summer home owners) voted by a margin of 143 to 79 against the plan—a margin wider than that of resident members.[103] According to one summer resident, the parkway would "adversely affect the value of my property and destroy its usefulness to me for reasons which are perfectly apparent. Evidently, Vermonters feel they are receiving a gift of ten million dollars and are willing to exchange this substantial, permanent and ever increasing summer population for a mob of cruising tourists from New York City, whose propensities are well known."[104]

Not everyone feared this "mob" and their "propensities," and many considered the proposal an appropriate way to open the mountains to all visitors—one that was in keeping with the democratic spirit of Vermont. James Taylor branded parkway opposition "a case of stacking up the special suppositions of a few against the possible good of the people as a whole."[105] And in David Howe's opinion, opponents were guided by "pure unadulterated self-satisfaction, or exclusiveness, or of the refined forms of selfishness that few care to see analyzed and exposed to the critical gaze of their neighbors."[106] Such criticisms appeared to have rolled easily off the backs of some like Roderic Olzendam. In a particularly melodramatic moment, he argued that he would like to "build a wall around the state and have but one gate. At this gate, if you came in an automobile, you would have to park your car outside and either take to horseback, go on foot, or in a carriage. Then, through the effort expended and the inevitable leisureliness of the trip, people would come to really know the glories of the Green Mountain state."[107]

The "glories of the Green Mountain state"—at least as defined by Olzendam and other parkway opponents—appeared to have triumphed in March 1935, when the Vermont State Legislature rejected the parkway proposal by a resounding margin of 126–111. Later that year, however, Governor Charles Smith asked the legislature to reconsider its vote, at which point they turned the issue over to the people to decide by a special vote on Town Meeting Day in March 1936. Both sides of the issue scrambled in the coming months to make their points and to shore up support for their causes. But despite the best efforts of supporters, Vermonters rejected the proposal by 43,176 to 31,101.[108]

Roadside beautification, the construction and management of the Long Trail, and debates about the Green Mountain Parkway all reveal defining characteristics of Vermont's early-twentieth-century tourist history. They reveal tourism's expanding economic role and the truly statewide scale at which tourism now operated. They reveal a willingness on the part of many different social groups to transform a state otherwise known as timeless and uncontrived in order to bring it in line with their perspectives on rural life. They reveal the growing power that non-residents were exercising over matters of land use, the growing power of leisure to shape the contours of landscape and identity in Vermont, and the continued importance that work-leisure relations played in discussions about the future of the state. Perhaps above all else, they reveal the escalating social tensions that by the late 1930s seemed endemic to tourist development in Vermont.

The noted Vermont author and parkway supporter William Hazlett Upson summarized the essence of those tensions in 1934 as he contemplated the future of the Green Mountain Parkway: "From a theoretical point of view," Upson wrote, "it might be desirable to leave our state exactly as it is. But from a practical point of view, it is impossible to leave things as they are." He continued: "Our problem, then, is not how to keep Vermont unchanged by futile attempts at isolating ourselves and discouraging visitors. We must accept the fact that we are going to have an ever increasing number of visitors. And the only thing we have to consider is the problem of how we can best handle this tourist traffic."[109] Upson's comments capture the challenges facing residents and visitors as they contemplated the reworking of rural Vermont — challenges that were both deep and serious. Vermonters would have to find ways to protect the attractive qualities of the rural landscape without compromising its economic value for the tourist trade. They would have to find ways to adjust their conceptions of a working rural landscape to fit the demands of a leisure economy. More than ever before, Vermont's rural landscape was positioned firmly at the intersection of diverse and competing demands: the demands of road and trail, visitor and resident, change and tradition. Together, Vermonters and their visitors would have to learn to navigate this middle ground.

As we shall see, issues like these would return with renewed vigor in the decades to come. But it was here in debates over roadside beautification, the Long Trail, and the Green Mountain Parkway that the first truly state-wide concerns about balancing growth and scenic beauty burst onto the scene. In this sense, debates about accessibility and unspoiled Vermont represented a turning point in the history of Vermont tourism. From the 1930s forward, many Vermonters would be as worried about protecting the state's landscape *from* tourism as they were about protecting it *for* tourism. As the *Rutland Daily Herald* put it in 1933, "If, in our ambition to attract visitors from the outside, we forget to keep the Green Mountains green, we are forever lost."[110] This was the lesson that many Vermonters took with them as they headed into the middle decades of the twentieth century. How to put that lesson into action became a lasting riddle in the reworking of rural Vermont.

CHAPTER 4

The Four-Season State

Creating a New Seasonal Cycle

In 1939 the Stephen Daye Press of Brattleboro, Vermont, published a contemporary daily journal entitled *New England Year*. Written by a Vermont farmwoman named Muriel Follett, the book provides an intimate look at the seasonal cycle of farm life for one Vermont family. In reading Follett's work, one is immediately struck by two impressions about farming in Vermont. First is just how hard rural families had to work in order to keep their farms going. Although Follett makes a point to discuss her family's leisure time, and although she takes time out to comment on the natural beauty of Vermont, her entries more often than not focus on the working routines that so strongly defined the lives of rural women and men. Second is the strong seasonal cycle that guided the Follett family's annual routines. From winter logging and ice cutting to spring sugaring and planting, from summer cultivating and haying to fall harvesting and food preservation, Follett recounts a world profoundly shaped by the seasonal round.[1]

Three years prior to the release of *New England Year*, the Stephen Daye Press had published another book on rural Vermont entitled *We Found a Farm*. Written by Charles Speare, an urban resident and owner of a Vermont summer home, *We Found a Farm* shares a similar goal with Follett's book in that it, too, explores the rhythms of nature and life in rural Vermont. But the books differ in two very telling ways. First, Spear adopts a more leisurely view of rural life than Follett — one shaped by his social position as a visitor to the state and by an aesthetic view of rural life rather than a personal experience with agricultural work. Unlike Follett, Speare does not engage rural labor directly, either in practice or in writing; when he does take the time at one point to comment on a working scene, he does so from a distant vantage point and from a detached, romantic perspective, likening the scene to "one of the Inness landscapes." Second, Speare differs from Follett in his engagement with the seasons. *We Found a Farm* says little or nothing about spring, winter, and fall. Instead, Speare's diary reflects the constrained seasonal limits of his vacation-based experience in Vermont.[2]

Each of these differences between the two books — their degree of engagement with work and their degree of engagement with the seasons — offers a point of entry into some of the defining characteristics of rural tourism in Vermont between the 1910s and the 1940s. At first glance, their different encounters with work suggest a strong dichotomy between visitors' leisure-based experiences of the rural landscape and residents' work-based experiences. This division certainly existed, and as we shall see, it defines a great deal of what follows in this chapter. Nonetheless, if this division is drawn too sharply it tends toward oversimplification. As we saw in earlier chapters, for instance, some visitors sought encounters with rural labor, trying (if with mixed results) to understand rural land and life through the context of work. What is more, residents' lives were never defined by work alone; like vacationers, residents encountered the rural landscape through leisure as well, even if their leisure time was often limited by comparison to the time they spent working. Therefore, I do not mean to suggest in this chapter that we compartmentalize vacationers' perspectives exclusively with leisure and residents' exclusively with work. What I do suggest is that we think of tourism at this time as being characterized by a deepening overlap between work and leisure. As visitors and residents reworked the rural landscape according to the dictates of the tourist economy, they continued to transform the nature of work-leisure relations, maintaining the centrality of those relations to ongoing discussions about landscape and identity in rural Vermont.

Equally important, I argue that this deepening overlap between work and leisure was now being played out on a widening seasonal frame. The expanding spatial scale of Vermont's tourist economy explored in the last chapter was matched at this time by an expanding temporal scale, as opportunities for vacationing spread into all seasons of the year. This brings us back to a second difference between Speare's and Follett's books. Throughout the twentieth century, summer remained the most popular time of year to travel to Vermont; Speare's limited seasonal focus on summer was indicative of that season's dominance in the state's tourist trade. Nonetheless, tourism's exclusive associations with summer were beginning to break down at precisely the same time that Speare wrote his book. A number of factors contributed to this, including transportation improvements, aggressive image-making campaigns, and the agency of nature itself, as new interpretations of seasonal weather and new kinds of physical environments opened up new opportunities for travel. Increasingly, visitors were encountering Vermont through the context of seasons other than summer. As they did, they encountered new kinds of seasonal work and new kinds of seasonal leisure, redefining them according to their new status within the state's tourist economy and fueling a powerful discourse of expansion, this time on a temporal as well as a spatial scale. That expansion placed the four seasons themselves in a middle ground where the meanings, spaces, and practices of work and leisure associated with each season were now all open to debate.

Work, Leisure, and Seasonality in Rural Vermont

Farming remained a cornerstone of Vermont's economy and cultural identity throughout the middle of the twentieth century, even as dairying—Vermont's dominant agricultural pursuit—entered into a period of profound change. As noted in chapter 2, Vermont farmers had made a transition from sheep to cows during the second half of the nineteenth century, increasing the state's production of butter and cheese for sale to regional markets. By the early decades of the twentieth century, though, another transition was underway. Improvements in refrigeration, storage, and shipping were making it easier for farmers to sell cream and fluid milk to regional urban markets, thereby redirecting the thrust of the state's dairy industry away from butter and cheese alone. This reorientation toward fresh milk inspired a cascade of changes. The number of farms in Vermont declined from 35,522 in 1880 to 29,075 in 1920, and then from 23,582 in 1940 to 19,043 in 1950. Between 1930 and 1950 alone, Vermont's farm population decreased by 27.5 percent, or 30,766 residents. Even as the state's number of farms declined, farm sizes increased, as farmers tried to adapt to new markets by increasing production. In 1920 the size of an average Vermont farm was 146 acres; by 1950 that number had risen to 185 acres. Herd sizes also rose, as did the output of milk from individual cows, thanks to new breeding and handling techniques. In 1935 the state's farms had an average herd size of 12.9 cows. By 1957 the average herd size had risen to 23.4 cows.[3] Taken together, figures like these demonstrate the profound changes occurring in Vermont's mid-century dairy economy. Yet despite such changes, dairying (and farming as a whole) remained a central feature of the state's economy and cultural identity.

A suite of daily working routines defined the lives of Vermont's farm families — routines such as cooking, cleaning, mending, and milking that no farm family could avoid. The same was true on an annual scale. Each season of the year was defined according to specific and repeating working routines. Men performed annual rituals such as wood cutting, ice harvesting, and logging in winter; harvesting, butchering, and plowing in fall; sugaring, manure spreading, planting, fence repairing, and stone picking in spring; and cultivating and haying in summer. Women often helped with these tasks in addition to gardening, preserving food, and cooking extra to feed the neighbors and hired hands who helped with larger seasonal jobs. In addition, rural Vermonters took on the never-ending work of running their towns. Governance and roadwork were civic duties before they were paid positions in rural towns, as were volunteer committees on health, education, or the town poor.

Although their free time was limited, rural people also participated in a variety of leisure and social activities. Many of these activities, such as skating, swimming, and evening dances, were seasonal in nature, based on natural conditons and on the schedule of farming activities. Other activities such as hunting and fishing were both a pleasurable way to spend time and a valuable part of the home economy. Socializing among rural neighbors also took place through the context of work bees — communal events when rural people tackled large jobs like cutting ice, building barns, and haying. Work bees gave farm families a chance to socialize while at the same time accomplishing an important task. That dual purpose was welcomed by those who regarded excessive free time with suspicion — a suspicion, we have seen, that often colored residents' opinions of vacationers. For such residents, work was at the core of what it meant to be "rural." Hard work, Vermont author Lewis Hill recalls, "merited great respect" among Vermonters and "was the means by which we could be a credit to our family, ancestors, and country." By contrast, he adds, "Relaxation and fun, we were made to understand at an early age, although not completely evil, came mighty close. A body was always made to feel guilty about relaxing unless everyone else was resting as well, which was a rare occurrence in our neighborhood, except on Sunday. Playing was even more suspect. On a Sunday it was sinful, and on a weekday it interfered with good, honest work."[4] Thus, while rural Vermonters defined leisure as part of their cultural identity, that identity was more often shaped by the work and seasonal routines that gave structure to their lives.

A range of image-makers also used the seasons and seasonal work to define a coherent identity for Vermonters, often by turning to romantic and stylized sentiments. "His clock being the sun's cycle, with the seasons for its ticks," one author wrote for *Atlantic Monthly*, the rural Yankee lived "by the deliberate pulsations of the universe."[5] Perspectives such as this often leaned towards nostalgia, with writers portraying rural Vermonters as the keepers of America's pre-industrial traditions. Even amidst the complexity of the post–World War II world, for instance, Vermont seemed to remain steadfast and unchanged—a place where, according to one 1946 perspective, "the spirit of America-when-she-was-new somehow survives even in an atomic and predigested age."[6] This interest in and romanticization of rural seasonality, the historian Michael Kammen has argued, persisted throughout the twentieth century, both in New England and beyond. For generations, urban Americans have turned to seasonal representations—whether in word or image—as means for framing celebratory impressions of rural life.[7]

This romanticization of the seasons also offered tourist promoters an opportunity to expand the temporal scale of vacationing in Vermont by redefining landscape and identity according to a wider annual frame. Advertising campaigns from seasonal brochures to 1930s films such as "Vermont around the Calendar" promoted a new range of seasonal activities including skiing, apple-picking, fall-foliage viewing, and maple sugaring.[8] This seasonal expansion in advertising found its consummate expression in *Vermont Life*, a quarterly general-interest magazine covering tourism, agriculture, industry, community, and the arts. Created in 1946 as the voice of the Vermont Development Commission—Vermont's latest state-funded promotional agency—*Vermont Life* became a powerful force behind the reworking of seasonal life statewide. The magazine's editors used color photography, folksy language, and the catchy phrase "Vermont is a way of life" to construct what have since become some of the most enduring images of Vermont's seasonal landscapes.[9] During the course of the subscription year, readers were treated to seasonal photographic essays and special commentaries on seasonal rural traditions. In the process, they were urged to think about rural identity, rural tourism, and the relationship between work and leisure in ways that extended beyond the confines of summer alone.

Although promoters typically defined the seasons through stylized and nostalgic lenses, the reality on the ground, of course, was often very different. As the British literary and social critic Raymond Williams argued in his book *The Country and the City*, this disconnect between representations of rural life and the often overlooked realities of rural workers' lives has a long history in Western culture.[10] The Vermont experience reflects and informs that history as well. Romantic, tourist-directed perspectives on seasonality frequently oversimplified the experience of rural Vermonters, all of whom worked harder than image-makers typically suggested, and many of whom embraced modern farming techniques and modern farm machinery whenever they could afford to do so. Indeed, any apparent continuity between residents' lives and old-fashioned farming was often the product of economic constraints rather than a desire to preserve a pre-industrial heritage. Many traditional farming practices persisted into the 1950s simply because Vermonters could not afford to change their techniques.[11]

Tourism added to this distortion, but it did so only to a point. For all their reliance on motifs such as timelessness and stability, tourist promoters also relied on discourses of expansion, growth, and development to enlarge the temporal focus of the state's tourist trade. There was nothing traditional about some of the ways in which rural Vermonters were now being asked to manage their affairs, nor was it tradition alone that attracted visitors to new kinds of seasonal activities. Residents and visitors alike were rethinking rural tourism entirely, expanding the kinds of experiences available to vacationers in an effort to expand tourism beyond summer alone. Their efforts transformed traditional patterns of work and leisure according to a new tourist logic, reworking landscape and identity across all seasons of the year.

Spring

Early spring has always been a slow time for tourism in Vermont. As the snow melts, the ground thaws, and spring rains fall from late March through April, Vermont enters its infamous "mud season"—a cold and sloppy time of year barely indistinguishable from winter, a time of year when dirt roads could become nearly or entirely impassable for weeks on end. Nevertheless, early spring was drawn into the orbit of tourism through its association with what is now one of Vermont's best-known working rituals: maple sugaring. By the middle decades of the twentieth century, maple products (which include maple cream, granulated maple sugar, maple candies, and the more familiar maple syrup) shared the stage with milk as Vermont's best-known agricultural commodity. Unlike milk, however, maple products were marketed to non-residents as something more than household necessities. Buying maple products, consumers were told, offered a way to connect with the state's unique regional identity. Most visitors did not make that connection by traveling to Vermont to purchase maple products during early spring when maple sugaring was underway, although many did. Instead, most purchased them by mail or at roadside stands and gift shops during other times of the year. In this way, maple products and the work involved in producing them were linked increasingly to the state's tourist trade, fueling a seasonally based discourse of expansion and transforming the identity of spring for residents and visitors alike.

"Sugaring" refers to the process of collecting raw sap from the trunks of maple trees and boiling it down to produce concentrated syrup or crystallized sugar. As air temperatures rise in late winter and early spring, maple sap begins to "run" (or move up from the roots through the tree's inner layers of bark), preparing the tree to bud. Sap runs best during a peak period in March and even into early April when air temperatures are above freezing by day and below freezing by night. Once trees eventually bud out, the chemical composition of their sap changes, making it useless for sugar production. Colonists in New England learned to make sugar from Native Americans, and from early on they cultivated stands of productive maple trees known as "sugarbushes" to increase their yields. What New Englanders commonly produced — even into the early twentieth century — was not the light, clear maple syrup we are familiar with today but a darker, more refined form of crystallized maple sugar. They used this product in place of expensive cane sugar.[12]

Sugaring is one of the state's most distinct seasonal rituals, and for Vermonters of all stripes it lends a powerful, work-based identity to the early weeks of spring. For early-twentieth-century farmers sugaring season began each year by tapping trees with spouts from which they hung buckets to catch the dripping sap. Depending on the weather that year, farmers typically tapped trees and hung buckets at a time when the snow still lay deep in the woods, requiring them to break out roads with oxen or horses and pack down additional trails with snowshoes. When the sap began to flow, family members and hired hands trudged each day through wet snow or cold mud to empty sap buckets into a larger vessel mounted on a sturdy sleigh. The sap was then transported to the "sugarhouse," where it was boiled down to the consistency of syrup or crystallized sugar in long wood- or oil-fired "evaporators." Men and women alike guided the precious sap through the evaporator's multiple chambers, keeping watch to prevent it from burning or boiling over. Then, as now, it takes roughly thirty-five to forty gallons of sap to make one gallon of syrup, and for the sake of quality the sap must be boiled soon after it is collected. That meant that on a day when sap was running strong, husbands and wives often took turns tending to the evaporator well into the night, drawing off gallons of finished syrup into bottles and adding fresh sap to the evaporator as needed.

Once the sap stopped running in April and the hectic, sleepless pace of gathering and boiling ended, farmers cleaned, dried, and stored their sugar-

FIGURE 4.1. Sugarhouse near Berlin, Vermont, c. 1900. Sugaring was often a time for socializing among neighbors, yet gathering and boiling sap was also hard work, requiring physical strength, great patience, and careful planning from one season to another. Photo by George R. Bosworth. Courtesy Vermont Historical Society.

ing equipment before taking stock of the year's output. First, there was the home supply to think of. As they had done for generations, many mid-century farmwomen continued to use crystallized maple sugar and liquid syrup in place of cane sugar. Beyond their daily uses as sweeteners in coffee or on donuts, maple products figured prominently in regional recipes for baked beans, maple ham, maple cakes, and maple cookies, to name only a few. Explore any number of Vermont cookbooks and recipe lists and you will quickly get a sense of the cultural and economic importance of maple sugar and maple syrup to farm households.[13]

Second, there was the market to consider. Not all farmers sugared for the market, but to many syrup was an important means of generating extra income during a season of the year that was otherwise relatively slow on the farm. In 1909 alone, sugaring generated a million dollars in revenue state-wide. Residents tapped five and a half million maple trees that year, producing nearly eight million pounds of sugar and over 400,000 gallons of syrup, which together were valued at 20 percent of the nation's maple crop. Vermont's sugar makers sold poorer quality syrup wholesale for flavoring in liquor or tobacco, or for blending with other sugars to produce cheaper grades of syrup. They saved their better quality sugar and syrup to sell at a higher price to neighbors, mail-order customers, stores, and eventually tourists. Muriel Follett recounts in *New England Year* how she and her husband made springtime sales trips to Connecticut, where southern New England retailers took her husband's "dry Vermont twang" as "proof that they are getting a genuine Vermont product."[14]

Sugaring was hard work, but it was also an opportunity for Vermonters to mix work with a bit of neighborly sociability, and even a bit of fun. Neighbors swapped stories from the long winter as they helped one another gather and boil sap. Early spring was also a time for popular "sugaring-off parties," when neighbors gathered to celebrate the tail end of the sugaring season. Partygoers ate donuts and "sugar on snow"—a concoction of thick syrup and snow or crushed ice—followed by sour pickles to cut the overwhelming sweetness of the syrup. These parties bound residents to a common heritage and to a common identity built through shared encounters with the work of sugaring.

Sugaring lent a sense of identity to rural Vermonters, as well as to the season of spring itself. According to Muriel Follett, "Maple sugar making is as long and steady in the lives of Vermonters as the state itself. Through the years Vermont farm families have waited, with hearts and minds attuned to the slightest [seasonal] change, for that warm morning when their maple trees would be ready to yield up their sap."[15] Others viewed sugaring as an almost instinctual inclination among Vermonters — or, as one put it, as "something that gets in your blood."[16] According to another, "You can tell when the weather feels like sugar weather. It's the air. It's an urge that tells you you'd better be getting up there [to the sugarbush]. I don't know. There's something about it that just puts that feeling right into you."[17] Feelings such as this tied springtime and the work of sugaring more closely to one another, making each central to the larger identity of rural Vermont.

But what of sugaring's connections to tourism? How was sugaring linked to the state's leisure economy? What effect did such links have on the identities and landscapes associated with sugaring and spring? Promoters began defining sugaring as something unique and marketable for tourism by the early years of the century. Speaking to the Vermont Maple Sugar Makers' Association in 1916, James Taylor (who was then with a precursor to the VSCC, the Greater Vermont Association) urged his audience to exploit the tourist potential offered by sugaring and maple products. Hotel owners should always use their best maple syrup for guests, he argued, and they should always make information available to visitors about local syrup producers in case visitors wished to purchase syrup directly from the source.[18]

By mid-century, promotions geared toward "maple tourism" were becoming more common. The Vermont Development Commission was now allocating new funds for maple sugar advertising, for instance, and maple retailers were now reporting years when over 60 percent of their sales were made to tourists.[19]

To make a case on behalf of maple sugaring, tourist promoters and agricultural officials often deployed nostalgic, stylized representations of old-time sugaring methods, reproducing familiar, idealized impressions of rural land and life.[20] Vermont author Noel Perrin later dubbed this romanticization of sugaring the "Wooden Bucket Principle." For Perrin, this principle refers to a broader "tendency to imagine almost anything in the country as simpler and more primitive and kind of nicer than it really is." Pictures of New England typically exclude mundane realities like gas stations and school buses, Perrin argued, just as they exclude the twentieth-century sugar maker's use of galvanized metal buckets in favor of the more rustic wooden buckets of old.[21] Packaging also became a crucial part of this romanticization. Crystallized maple sugar had traditionally been sold to consumers in plain, functional paper wrappers. Now, however, marketers were making a more concerted effort to package maple products in ways that reinforced romantic discourses about rural life. Maple Grove Candies in St. Johnsbury, Vermont, for instance, packed their maple products in a "unique 'Sap Bucket' of natural wood."[22] And syrup cans were now being painted with scenes depicting traditional sugaring techniques, some of which had gone out of fashion long before.

General notions of purity or cleanliness informed tourist-directed promotions of maple sugaring as well, suggesting to consumers a symbolic link between maple products and the essence of rural life.[23] The state's publicity director, Walter H. Crockett, suggested this point to an audience of sugar producers in 1921. Maple producers should cultivate a class-based identity of taste and elegance for their products, Crockett argued, treating maple syrup as an "aristocratic commodity" capable of appealing to "city physicians or bankers," to "high grade automobile owners" or any visitor willing and able "to pay good prices for gilt edged goods." To do so, they should emphasize maple products' associations with "cleanliness, the sweetness and beauty of the forest, and nothing unclean or repulsive." That meant convincing the public "that the light colored maple products are pure and that the dark products are impure."[24] Suggestions like these dovetailed with both a tourist-based approach to marketing and a more general transition occurring in the sugar industry itself. With national prices for cane sugar falling and production, packaging, and transportation techniques

improving, Vermont's sugar makers were increasingly producing clear, deli-
cately flavored syrup rather than the dark, strong-flavored sugar many had
traditionally produced. That trend, of course, affects how we think about
maple products today.[25] The production of lighter-colored syrup, coupled
with aggressive marketing to non-residents, Crockett and others suggested,
would make syrup a hallmark of rural identity in Vermont. Light not dark,
clean not dirty, maple syrup symbolized the best of rural American life.

Arguments like Crockett's point to the larger redefinition of sugaring
and of early spring that accompanied the seasonal expansion of tourism. To
make that redefinition possible, visitors had to be convinced that spring
and the work of sugaring offered meaningful ways to define and encounter
rural life. Visitors were encouraged to make this connection in a couple of
different ways. On the one hand, they could travel to Vermont at the dawn
of springtime to witness sugaring first hand. The owners of the Harlow
Sugar House in Putney, for instance, encouraged visitors to come to Vermont in March and early April, when they could "ride on our horse drawn
sled, eat sugar on snow and watch us make maple syrup."[26] It is impossible to
know how many visitors braved bad roads and volatile weather to see sugaring in action, but their numbers were certainly far lower than those who
visited the state during the height of summer. Yet there was enough interest in seeing sugaring first hand for farmers and the state to continue allocating money to advertising campaigns. On the other hand, visitors could
encounter sugaring and the arrival of spring in Vermont through the act of
consumption. It is entirely common for Americans to create group- and
place-based identities through the products they buy.[27] Recognizing this,
promoters suggested that non-residents could connect intimately to rural Vermont through the purchase and very literal consumption of maple products. If you could not go to Vermont in early spring, the season and its rural
traditions could come directly to your kitchen table.

From the perspective of maple promoters, Vermont's producers and retailers had to be trained to think about sugaring in ways that fit its associations with tourism. Walter Crockett's advice to the Vermont Maple Sugar
Association about the purity of their products offers a good example. Rather
than treat maple syrup as a traditional household commodity, rural Vermonters were now being encouraged to see it as something that evoked
nostalgic sentiment among non-residents, something that should be sold
not as a cheap substitute for cane sugar but as an expensive specialty item.
Maple products, Vermonters were told, should be thought of as luxury
goods worthy of gift status among discerning, even wealthy, shoppers. That
required Vermonters to think of maple products more as a tourist might

FIGURE 4.2. Sugarhouse and welcome sign, no date. Not all Vermonters would have wanted tourists to drop by their sugarhouses. Yet a sign like this was both an invitation to visit and an educational marker for visitors who might not otherwise know how to identify a sugarhouse from other farm buildings. Photo by the Vermont Travel Division. Courtesy Vermont Historical Society.

think of them, rather than as rural Vermonters had thought of them for generations.

Sugaring's sentimental, tourist-directed identity did not always match the producer's actual working experiences. Despite the old-time imagery adorning syrup bottles, mid-century sugar production was moving in decidedly more modern directions, particularly as producers looked for ways to improve yields through scientific research and the use of laborsaving technologies. Tractors were replacing the horses and oxen used for collecting sap. State-mandated specifications about the purity and density of syrup were forcing farmers to learn new techniques. And while most mid-century farmers continued to collect sap in galvanized buckets (wooden ones had long since been burned, piled out of the way in the barn, or sold to tourists), plastic tubing was now being used by some to deliver sap directly from trees to the sugarhouse.[28]

That disconnect between image and reality, between the consumer's view and the producer's, suggests a deeper masking of the labor involved in sugaring and in farming more generally. As the geographer Don Mitchell has argued, landscapes typically disguise the true nature of their production — the work, the labor, the relationships that go into their making.[29] Likewise, sugaring's spaces of production — the sugarbush and the sugarhouse — were commonly represented to outsiders as sites of "colorful activity" and "fun" rather than hard physical labor.[30] Of course, non-residents were often told that sugaring was hard work. Yet popular representations more often than not downplayed the cold fingers, wet feet, sore backs, long hours, and worry that also defined sugaring. Indeed, if sugaring was so fun, why did sugar makers like the Harlow family of Putney invite visitors to "watch" the family sugar rather than join in to help? The answer, of course, is that sugaring was hard work, and although there may have been a certain sense of fun associated with it, sugaring was necessarily something more than a quaint pastime. A growing number of tourists may have defined spring through the context of sugaring. But that did not automatically mean they defined sugaring through the context of work.

Tourism's tendency to mask the reality of rural work extended to other seasons of the year and to other farm-related activities, suggesting a deepening disconnect between work and leisure even as the two grew more intertwined. One mid-century writer expressed this irony neatly when he wrote that "Late summer and fall, here in Vermont, are peaceful times, times of drowsiness and surcease." But for whom was this so? After all, just a few short paragraphs later this same author contradicted his claims about "drowsiness" by noting, "The second crop of grass, the short rowen is being

mowed and pressed into compact bundles in the fields, redolent with red
clover and alfalfa, to fill the long barns, next January, with the odor and feel
of summer."[31] By failing to mention the energy, the effort, the long days,
the extra cooking, and the dusty, heavy labor that went into haying, this
author, for one, reinforced a fundamental disconnect between visitors'
leisure-based conceptions of rural life and the reality of residents' working
lives. Similarly, journalist Philip DuVal expressed concern that he and his
wife would have to work should they decide to take a vacation on a Vermont farm. His wife quieted his worries, and perhaps those of his readers,
by reassuring him, "No, you just sleep, eat, and watch other people work."[32]
How, in light of such comments, could visitors and residents ever see eye
to eye? How, in light of tourism's growing influence over sugaring itself,
could residents retain a meaningful sense of control over the definitions
and practices associated with rural work?

Winter

Spring tourism drew aspects of traditional rural work into its orbit, redefining them through their convergence with leisure. The origins of winter tourism in Vermont tell a slightly different story, although it, too, is a
story of tradition and tourism coming together. Rather than begin defined
as a convergence of rural work and leisure, winter tourism's origins in the
1910s and 1920s were defined by a convergence between traditional types of
rural *recreation* and the expanding tourist economy. Winter tourism's earliest incarnations emerged out of residents' pre-existing leisure activities
and leisure spaces, including those used for skating, sliding (or sledding),
tobogganing, and snowshoeing. Promoters and entrepreneurs eventually
added skiing to these, capitalizing as best as they could on the unique, seasonal opportunities offered by the nature of winter weather. In the process,
they redefined the identity of winter in Vermont for visitors and residents
alike, building the foundation for a successful winter-tourist economy.

Regional commentators have often described winter as a time when
rural Vermonters hole up around woodstoves to play checkers and wait in
unspeaking silence for spring. Compared to other seasons of the year,
many mid-century Vermonters did find themselves with extra time on their
hands during winter, but winter remained a busy time of year nonetheless.
Farmers cut ice from local ponds for home iceboxes or to keep milk cool
during summer until the milk truck arrived—an activity that remained
important into the 1930s in areas not yet reached by rural electrification.[33]

Some worked as loggers for lumber companies or harvested timber on their own property, both for fuel and for sale to local mills. And of course routine activities such as milking, cooking, and cleaning continued throughout the winter, just as they did the rest of the year, the only difference being that they were made more difficult by snow, cold temperatures, and darkness. Many of us today could not even imagine waking up in a dark, poorly heated home on a sub-zero day to milk a herd of twenty cows before breakfast.

Snow and cold also made it more difficult to leave the farm, lending some truth to the image of winter as a season of isolation. Radios and telephones reduced that sense of isolation to a degree, but the chance of being snowed in for more than a day at a time was still quite high for many mid-century families. Those who lived near villages or along main roads and established milk-truck routes typically had the benefits of snowplowing and winter automobile travel, but those who lived along more isolated back roads often had a more difficult time getting to town more than a few times each winter. By the early 1930s state and town crews were plowing snow on over 2,500 miles of Vermont's 3,500 miles of improved roadways. Many towns continued to roll their back roads, packing them down to make it easier to travel by sleigh than by car. This process of "snowrolling" worked well enough until warm spells softened the snow, making travel sloppy if not impossible for weeks at a time.[34]

Despite limitations to travel, winter was more than a time of hibernation broken only by the first run of sap in the spring. Winter's slower pace also meant that it brought opportunities for socializing and fun. Meetings of the Grange and the Ladies' Aid Society, educational forums, movie nights, and school plays were all common forms of winter entertainment in rural communities.[35] In addition, neighbors often came together on winter evenings for informal gatherings, sometimes called "kitchen junkets." Guests would move furniture out of the host's kitchen or some other large, warm room to make way for music, dancing, and socializing. Visitors would eat baked goods, drink coffee (and sometimes hard cider), take up a collection for the band, and dance into the night. In some cases, residents recall these parties happening just about every Saturday night.[36]

The nature of winter weather also offered Vermonters of all ages the opportunity to participate in a variety of outdoor recreational activities. Most important among these were what Vermont author Charles Edward Crane affectionately called "the three S's — skating, sliding, and sleigh-riding."[37] Before the birth of resort-based ski tourism in mid-century Vermont (see chapter 5), residents of all ages pursued winter activities such as these on just about any millpond, hillside pasture, or rolled country road. Others

pursued winter sports such as ice boating, tobogganing, and sliding with
"traverses" — large, fast-moving sleds with runners and side rails that were
capable of holding up to a dozen riders at a time. Turn-of-the-century Vermonters also founded winter sports clubs to coordinate recreational activities in larger towns and smaller rural communities. Although these clubs
waxed and waned in popularity, winter recreation remained an important aspect of village and farm life from the end of the nineteenth century forward.[38]

Winter sports clubs turned out to be pivotal to the state's earliest winter-
tourist economy. Turn-of-the-century clubs in towns from Burlington to
Woodstock to Stowe periodically organized winter carnivals to win new
converts to winter sports, boost civic pride, and attract out-of-state visitors
to Vermont during a decidedly slow time of year for tourist travel. These
winter carnivals may not have drawn tremendous numbers of vacationers,
but they drew enough to put winter on the adventurous traveler's mental
map and to convince tourist entrepreneurs to begin thinking seriously
about the economic potential of winter sports.[39] As one promoter wrote in
1907, "To the weary people of the crowded cities that would like the experience of sleigh-rides, coasting, tobogganing, snowshoeing, skating, ice-
boating, in the snappy bracing air of a genuine Vermont winter, a weekend visit to some comfortable village in the state will afford rare delights
that maybe have only been read of in story books."[40] Promotional outlets
like *The Vermonter* reinforced sentiments like this, publishing notices and
feature articles on winter recreation. By the 1910s it was becoming clear
that many visitors were eager to participate in winter sports, and that traditional patterns of rural recreation could be put into the service of a modest
winter-tourist economy. That realization opened the door a bit wider on
the seasonality of tourism.

By the 1920s skiing's popularity and its importance to Vermont's winter-
tourist economy was beginning to match that of traditional activities like
skating and tobogganing. Scandinavian immigrants first brought skiing to
communities in New England and the upper Midwest during the second
half of the nineteenth century. Immigrants typically skied as much for
transportation as for fun, but they also organized recreational ski clubs and
ski tournaments, both of which caught the attention of native-born Americans. As a recreational term, "skiing" was first used to describe a number
of different activities. One of the most important of these was ski *jumping*,
which by the 1910s and 1920s had become a wildly popular event at winter
carnivals across the northern United States. Citizens in towns from Duluth,
Minnesota, to Berlin, New Hampshire, to Brattleboro, Vermont, built ski
jumps and held regional and national tournaments that drew thousands of

WINTER SPORTS

are "Regular Routine" in

VERMONT

The Green Mountain State is each year welcoming more guests who are attracted by the unusual facilities this region offers for the enjoyment of winter sports.

The scenic attractions of Vermont in winter are superb. Snow-blanketed hills, white-burdened evergreens, frozen brooks and splendid fresh-crystalled landscapes lure the sports lover to this invigorating winter atmosphere. Snow-shoeing, skiing, tobogganing, skating, tramping, sleigh- and straw-rides are all a part of Vermont winter life.

The Vermont hotels are for the most part all-the-year hostleries—offering a warmth and cheer that is complete and unique.

To anyone interested in winter sports we will gladly send our free book on "Hotels and Boarding Houses in Vermont," and such other specific information as may be requested.

VERMONT PUBLICITY BUREAU, Montpelier, Vt.
Harry A. Black   Secretary of State

FIGURE 4.3. Winter sports advertisement, 1922. Advertisements like this, which appeared in *Outing* magazine, helped to expand Vermont's tourist appeal beyond summer alone, laying the groundwork for an early winter-tourist industry based on snowshoeing, skiing, tobogganing, skating, tramping, and sleigh- and straw-rides.

spectators. In addition to jumping, the term skiing also referred to a mix of what we would now call "downhill" and "cross-country" skiing. In an era before mechanized ski lifts, ski "running," as it was sometimes called, necessarily included both the uphill climb and the downhill run. And like activities such as snowshoeing or tobogganing, skiing could be done on any hillside pasture or forest trail.[41]

The historical record suggests that some native-born Vermonters began skiing for fun as early as 1900. By the 1910s skiing was earning wider coverage in the state's popular press, both as an activity of interest to residents and as an addition to the state's winter-tourist trade. Visitors by this time were taking to skiing in growing numbers, particularly by comparison to Vermont's more traditional wintertime activities, such as sliding, snowshoeing, and skating. That break with tradition was by no means instantaneous; visitors continued to embrace traditional activities like skating and snowshoeing, as did the promoters who catered to them. As one commentator noted in 1929, "It is no uncommon sight to see merry parties at the Grand Central in New York or the North station in Boston on a week end [sic] carrying snowshoes, skis and skates, and all ready for a few days on Vermont's snow clad hills and frozen waters."[42] The balance between skis and skates or snowshoes was shifting nonetheless, and in the years to come skis would become a far more common sight at places like Grand Central Station.

That shift was accompanied by broader, concurrent shifts in the meaning of winter and seasonal tourism in Vermont. One of the architects of this transition was a Vermonter by the name of Fred Harris. Harris was born in Brattleboro in 1887, and he learned to ski and make ski equipment at a young age. By the time he enrolled as a freshman at nearby Dartmouth College in 1907, Harris was already an accomplished skier and ski jumper. While at Dartmouth, Harris founded the now famous Dartmouth Outing Club, making a name for himself as a leading proponent of winter sports.[43] Upon graduation, Harris took his zeal for winter sports back home to Brattleboro, where he wrote widely about skiing and helped move it firmly into the state's tourist economy. In a 1912 essay entitled "Skiing and Winter Sports in Vermont," Harris outlined some of the broader trends associated with this transition toward tourism. In particular, he appealed to local farm boys to take up skiing as a fun and effective means of winter travel, and as a way to infuse the sport into the state's seasonal identity, not only among residents but among visitors as well. The state's economy would only benefit, he wrote, if the name of Vermont "became synonymous with winter sports."[44] Whether farm boys ever saw themselves as the ambassadors of winter tourism that Harris seemed to have in mind is impossible to know. What is clear is that Harris saw an opportunity to introduce a new leisure activity into the local culture and eventually to merge that activity into the tourist economy.

The success of winter tourism in Vermont depended on the ability of promoters like Harris to transform the meaning of winter along new, tourist-based lines. First, promoters needed to overcome any hesitations visitors might have had about the safety and comfort of winter travel. As a concerned Walter Crockett told Dorothy Canfield Fisher, too many urban New Yorkers "think one is taking his life in his hands to venture into Vermont in winter . . . I would like to know how the general public can be educated out of the idea that Vermont lies very close to the Arctic Circle."[45] The most obvious way to reeducate visitors about winter safety was to emphasize the modernity of Vermont's roads and the care now being taken to maintain them during winter. From Charles Edward Crane's optimistic perspective, "There's scarcely a day in winter when you couldn't traverse the state fairly comfortably in your heated car all the way from the Massachusetts state line to the Canadian boarder . . . We have well-plowed roads now, and when they happen to be at their best in winter, they make an unusually smooth surface." Others echoed Crane, citing paved roads and diligent snowplowing as evidence not of winter's isolation and hardships but of its safety, comfort, and accessibility.[46]

Second, winter tourism's success depended on the visitor's ability to understand and embrace the very essence of winter in Vermont. Promoters felt compelled to teach visitors to appreciate winter's moods, beauty, and charm, making them familiar with winter in the same ways that they were familiar with summer. Charles Edward Crane's 1941 book *Winter in Vermont* was the most thorough attempt to do this. It touches on a surprising array of topics, from weather and scenery to folklore and recreation, emphasizing to the reader the uniqueness and charm of this underappreciated time of year. Crane even went so far as to take other Vermont authors to task for ignoring winter. Wallace Nutting, for one, "apparently took flight with most of the other city folk and didn't stay to see even the first snowfall."[47] The success of winter promoters like Crane depended on their ability to construct a scenic identity for winter that rivaled summer — an identity that defined Vermont as the "Switzerland of America." According to Walter Crockett, "Vermont is as beautiful in winter as it is in summer. The blue and white effects that one gets by looking across Lake Champlain at the Adirondack Mountains, and the rose tints that one gets on Mansfield from [the University of Vermont], as the setting sun of winter shines on the whiteness of the mountains, are glimpses of beauty to be cherished in the memory as long as life lasts."[48] Beauty such as this — the dazzling sun and the polished whiteness of the landscape — was available only in winter; according to promoters, visitors needed only to get in their warm cars and drive in comfort to see it.

Promoters also depicted winter as a time of exhilarating action, vigor, and health — a time when exposure to the outdoors made one stronger, both in body and in character. Vermonters became the embodiment of this message. "When snow falls on Vermont," Crane argued, "I like to think not only that it is remolding the landscape, but that it is also reshaping our character, to the extent at least that it renews our spunk."[49] That spunk was not reserved for Vermonters alone, however. Rather, promoters suggested, it was available to all who embraced winter recreation. As Walter Crockett stated, "The exercise obtained from mountain climbing, snowshoeing, coasting, skating, skiing, etc., make the blood run more swiftly, give a healthy glow to the body, induce hunger and benefit the office worker materially, sending him back to his task, with a stronger body and a more contented mind."[50] Another went so far as to suggest that winter sports enthusiasts were "the reincarnation of New England ancestry" because of their rugged strength and ability to endure challenging winter conditions.[51]

Claims like this applied to both men and women, for although skiing and winter sports were often couched in masculine terms, young women often took to skiing as eagerly as men.[52] In one writer's view, men and women who visited Vermont in winter were both "well tanned and seem inured to weather. Some are bronzed veterans of the open. All make a business of being in condition. Cold, keen winds trouble them not at all."[53] Through assurances such as these, then, winter sports seemed to offer health and fun, as well as a cultural connection to the best of winter in Vermont.

Promoters needed to do more than convince visitors to come to Vermont in winter; they needed to convince residents to think about winter in entirely new ways as well. I explore this trend in greater detail in chapter 5, but it is worth introducing briefly here. Winter weather had actually been an important part of Vermont's working routines for generations. Snow and freezing temperatures, for example, made it possible to harvest ice and to skid logs out of forests more easily. But as the popularity of winter sports grew, residents were encouraged to re-envision winter in terms of its economic potential on the tourist stage. Winter weather took on an entirely new economic value through its associations with tourism; snow had now become "white gold," one commentator later noted, taking on all the trappings of a "commodity that can be enjoyed or sold to advantage."[54] Snow had been "rediscovered" by Vermonters, another added, and those who had once endured winter with "resignation and despair" now felt "more kindly towards the season their forefathers dreaded."[55]

Whether this claim was entirely accurate or not, winter sports had, in fact, added a new leisure-based value to phenomena like snow and cold, redefining their economic potential and in time transforming the nature of winter work in rural communities statewide. Winter recreation had once been an escape from work for rural residents, but now it was increasingly being absorbed into an expanding seasonal tourist economy. As that happened, hotel owners and local entrepreneurs began to dream up new ways to capitalize on snow's economic potential, redefining the meaning and value of winter, and expanding tourism's influence across a growing seasonal frame.

Fall

Vermont has always been associated in the popular mind with the color green—a trend that, in many respects, reflects the dominance of summer tourism in the "Green Mountain State." During much of the year, however, Vermont is actually far less green than its nickname suggests. For many months, it is white with snow or brown with bare vegetation and dead grass. And for a few weeks each fall, the state is bright with the color of changing leaves. One can find references in Vermont's tourist literature to the beauty of fall as early as the 1880s and 1890s, but it was not until the 1930s and 1940s that autumn was truly absorbed into the state's seasonally expanding tourist economy. That expansion hinged on the popularity of fall foliage and the fall harvest, both of which, I argue here, were shaped by their associations with the context of rural work. As fall's popularity grew among tourists, its working traditions were brought more firmly into the orbit of leisure, strengthening an expansionist discourse among tourist promoters and entrepreneurs. Tourism by mid-century had therefore made inroads into each of the four seasons of the year, redefining landscape and identity across the entire seasonal frame and, as we shall see in a moment, redefining the very nature of that frame itself.

Three historical-geographic patterns converged to make fall foliage touring popular among visitors across northern New England. First was the rising popularity of automobile touring and the new travel schedule that it made possible. Because automobiles made travel quicker and easier by comparison to trains, they allowed visitors from regional cities to take shorter, day or weekend trips to Vermont, New Hampshire, or Maine. This was a benefit for families with children who were in school during the fall, or for academics who could not get away from teaching duties once the fall semester began. A range of industry observers recognized this promising coupling of automobile tourism and fall. The National Park Service, to take one example, appealed to fall touring as a means for boosting the Green Mountain Parkway, arguing that fall offered "the possibility and probability of a considerable extension in the length of the recreational season." Indeed, national park planners predicted, "it may well be that the viewing of the autumn foliage may in time come to be one of the really outstanding nature pilgrimages of America."[56]

The park service's prediction has proven true, of course, in part because of rising rates of automobile touring, and in part because of aggressive new promotions defining fall as one of Vermont's most desirable seasons for travel. As weekend getaways became easier for middle-class easterners, a growing number of travel guides focused their attention on fall touring.[57] The success of such promotions rested, in large part, on their ability to portray Vermont's beauty as the product of more than the color green. Advances in color photography and color printing in the 1930s and 1940s helped enormously with this, making possible new color brochures and popular spreads such as those that annually graced the fall editions of *Vermont Life*. Increased state expenditures on fall tourism between the 1930s and the 1950s helped as well. From the Vermont Development Commission's new "Fall Tour Desk" came radio reports on foliage conditions, advertisements in regional newspapers, and updates sent directly to Radio City Music Hall's Vermont Information Center. As in the cases of winter and spring, state officials were now doing all they could to tie fall more firmly into Vermont's tourist economy.[58]

None of this would have been possible, however, without the concurrent ecological changes that made Vermont a land of brilliant fall color—changes rooted in the environmental history of farm work in Vermont.[59] Some of the literature's earliest references to fall coloring—such as an 1897 gesture to the "flaming foliage of the maple orchards" in Newport—were tied directly to Vermont's working landscape.[60] The fact that sugarbushes (rather than entire mountain vistas) were cited specifically as sites for some of Vermont's best fall coloring should remind us of the relative openness of the state's turn-of-the-century agricultural landscape. Even a quick glance at landscape photographs from the 1890s confirms the patchy nature of Vermont's forests and the extensiveness of its open fields, meadows, and pastures. Early autumnal color was therefore associated with smaller stands of trees, such as woodlots, riparian zones, and, in the case of Newport, sugarbushes, rather than expansive hardwood forests.

By the start of the twentieth century, however, the patchiness of Vermont's forests was giving way to a more extensive and contiguous cover. As logging declined, as farm abandonment continued, and as summer home owners took former farmland out of commission, the state's once open landscape began to grow back to trees. More importantly, it grew back with

FIGURE 4.4. Fall touring promotion, 1949. As the seasonal tourist trade expanded in Vermont, visitors were offered new ways to enjoy traditionalized scenes. Although this image, like many early fall promotions, was printed in black and white, it left little doubt that Vermont's "blazing colors" were sure to please the fall auto tourist. © and courtesy of *Vermont Life* magazine.

the kinds of trees that turned attractive colors in fall. In most places the pattern of ecological succession associated with this transformation favored deciduous, colorful tree species such as beech, ash, and especially maple, as opposed to evergreens such as white pine or spruce.[61] As early as 1892 one fall travel writer in New England took note of the "many abandoned fields gone to brush, mauve, maroon, crimson, and purple-colored with their dense growth of bushes, scarlet-lined along the fences by rows of sumac."[62] By the 1920s and 1930s, that brush had given way to trees across hundreds of thousands of acres of former farmland.

What this meant, of course, is that the success of fall tourism was inversely related to the success of agricultural and industrial work in the state: It was the very *lack* of work on many Vermont farms that allowed forests to return to the state's hillsides, thereby creating the opportunity for a new tourist activity. Even passing comments about sumac along the state's stone walls (such as that above) indicated that farmers were no longer cutting that pioneering species away from their fence lines.[63] The ecological changes that followed the transformation of rural work in Vermont therefore added nature's agency once again to the list of causal factors shaping the state's seasonal tourist trade.

We will return to fall foliage in a moment, but only after exploring another aspect of the season's tourist trade — the fall apple harvest. Generations of Vermont farmers grew apples in small orchards, primarily to meet domestic needs. During the early decades of the twentieth century, however, improvements in growing, transportation, and storage boosted the market potential of apples, prompting some Vermonters to plant larger, more specialized orchards. Vermonters planted more than 285,000 apple trees across 4,000 acres between the early 1910s and the mid-1920s alone.[64] Most of these trees were planted along the shores of Lake Champlain in towns such as South Hero, Shoreham, Ferrisburg, and Shelburne, and along the Connecticut River Valley in towns such as Putney and Brattleboro. Here, in proximity to the beneficial micro-climatic effects of water bodies, Vermonters grew varieties such as Baldwin, Northern Spy, and eventually McIntosh, each of which had unique benefits based on size, taste, or storability. During the early decades of the century, local men and women harvested apples at the state's larger orchards, earning a bit of extra income for themselves and their families.[65]

As in the case of maple syrup, promoters deployed the apple as a marketable tourist symbol, linking it in visitors' minds to the very essence of fall. They did so, in part, by defining commodities such as apple pies and apple cider, and activities such as apple picking and harvest dances as ro-

mantic icons of regional identity. Perhaps not surprisingly, that contradicted the experience of many farmers. Mid-century apple growers, *Vermont Life* suggested in a feature on Badlam Orchards in Ferrisburg, were
often quite modern and scientific in their approach to farming. That did
not mean, however, that magazines like *Vermont Life* hesitated to link
apple picking to the state's romanticized agricultural heritage.[66] As in the
case of sugaring, tourists were told they could connect to that heritage either by traveling to Vermont to pick a few apples themselves or by purchasing apples directly from local farmers. A growing number of farmers
began encouraging visitors to pick their own apples and to stop at roadside
farm stands, where they now sold at least some of their produce at higher,
retail prices. They got help from the Vermont Agricultural Experiment
Station, which took an active role in advising Vermont's farmers about how
to make roadside stands more appealing to passing travelers. By the mid-
1940s, the agency reported, 45 percent of those who stopped at roadside
markets were tourists.[67]

Vermont's mid-twentieth-century apple farmers sold their product through
the mail as well. The owners of Windy Wood Farm in Barre, for example,
began selling and shipping fancy, gift-wrapped apple packages during the
1930s. They expanded their efforts in the 1940s by shipping a wider array of
apple, maple, and cheese products from the Thanksgiving holiday through
the first of the year. And, of course, the farm's advertising brochures encouraged guests to stop in for a visit, see the apple harvest in action, and
watch apple cider being made on the farm's cider press.[68] Efforts like these,
coupled with the sale of apples at roadside stands throughout Vermont, encouraged travelers to identify the apple harvest as yet another reason to take
a trip to Vermont on a crisp fall weekend.

Fall and the annual harvest, of course, had long-standing identities
among local residents that were independent of tourism — identities that
were typically far less romantic and far more rooted in the practicalities of
work. Fall was a season of hard work and diligent preparation for the winter months ahead. That fact remained true for many Vermonters, even
after rural electrification and automobile travel began to ease many of the
challenges associated with rural life. Indeed, many mid-century families
continued to rely, at least in part, on generations-old methods of food
preparation and storage. Many continued to butcher animals in the fall,
salting or smoking the meat to preserve it. They gathered fruits and vegetables in late summer and early fall, preserving their crops for the winter
by drying, canning, root cellaring, or burying them in sawdust or earthen
pits. Fall meant making sure that the house's foundation was banked

FIGURE 4.5. Roadside stand near Bennington, Vermont, fall 1939. Even modest roadside operations like this allowed farmers to market produce at higher retail prices. Roadside stands also gave
auto tourists an opportunity to connect with the harvest on their autumn travels in Vermont. Photograph by Russell Lee, Library of Congress, Prints & Photographs Division, FSA-OWI Collection.

against the winter cold, and that the woodshed was fully stocked. Fall was
a season where the long-term benefits of work were measured across time
in terms of both food and warmth.

Apples played a key role in meeting these basic needs, transcending the
season of fall and linking the work of one season to that of another. To
understand how, we need to consider the many different methods Vermonters used to preserve apples for use later in the year. Apple preservation may seem like an odd tangent to take, but it turns out to matter deeply
to our understanding of seasonality and tourism in mid-century Vermont.

Apples store fairly well by comparison to other fruits, but they will spoil
in a matter of weeks without some form of preservation. Mid-century improvements in refrigeration certainly made it easier to store apples for home
use and retail sale over a longer period, but even so, many women continued to preserve apples through a variety of traditional means. Some dried
apple slices on racks placed on wood stoves, hanging those slices in bags out
of the reach of mice and rehydrating them later for use in pies or other
recipes. Others canned apples or ready-made apple-pie filling for future use.

By whatever means of preservation, the ability of farmwomen to store apples for future use was central to their families' nutritional needs as well as their sense of domestic self-esteem. As one woman later remembered, guests and hired hands often expected apple pie with their meals, and they often judged a woman's ability as a housekeeper by the quality of her apple recipes.[69]

Men preserved apples in an additional way as well. Each year thousands of Vermonters crushed and strained apples to produce cider, which they then preserved by converting it into a "hard," or alcoholic, drink. For contemporary Americans, apple cider is a sweet, refreshing beverage most often found on supermarket shelves in fall but, thanks to modern refrigeration, now available at other times of the year as well. Rural Vermonters have always drunk sweet cider too, but because of its short shelf life, they did so only during fall. Consequently, many converted sweet cider into hard cider, which resisted spoiling and when stored properly could last for well over six months. Hard cider is produced by fermentation — a process by which yeast is used to convert sugar into alcohol. It is cheap and easy to produce, and it remained a popular drink in Vermont into the middle of the twentieth century. For many, cider making was an annual ritual — one that could yield one or two barrels each year.[70] As one Vermont farmer recalled of his mid-century boyhood, "Everybody considered [cider making] the proper thing to do . . . It was [consumed] more or less on special occasions, like haying, or some party, celebration."[71] Cider therefore remained important to the cultural identity of early- to mid-twentieth-century Vermonters, informing work and leisure beyond the season of fall itself. Men who made cider — like their wives who preserved apples for cooking — preserved the bounty of one season in order to serve the needs of another.

Hard apple cider will eventually spoil, of course, but ironically "spoiled" cider was actually quite valuable to food preservation. When stored in open containers for over a year, apple cider will turn to vinegar, an essential ingredient in preserving food. Vinegar was traditionally used to pickle — and thus to preserve — a variety of farm products including meat, eggs, and vegetables. The most notable of these were cucumbers, or what most people simply call "pickles." Pickles were an important nutrient source during times of the year when fresh vegetables were scarce, and they were an essential part of sugaring-off parties in spring, where, as mentioned earlier, their sour flavor counterbalanced the intense sweetness of sugar-on-snow. In this way vinegar, like hard cider or dried apples, transcended the season from which it came, again preserving the bounty of one season to meet the needs of another.

Whether this seasonal transcendence was achieved through cider, vinegar, or dried apples, it should remind us that the lives of many mid-century

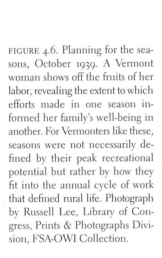

FIGURE 4.6. Planning for the seasons, October 1939. A Vermont woman shows off the fruits of her labor, revealing the extent to which efforts made in one season informed her family's well-being in another. For Vermonters like these, seasons were not necessarily defined by their peak recreational potential but rather by how they fit into the annual cycle of work that defined rural life. Photograph by Russell Lee, Library of Congress, Prints & Photographs Division, FSA-OWI Collection.

rural residents were still characterized — as they had always been — by close interconnections between the work of one season and that of another. One did not harvest apples, dry them, or turn them into cider or vinegar for immediate use; one did so to provide for other times of the year. One did not cut wood in winter for immediate use; one did so to heat homes the next winter, to make sugar in spring, and to cook food all year long. And one did not make maple sugar to last only through the spring; one made it to supply one's home and the homes of one's neighbors for months to come. The seasonal cycle of work in rural Vermont was just that, a cycle — *an interconnected series of events.* It was a progression of time with no clear beginning or end, a progression of time in which the activities in one season contributed very clearly to the quality of life, or even survival, in another. We can talk about this cycle without idealizing it and without romanticizing it as an indication of a rural culture rooted wisely in the rhythms of nature. Indeed, it is doubtful that very many rural Vermonters saw their lives in such romantic terms. Rather, the seasonal cycle was indicative of practical adaptations merged with tradition. It was, above all else, a *work-based* context through which Vermonters encountered the rural landscape.

Tourism's spread into all four seasons of the year introduced a new,

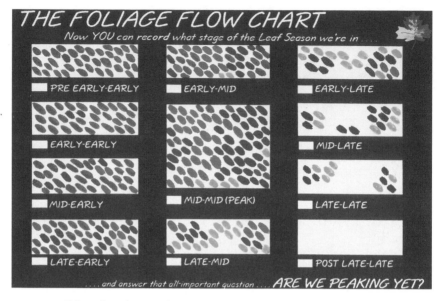

FIGURE 4.7. Foliage flow chart, no date. Many "leaf-peepers" will tell you that Vermont's fall colors go through a number of different peaks annually, as different species of trees show their colors at different times. But as suggested by this post–World War II postcard (originally produced in color), discerning travelers should be able to determine just when fall's seasonal offering was at its best. Copyright Leisure Years Unlimited. Courtesy Vermont Historical Society.

leisure-based context through which Vermonters and the guests would now encounter rural seasonality. Here we return to the case of fall foliage. According to the calendar, fall, like all seasons, is three months long. But naturally any three-month period in Vermont is characterized by a great deal of variation in weather, vegetation, scenery, and, by extension, leisure activities. Although mid-century promoters like Vermont Governor Stanley Wilson assured foliage viewers that it was "reasonably safe to take a trip" to Vermont from September 15 to October 15, fall colors are always better during some parts of this period than others. After all, deciduous tree species change color at different times and different rates, which themselves vary depending on factors such as latitude, elevation, and the weather in any given year. Consequently, visitors began to think of foliage as having a "peak" — a time when colors were at their very best and when, according to the traveler's perspective, it was most worth visiting Vermont.[72] As early as the 1930s, New England publicists used airplanes to monitor color changes and to provide up-to-date press releases for newspapers and radios about peak leaf conditions in northern New England.[73] The peak was not to be missed, they suggested, so visitors had better listen close.

By organizing travel plans around the foliage peak, visitors rejected the value of the long, subtle, and admittedly often dreary transition periods between summer, fall, and winter. The whole idea behind the foliage peak was that it contrasted sharply with other times of the fall season, when, presumably, rural Vermont was less beautiful and less accommodating to visitors' desires. The logic of the peak, moreover, applied to seasons other than fall as well. Just because it is "winter" does not mean that snow conditions are right for winter sports. Just because it is "summer" does not mean that conditions are ideal for outdoor activities like swimming. And just because it is "spring" does not mean that maple sugaring is underway or that the state's forests and fields are bursting with new growth. By learning to pick and chose from those parts of the seasonal cycle when conditions were best for certain activities, visitors bestowed a seasonal identity on rural Vermont based not on an interconnected cycle but on a *segmented sequence of episodes*, each of which represented a season, but each of which remained unconnected and often very distinct from the others. Unlike rural residents, who coped with Vermont's seasons in all of their moods, visitors who did not find conditions at their peak could always pick up and go home. Visitors could choose to control their experience of the rural landscape, fueling a sense of entitlement to a rural vacation that was "just right"—no matter what the time of year.

Before Vermont became a four-season state for visitors, it was a four-season state for residents, who shaped rural landscapes and defined rural identity according to a range of traditional patterns relating to work and leisure. As Vermont's tourist economy spread into all seasons of the year, however, new elements of rural work and rural leisure were increasingly folded into the tourist trade. Those participating in that economy were now reworking landscape and identity across a wider annual frame, yielding an expanding middle landscape situated at the confluence of changing networks of seasonal work and seasonal leisure. Whether through the context of winter, spring, or fall, tourist promoters and visitors adopted new kinds of imagery, activities, and places into their conception of rural land and life. New ideas about the value of nature—about its scenery, weather, and seasonal moods—coupled with changes in the natural environment itself generated new opportunities for visitors and residents alike.

Between the 1910s and the 1940s, then, residents' seasonal patterns of work and leisure were being reinterpreted and renegotiated within the framework of an expanding tourist economy. As that process unfolded, control

over the future of rural Vermont was also being reinterpreted and renegotiated. Tourism's expansion beyond its traditional summertime focus meant
that visitors now traveled to Vermont during seasons that had once been
the exclusive domain of residents. As tourism's reach spread into new seasons of the year, residents found themselves having to deal with tourists in
greater numbers and with greater regularity. They found themselves faced
with new choices about seasonal employment and new decisions about the
economic value of their state's rural landscape. And in time they would
find themselves facing a degree of tourist development and social conflict
unlike anything they had faced before.

CHAPTER 5

New Paces for Old Places

Creating Vermont's Mid-Century
Ski Landscape

If you thumb through the pages of coffee-table books and other regional portraits of New England from the 1960s you will find a steady supply of sentimental rural imagery. Take, for example, the 1967 Time-Life book *New England.* The book's chapter on "The Yankee Character" features the picture-perfect town of Peacham, Vermont, and reproduces faithfully the nostalgia and romanticism that generations of tourists had used to make sense of the state's rural landscape. Peacham's residents still lived a "quiet, unchanging way of life," the book claimed, one characterized by steadfast tradition and a "deep and abiding intimacy with nature." The portrait went on: "Here, in miniature, is another era — a community of self-reliant farmers, skilled artisans and small merchants . . . Here Thomas Jefferson's dream of a nation of small landholders and craftsmen is still alive. What for most Americans is but a dim ancestral memory of quiet and peaceful country pleasures is for Peacham a continuing reality."[1] Representations of rural Vermont such as these were firmly entrenched in the popular imagination by the middle decades of the twentieth century, and by this point in our story they are not new to us either. Indeed, what makes a passage such as this so important to this chapter is not so much what it says, but what it fails to mention.

If you traveled to Peacham in 1967 from any city or suburb you would have no doubt encountered a place that felt very different to you — a place where you might even imagine yourself standing in the midst of Jefferson's centuries-old vision for America. That might very well have been the kind of "rural" you were after: one where landscape and identity were defined by tradition, simplicity, and charm. But if you got in your car and drove a half hour outside of Peacham you would be in the heart of Vermont's ski country. There you would find a very different kind of landscape — one in which Vermonters and vacationing skiers were faithfully reproducing a discourse of modern technological achievement so widespread in post–World War II American culture. There you would find people who were celebrating and even encouraging changes associated with modernity and a "space-age" future. You might even wonder wistfully what happened to the Jeffersonian "Yankee Character" you read about in Time-Life's *New England*.

Then again, you might have found exactly what you came to Vermont for in the first place. And in that case you would have certainly not been alone. Just as mid-century publications like *New England* were flooding the market with sentimental imagery, tourist promoters and tourist entrepreneurs were also busy reworking rural Vermont along markedly different lines. This chapter and the next explore the nature of that transformation, the ways in which residents and non-residents responded to it, and the changing patterns of work and leisure that defined it.

By the 1950s skiing had become a primary tourist activity in Vermont as well as in neighboring New Hampshire, where many of the same kinds of trends explored in this chapter played out. Boosted by slick promotions that dubbed Vermont "America's First" in skiing and encouraged visitors to "Ski Vermont, the Beckoning Country," the ski industry captured a growing share of the state's promotional budget and tourist revenue. Resort owners and operators used radio, magazines, newspapers, and brochures to advertise annual improvements to trails, ski lifts, and other services. Skiers could read daily snow reports in regional newspapers. Or they could get all of the latest information on ski conditions by visiting the Vermont Information Centers scattered around the northeast.[2] Vermont was "beckoning," indeed. To sound its call, the state's ski promoters and ski-area developers highlighted their commitment not to a nostalgic vision of the past but to a modern, even futuristic vision of what Vermont might yet become. Skiers who came to postwar Vermont still wanted to find old barns and church steeples, and they still could. But they also wanted to find a highly modern and technologically sophisticated ski landscape.

Promoters and ski-area developers gave them what they wanted, turning to technology not only to boost the state's popularity among skiers but to manage the activities of ever increasing crowds, control the uncertainty of nature, and insure the future of their investments. In the process, ski-area developers redefined rural leisure, rural work, and the relations between

them, magnifying skiing's influence over landscape and identity in the state
and fashioning a rural middle landscape where residents and visitors would
be forced to think more critically than ever before about tourism's role in
the future of rural Vermont.

Transportation and the Consolidation of Skiing in Vermont

Skiing's popularity first took root in the 1920s and 1930s among adventurous travelers from New England in the east to the Rocky Mountains in the
west. But the sport's popularity boomed in the decades following World
War II, when middle-class Americans embraced skiing both as a recreational outlet and as a forum for economic self-expression during a period
of national economic prosperity.[3] Vermont emerged quickly as a regional
and national leader in skiing, although during the 1930s the state had nothing remotely like the large, full-service ski resorts that many of us associate
with skiing today. Early skiing in Vermont took place on an unorganized
and dispersed network of logging roads and pastures, fitting relatively unobtrusively into the state's preexisting rural structure. Some towns boasted
a few popular hillsides where skiers might buy lunch or hot chocolate at a
local farmhouse, fueling up for a day of climbing up hills and skiing back
down. Others offered little more than old logging roads where daring skiers
could plunge through the woods on trails that once served the timber economy. As one resident later recalled, Vermont's "farm and village hills were
laced with ski runs." Local children skied for free wherever they pleased,
while Vermont's earliest ski tourists paid little, if anything, for the privilege
of gliding through pastures and woods.[4]

From the perspective of entrepreneurs, the dispersed and unregulated
nature of skiing in the 1930s was a problem, as it made it difficult to create
a cohesive ski economy from which they could make money. To remedy
that, Vermonters began assembling a recreational landscape that consolidated skiing geographically into specific towns and more clearly defined
ski areas. By 1939 roughly forty Vermont towns claimed some sort of organized facilities for skiing, whether open trails, forested ski runs, mechanical ski tows, or ski jumps. In time, many of these facilities would provide
the nucleus for the state's larger resorts.[5]

Transportation played a vital role in this process of consolidation. Like
summer tourism, Vermont's winter-sports economy was built initially on
the railroad. Railroad companies capitalized on the expanding seasonality
of tourism in the 1930s by running special "snow trains" from cities such as

Boston, New York, and Hartford to ski destinations across northern New England. Some snow trains offered Sunday excursions only. Others set out after work on Friday evenings, dropping their passengers off on Saturday morning in the White or Green Mountains and returning them home in time for work on Monday. By the middle of the 1930s tens of thousands of young, middle-class urban residents were riding snow trains north each winter in search of healthy, sociable, and challenging recreation. The popularity of skiing among this generation of young travelers helped carry the sport successfully through the Great Depression and World War II, sustaining the interest of investors and ski-area developers.[6]

New England's snow trains helped to consolidate skiing in Vermont by encouraging ski development near railroad depots. As growing numbers of young people took trains north to ski, private ski clubs, railroad companies, and local entrepreneurs began investing in skier services in and around train stops such as Manchester, Brattleboro, and Waterbury. Skiing in such places also benefited from public funding. The Civilian Conservation Corps — a federal, depression-era relief agency — built and maintained many of Vermont's early ski trails, most of which were little more than narrow tracks through the forest, but some of which evolved into established resorts such as Big Bromley and Mount Ascutney. Most notable among the state's early trail networks were those on Mount Mansfield in the northern Vermont town of Stowe. Mansfield's popular trails and proximity to a rail stop in Waterbury quickly made it one of the region's best-known ski destinations.[7]

Snow train service ended in New England after World War II as rates of automobile ownership increased and growing numbers of skiers chose to drive to Vermont instead. Nonetheless, postwar skiing continued its course toward consolidation, only on a larger scale. Rising rates of automobile travel gave resort developers the confidence to invest in regions of the state not serviced directly by passenger rail, thereby expanding skiing into new parts of the state. In particular, skiing spread dramatically in the postwar years along the high, central spine of the Green Mountains, where skiers could find the state's highest rates of snowfall and most challenging terrain. Publicly financed improvements to central Vermont's north-south artery, Route 100, also gave motorists better access to the state's highest mountains, encouraging investors to open new resorts or expand preexisting ones along this emerging corridor of ski development.[8] The construction of Vermont's interstate highway system during the 1960s, coupled with the improvement of state roads linking the interstate to Vermont's ski areas, only accelerated this trend, as did the state-funded construction of "access roads" connecting resort parking lots to main highways. By the mid-1960s the Vermont legislature had spent roughly one million dollars for new access roads into some of Vermont's largest ski areas.[9]

Improvements in winter road maintenance also contributed to skiing's consolidation and formal organization. Unlike most other types of outdoor recreation, skiing is at its best precisely when the weather is at its worst for driving. That fact posed a significant problem for skiers as well as for developers interested in opening or expanding a ski resort. Both parties registered complaints with the state, noting that ski areas located off Vermont's main highways were often inaccessible after storms. State highway officials responded over the years by stepping up their commitment to winter road maintenance, and state promoters responded by making bold promises about the state's ability to keep roads open to skiers throughout winter months.[10] Local officials responded as well (often begrudgingly and with considerable debate at annual town meetings) by raising taxes and allocating larger percentages of their annual budgets to winter road maintenance. What was good for skiing, some were willing to argue, was good for the state as a whole.

Skiing and the Technological Turn in Mid-Century Vermont

Transportation systems were crucial to the consolidation and success of resort-based skiing in Vermont, but they would have meant little without the ski technologies that were being deployed simultaneously at ski areas from Vermont to New Hampshire, from Idaho to Colorado. From the middle of the 1930s forward, engineers and ski entrepreneurs worldwide manufactured an astounding array of ski lifts, trail-grooming technologies, and snowmaking systems, all of which allowed them to create and expand upon a decidedly technological leisure landscape.[11] Ski technologies did not necessarily determine the contours of that landscape outright, but they did allow resort owners and operators to exert at least some measure of control over skiers, over nature, and over the future of their investments.[12] Consequently, they allowed resort owners to transform landscape and identity in Vermont on new, dramatic scales, making technology a powerful tool in the reworking of rural Vermont.

Historians often trace the birth of modern skiing in the United States to a January 1934 conversation at the White Cupboard Inn in Woodstock, Vermont. As a group of tired New York skiers replayed the day's activities and nursed their aching muscles, the topic of conversation turned to skiing's most inescapable challenge: the uphill climb. Train service, trail networks, and good roads were one thing, but without any form of mechanized up-hill transport, skiers in the early 1930s still had to trudge uphill before being able to ski back down. This uphill climb meant the loss of precious energy and precious hours of winter daylight, both of which might otherwise be devoted to the downhill run. So, with $75 as a start-up incentive, the guests at the White Cupboard Inn challenged the inn's owners, Robert and Eliza-beth Royce, to develop a mechanical solution to the uphill climb. The Royces quickly secured information about an obscure "rope tow" used by skiers in Canada, and with the help of a local inventor they managed to have their own rope tow up and running a few weeks later on a hillside pas-ture owned by a farmer named Clinton Gilbert. The tow was a modest contraption. Its 1,800 feet of continuous rope wound its way through a se-ries of sheaves powered by an old Model T Ford. As modest and clumsy as it was, however, the tow gave Vermonters a proud place in the history of skiing, and it gave ski promoters an opportunity to dub Woodstock the "Cradle of Winter Sports."[13]

Other entrepreneurs followed suit, in Vermont and elsewhere, experi-menting with a range of ski lift designs, many of which are still in use today. These ski lifts typically fall into three classes. First, "surface lifts" (such as rope tows, t-bars, j-sticks, and poma lifts) pull skiers uphill with their skis on the ground, either by providing a continuous rope to hold on to, or by providing bars or platters against which skiers lean for support as they are pulled along. Second, "chairlifts" carry skiers uphill in chairs attached to a cable and suspended above the ground by a series of towers. Between the 1940s and the 1960s chairlifts seated either one or two passengers at a time and were commonly regarded as the most efficient and cost-effective method of moving skiers uphill. Finally, "gondolas" and "tramways" carry skiers uphill in enclosed cars capable of holding a handful of passengers (in the case of gondolas) to as many as sixty or seventy at a time (in the case of tramways).

Rope tows were the most popular among these during the 1930s and 1940s. By 1948 rope tows outnumbered chairlifts and all other surface lifts at Vermont's fifty-four mechanized ski areas by seventy-nine to ten.[14] Rope tows had a number of important advantages for Vermont's early ski entre-preneurs. In particular, they were small, low to the ground, and relatively easy to build—all of which placed them within the means of private ski clubs, municipal parks, and the dozens of individual landowners who opened small-scale ski operations on their land. Rope tows also fit neatly into the kind of ski terrain favored by many early skiers. Compared to difficult trail systems cut in forested, high-elevation terrain (such as the trails on Mount

Mansfield), ski trails serviced by rope tows were often located on the hillside pastures to which many skiers in the 1930s were accustomed. The rope
tow simply made it possible for skiers to make more downhill runs each day.

Like transportation systems, rope tows helped to consolidate skiing in
specific places where entrepreneurs could now charge money for the services they provided. Vermonters built new clusters of trails around rope tows
in towns throughout the state. These tows and the trails that surrounded
them were either operated by the landowners themselves or managed by
leaseholders. In either case thousands of skiers were soon paying about a
dollar a day for the privilege of using them.[15] Rather than encouraging vacationers to ski on any old pasture or logging road, then, rope tows helped
draw people to very specific places and then keep them there all day. Such
places now had names and signs out front. They could be mapped, advertised, and improved upon. They took on specific identities based on the
nature of their trails and tows. In short, they became identifiable "ski areas,"
where land owners and investors could bring skier activities and skier profits more firmly under their control.

That control proved easier to achieve on smaller hillside pastures like
Clinton Gilbert's in Woodstock than on more expansive, higher-elevation
trail systems like those on Mount Mansfield in Stowe. Such systems were
often too long or the terrain around them too rough to pull people uphill
along the surface with rope tows. To reach the top of a place like Mount
Mansfield, skiers in the 1930s still had to walk, and that meant that they
could look forward to only one or two downhill runs per day. Nevertheless,
the allure of high-mountain skiing was strong enough to encourage investors
and inventors from the United States to Europe to pursue new kinds of uphill transport. What they came up with, of course, was the chairlift.

The nation's first chairlift was installed at Sun Valley, Idaho, in 1936,
and its popularity and cachet quickly caught the attention of investors back
east, some of whom saw Mount Mansfield as the region's most promising
candidate for a chairlift. One of these investors was Roland Palmedo, a
Wall Street banker and president of the Amateur Ski Club of New York. In
the late 1930s Palmedo and roughly fifty investors established the Mt. Mansfield Lift Company to finance the cost of constructing a chairlift, manage
it once it was running, and collect profits earned from selling rides. Their
lift was completed in 1940 at a cost of $100,000, and at over a mile in length
it was the longest in the world. It cost skiers about a dollar to make the
twelve-minute journey to the summit, and despite a slow period during
World War II Mansfield's lift was soon providing returns to its investors.[16]
The lift carried 105,506 riders during the 1946–47 ski season alone, nearly

a 100 percent increase from the previous year. And by 1953 the lift's managers were celebrating its millionth rider.[17]

Palmedo's Mt. Mansfield Lift Company reflected the expanding scale of operations, the intensification of outside investment, and the application of professional management that would come to characterize the national ski industry in the ensuing decades. In Vermont Palmedo's success encouraged others to construct new chairlifts and new, more sophisticated surface lifts at resorts like Pico and Big Bromley (see figure 5.1). Mechanized uphill transport quickly became an essential part of Vermont skiing, and within a decade examples of skiing's technological turn could be found in just about every corner of the state. Although lift construction and resort expansion slowed considerably during World War II, they picked up immediately following the war as investors cashed in on increasing rates of travel among the American middle class. Resort owners hired engineers from chairlift companies such as Hall, Poma, and Riblet, who pieced together scores of new chairlifts statewide during the 1950s and 1960s. In 1950 Vermont had ninety lifts (seventy-nine of which were rope tows) at forty-nine ski areas. By 1969 the state boasted 205 lifts (only fifty-five of which were rope tows) at seventy-nine ski areas.[18]

Postwar skiers from Massachusetts, Connecticut, New Jersey, and New York flooded to established ski areas such as the Stowe resort on Mount Mansfield, and they flooded to entirely new operations such as Mad River Glen, Madonna Mountain, Glen Ellen, and Sugarbush. The total number of skier days in Vermont (defined as "one skier, one day") rose steadily from 400,000 in 1949 to over one million ten years later. Meanwhile, the Vermont Development Department proudly listed skiing's gross statewide income at $19.2 million for the 1959–60 season, up $13 million in ten years.[19] Ski lifts were given a great deal of credit in the press for these kinds of numbers, both as they applied to Vermont and to other states. As one journalist writing for *Travel* magazine argued, the modern ski lift had been a central force behind skiing's postwar boom: "More lifts mean more mountains transformed into winter playgrounds, easier ways to get to them, more vacationers captured by the lure of skiing, more lifts consequently needed, and so grows the cycle."[20]

That cycle of construction and expansion had important consequences for the physical and economic geography of skiing and ski resorts in Vermont. As ski lifts grew longer and reached into higher elevations, so did the trails they serviced.[21] This meant that skiing relocated upward from valley farms and hillside pastures into higher mountain environments where snow was more reliable and trails were longer and steeper. It also meant that the costs involved in building and running a ski area in Vermont were reach-

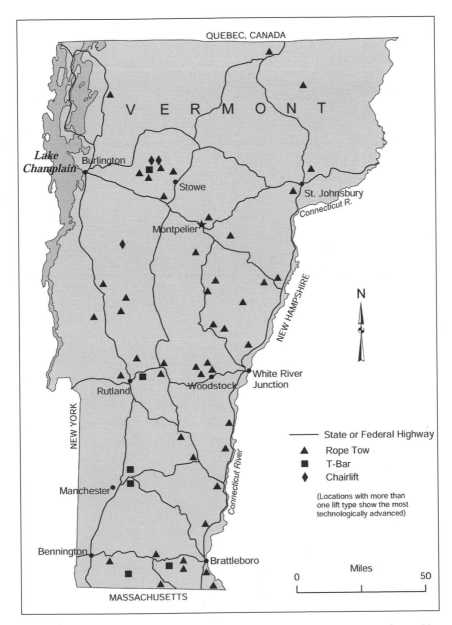

FIGURE 5.1. Mechanized skiing in Vermont, 1948. Ski lift technology was commonplace in Vermont by 1948, with rope tows being the most numerous. This map shows a wide distribution of ski lifts at large and small resorts scattered across the state. Within a couple of decades, however, the state's largest resorts were far more concentrated along the high, central spine of the Green Mountains (see figure 5.2). Adapted from Research Division of the Vermont Development Commission, *Vermont Ski Facilities* (1948).

ing new heights, and that owners and operators were increasingly being locked into a competitive cycle mediated by expensive, technologically sophisticated ski landscapes. The result was an economic climate in which it became increasingly difficult for any but the largest, highest, and most technologically advanced resorts to compete.

That trend was expressed clearly during the 1950s and 1960s by the large and powerful new resorts that were coming to dominate Vermont's ski industry. Many of these resorts were located along the high, central spine of the Green Mountains, where they were aided not only by chairlifts but by improvements to Vermont's Route 100 as well (see figure 5.2). Some, such as Stowe and Big Bromley, were privately owned ventures led by powerful individuals who brought order and a top-down approach to the management of housing, lifts, roads, and trails. By the start of the 1960s, however, a majority were public corporations beholden to stockholders and committed to the bottom line.[22] Whether public or private, though, the stakes for Vermont's resort owners and operators were rising, and the landscapes they produced to meet those stakes were necessarily growing more elaborate and more extensive.

Ski industry officials tried to mitigate their risks by reworking the landscape in ways that encouraged skiers to return week after week, year after year. One concern they had was how best to "alleviate the present overcrowding" that began plaguing some ski towns like Stowe as early as 1945.[23] Quite simply, Vermonters had never had to deal with such massive concentrations of tourists as the two or three thousand visitors that might descend on a small ski town in a single weekend. Consequently, crowd control became an issue in Vermont in a way it had never been before. Developers now had to plan for parking, for instance, and their resorts had to have just enough lifts to keep lines relatively short but not so many that the resort could not pay for their operation and upkeep. Trail design offered an additional measure of crowd control. Funneling large numbers of skiers downhill in safe, interesting, and challenging ways became a key goal of resort developers, who carved complex trail networks where skiers merged, flowed, and granted rights-of-way to one another.[24] Vermont's best-known trail designer was Fred Pabst, owner of the Big Bromley resort and heir to the Pabst brewing fortune of Milwaukee, Wisconsin. Thanks to the Civilian Conservation Corps and private trail builders, Bromley Mountain already had a handful of trails in place by the time Pabst purchased the mountain in 1935 and began developing a massive new system of trails. That system became so large, Pabst boasted, that novices and experts alike could fan out over one thousand acres of skiable terrain.[25]

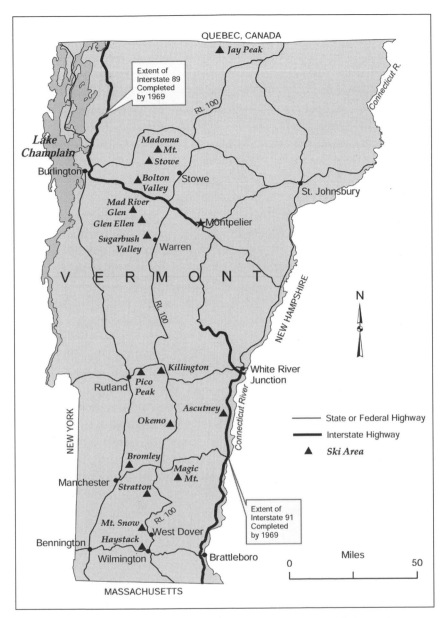

FIGURE 5.2. Major ski areas in Vermont, 1969. By the late 1960s, Vermont's largest and most successful resorts trended north-south along the state's highest peaks. Advanced ski lift technologies played a significant role in creating this pattern by allowing ski trails to reach into higher and more rugged terrain.

They could only do so, of course, if there was enough snow to cover the trails. No matter how many lifts or trails resort owners and operators had in place, their success depended ultimately on the weather. All were acutely aware of that fact, and of what Pabst described as the "complaint on the part of the skier that Vermont has a short season."[26] Chairlifts had helped to mitigate the problem of snow coverage a bit by allowing resorts to open in higher mountain terrain where snowfall was somewhat more reliable and where the snowpack lasted a little further into spring. But despite more favorable climatic averages at higher elevations, and despite assurances about the reliability of snowfall in Vermont's often-cited "120-inch snow-belt" along the crest of the Green Mountains, none of the state's ski areas was immune from drought or warm winter temperatures. The average number of skiing days statewide in December of 1937 was eleven, for example, and in February of 1938 only nine and a half.[27] From the standpoint of building a reliable, profitable ski industry, these were not encouraging numbers. Ideally, then, what industry officials needed were ways to match unpredictable variations in the weather to the predictable periods of high skier demand, such as weekends and holidays. It was easy to forget about bad winters when the snow was good, but when slopes lay bare for too many weekends in a row, or when it rained through the busy Christmas season, people who depended on skiing for their livelihood began getting nervous.

As a partial solution, resort owners and operators designed their trails carefully, both to make the most of what snow they did have and to make the snowpack last as long as possible. Such efforts granted at least some degree of control over weather's uncooperative periods. Here again, Fred Pabst led the way. Unlike most ski areas, which face east or north, Big Bromley faced directly south. This meant that trails on Bromley Mountain were exposed to more intense solar radiation than trails facing other directions, and that Pabst was at an even greater risk of losing his most precious commodity when temperatures rose to the melting point or when the sun rose higher in the sky during late February and March. To make up for this disadvantage, Pabst developed a program of trail construction designed to allow Big Bromley to open its trails to skiers with only a few inches of snow cover. As he described it, "The idea was to grade all our skiing areas — take out the rocks, stumps, and everything else that would get in the way — so a minimum of snow would cover everything." He planted grasses such as redtop and timothy that smoothed the surface, and that for a time provided a harvestable crop of hay during the summer. According to Pabst's optimistic account, his efforts paid off: In an era when most ski trails required at least ten inches of snow for safe skiing, he claimed to have gotten his mountain "smoothed out so thoroughly that four inches — just four little inches — of packed snow would cover it completely."[28]

In addition to trail design, resort owners and operators turned to a host of new grooming, packing, and crust-breaking technologies designed to increase the longevity, safety, and appeal of their snow-covered trails. Packing fresh snowfalls makes for easier skiing and for a denser snowpack, which in turn decreases rates of melting. Some of Vermont's ski area owners had been packing their trails since the 1930s, when local kids sidestepped down trails with their skis on in exchange for free skiing.[29] By the 1950s, however, resort operators had adopted groomers (or "snow cats") designed, as some put it, to "cultivate" snow for greater longevity and better coverage of bare spots. Although the first groomers were little more than tractors mounted on tank treads, new models from the 1950s and 1960s added increased power, mobility, and stability on steep slopes, as well as a host of front-end plows and pull-behind attachments designed to push, pulverize, and smooth the snow surface. A quick scan of any trade journal or ski-resort publication from the 1960s reveals a tremendous number of advertisements and articles praising the benefits of grooming equipment. Machines such as the "Tucker Sno-cat," for example, provided "safer, better-groomed, longer-lasting ski slopes," while attachments like the Spryte "hydraulic mogul cutter" leveled "unwanted mounds of hard snow for smooth follow up packing."[30]

New machines now prowled Vermont's slopes by day and night, breaking the forested winter silence with the roar of their engines and shaping and reshaping an ephemeral snow surface to meet the challenges of nature and the demands of skiers alike. Groomers standardized snow surfaces, making it easier for skiers to navigate the downhill run. They increased the number of skier days by packing snow, by minimizing hazardous conditions, and by redistributing snow from drifted areas to bare patches. In the process, they raised skiers' expectations about the quality of snow surfaces, which in turn saddled owners and operators with added costs for equipment and personnel. As they had done with ski lifts, owners and operators expanded their grooming fleets to meet the rising bar of skier expectations and to keep pace with their competitors. The more that owners tried to insure their success by controlling the ski landscape, the more costly their technological solutions became.

Snowmaking became the consummate expression of this equation. Although grooming technologies helped protect and improve snow reserves, they did nothing to address the fact that prolonged periods of drought or rain could lead to financial ruin. As capital investments in Vermont's postwar ski industry grew, owners searched for a means of climate control that would guard more decisively against high points in the seasonal tempera-

FIGURE 5.3. Sugarbush ski resort from the air, 1971. Building ski landscapes on this scale required sizable investments and a generous use of ski technologies. And while some Vermonters looked at scenes like this with pride and hope, others saw them as scars on an otherwise attractive landscape. Photograph by the Vermont Travel Division, courtesy of the Vermont State Archives.

ture curve and their corresponding low points in the seasonal profit curve. To meet that challenge, a handful of pioneering entrepreneurs in southern New England began experimenting in the 1950s with "artificial snow." Credit for New England's earliest efforts at artificial snow goes to a man named Walt Schoenknecht, manager of Mohawk Mountain ski area in Connecticut, who used 450 tons of crushed ice in early 1950 to patch bare sports on his mountain's trails.[31] The technique was costly and clumsy, but it met with just enough success to get other ski area owners thinking more seriously about artificial snow. They got an unexpected break on a cold day in late 1950, when an irrigation experiment at the Larchmont Farms Company in Lexington, Massachusetts, accidentally yielded a pile of manmade snow. It was a fluke, but it proved that the right mix of water, compressed air, and cold temperatures could actually produce snow. Larchmont began marketing snowmaking systems right away. Schoenknecht purchased one himself at a cost of $25,000, and other resort owners soon followed suit.[32]

Snowmaking operates on essentially the same principles as agricultural irrigation, except that snowmaking systems must move water during winter over terrain far more rugged than the average farm field. This is not cheap to do. Such systems require networks of permanent piping, movable hoses and nozzles (or "snow guns"), and large air compressors capable of pumping water uphill and of firing a mist ten to twenty feet into the air that then freezes into snow on its way back to the ground. They require reliable water sources—which in many cases means constructing reservoirs and diverting streams—and they require extra staff to move snow guns periodically and to redistribute and smooth over fresh mounds of manmade snow.

The costs involved in all of this made many resort owners initially hesitant to install snowmaking systems. But because snowmaking increased precipitation totals and allowed operators to stockpile snow in reserve against Vermont's inevitable wintertime warm spells, few could resist its attractiveness for long. A handful of Vermont's resort owners began installing systems in the late 1950s, spurred on by investment worries, changing skier expectations, and enticing trade-journal advertisements. The Mount Ascutney resort, for one, had a system in operation by 1957 capable of producing an inch of snow per hour over two thousand square feet. Others resorts started small but expanded their systems dramatically in 1965 after a debilitating drought cut into crucial holiday season profits. By the late 1960s resorts such as Sugarbush and Stratton were advertising snowmaking coverage for a mile or two of trails, while Big Bromley, with its south-facing slopes, had installed coverage for 77 percent of its skiable terrain.[33]

As owners and operators expanded their snowmaking systems, they added yet another layer of technology to the Vermont landscape—one intended to protect their growing investments by providing a degree of control over the uncertainty of winter weather. Snowmaking made it possible to predict profits and expenditures with greater certainty, and to open resorts earlier and close them later than ever before. "Modern snowmaking at many resorts," one promoter stated with confidence, "assures every skier delightful schussing."[34] Together with lifts and trail-grooming technologies, then, snowmaking bred a measure of confidence among ski industry officials, who saw in technology an ability to solve problems and create new opportunities.

A New Pace for Leisure

Vermont's traditionalized rural landscapes, Anglo heritage, and reputation for patriotism remained as popular as ever through World War II and the

cold war, and the state continued to be associated in the popular mind with
ideas about liberty, community, family, and faith considered by many to be
at the core of American national identity.[35] That reputation became naturalized in the national imagination through a number of cultural outlets.
Norman Rockwell's stylized and ethnically homogenous paintings (many
of which were based on his neighbors in Arlington, Vermont) became im-
mensely popular at mid-century, for instance, as did the 1954 Bing Crosby
film *White Christmas*, which was set in Vermont and, as one Vermont historian put it, offered "a perfect recipe for sentiment and nostalgia, for evok-
ing memories of the simple 'good old days' of long ago."[36] *Vermont Life*
magazine also maintained its status as a showcase for tradition. In 1969
the magazine's editors put together a book of essays and color photographs
drawn from their archives entitled *Vermont: A Special World*. The book captured the persistent sentimentality by which many Americans approached
Vermont during the second half of the century, privileging the endurance
of Vermont's traditionalized rural charms rather than exploring the changes
its authors admitted were regrettably at work in the state. What made Vermont "special," they suggested, was not that which was new, but rather,
that which was old.[37]

Of course, none of this hurt the state's tourist trade, which still relied as
it had for generations on idealized, sentimental conceptions of rural land
and life. But a sentimental identity and timeless landscape alone were no
longer enough to keep all segments of Vermont's tourist trade going. Ski
technology set in motion a transformation of many rural communities along
new and decidedly modern lines. Investments in skiing yielded a new
leisure landscape that reflected and reproduced a national embrace of technological achievement, modern design, and new social conventions. That
which was exciting, modern, and new was selling among mid-century tourists
as never before, setting a new pace for leisure that challenged but did not
supersede the old. I explore that shift here by examining technological encounters with recreation, the advent of new architectural styles in Vermont,
and a range of changing social patterns associated with class, gender, and
sexuality.

When tourist promoters and industry observers talked about ski technology, they often did so in a language that gave new meaning to the high-
mountain environments now being opened up by chairlifts. As Charles
Edward Crane explained in *Winter in Vermont*, Stowe's first lift was very
much like "taking a rocket to the moon. Not that it moves fast, but it takes
you up and up into what seems like another world . . . such as you might
expect on another planet."[38] Venturing into that otherworldly terrain de-

manded that skiers develop a basic understanding of topography and map reading. While the open slopes of smaller ski areas were often entirely visible from a single vantage point at the base of the rope tow, higher, longer slopes were not. Consequently, resort owners now had to furnish skiers with trail maps, and skiers now had to be able to use those maps to navigate their way back down safely. Vermont's mountains were no longer a backdrop to scenic rural vistas, nor were they the exclusive domain of adventurous hikers. Thanks to ski lifts, their mysteries and their dangers were open to greater numbers of people than ever before.

Ski lifts also reconditioned the visitor's experience with gravity and the familiarity with the landscape one gains by climbing a mountain on foot. As the environmental historian Richard White has argued, we develop an intimate understanding of nature through the human energy we expend to overcome nature's challenges and limitations.[39] Early skiers, for example, had no illusions about the powerful roles that gravity and topography played in their sport. As they tramped to the tops of trails through the snow, they gained a degree of familiarity with nature that the next generation of skiers would not have.[40] By eliminating the uphill climb, ski lifts masked the power of gravity and set in motion enduring expectations that skiing should demand only as much human energy as needed for the downhill run. This was particularly true of chairlifts, which gave skiers opportunities to sit down, rest, and save energy for a greater number of ski runs per day.

Vermont's ski lifts were only one example of the growing reliance on mechanical technology so prevalent in America's postwar recreational culture. Whether in the form of motorbikes, snowmobiles, motorboats, or chairlifts, mechanical power created new expectations about speed, action, and the labor involved in outdoor recreation. Vermonters helped spearhead this trend by making engines an integral part of the skier's experience. Skiing is a physically challenging sport, but few by mid-century expected to have to work at it. Instead, skiers wanted fast-paced downhill runs and comfortable, mechanized uphill rides as they chatted with friends and family along the way. Skiing nonetheless assumed a reputation for being a fast-paced, aggressive, and daring way to spend one's free time. Even ski trails conveyed a thrilling sense of adventure and danger. A trip down trails with names like Riperoo, Shincracker, and Jaws of Death was no sleigh ride over the river and through the woods. That impression was reinforced by the visitor's encounter with technology, which, if anything, made many skiers more impatient even as it made it possible for them to ski more runs each day. Skiers did not come to Vermont to wait in line, they came to ride up mountains and ski back down them with promptness and efficiency. The faster lift technology became, the stronger the visitor's sense of entitlement to speed.[41]

Tied up in all of this was the more general sense of newness, activity, and change now informing Vermont's entire resort landscape. Although some ski areas in postwar Vermont remained small, even rustic, family affairs, the state's largest, newest, and most innovative resorts were unquestionably its most popular. Places like the Mount Snow resort, for instance, were simply unlike anything skiers had seen before, either in Vermont or elsewhere. When Mount Snow opened in 1954 on the pastures and wooded slopes of the former Snow family farm in West Dover, it did so not with a couple of rope tows or a small single chair. Instead, it opened with two uniquely designed double chairlifts, a large base lodge, and more lifts on the way. Under the direction of its charismatic president, Walt Schoenknecht (the same Connecticut ski manager who had pioneered snowmaking), Mount Snow developed the capacity to move more people uphill than any resort in the state, giving the resort a reputation for innovation. For many industry observers, Mount Snow typified a new kind of modernist identity now taking hold in ski communities — one based less on the marketing of old-time rural traditions than on representations of brightly colored plastics, snow-irrigated hillsides, and space-age lifts. Schoenknecht's gimmicks and technological feats (including a two-hundred-foot ice fountain and a year-round outdoor swimming pool), and his grandiose (yet failed) plans to create lift connections between mountains and villages across Windham County, imparted an ultramodern identity to this formerly quiet farm community.[42] That identity became wildly popular among many skiers and among promoters, who used places like Mount Snow to redefine Vermont according to an unabashedly modern image. In a postwar culture defined by Disneyland, fast food, fast cars, and space exploration, even rural Vermont now seemed "zany," "flashy," and just plain fun. By the 1950s it was clear that Vermont was changing fast, and among many ski promoters and ski tourists that was a positive thing indeed.

New architectural styles compounded this impression of modernity and change, both at Vermont's ski resorts themselves and in the towns that surrounded them. This played out in two ways. First, postwar architects began designing vacation homes for skiers with a more fashionable, modernist feel — homes with vaulted ceilings, floor-to-ceiling windows, all the latest amenities, and in some cases the ability to ski directly to the base of a ski lift.[43] Vermont became something of a proving ground for new architectural styles during mid-century, rather than a place for leaning barns and quaint farmhouses alone. Second, postwar ski architects across the United

the
MT. SNOW *Spirit*

What makes a ski resort great? Great mountain, great snow, naturally. But there's more. **Call it spirit.** At Mt. Snow there's a refreshing spirit of new ideas always working to widen your pleasure.

Where else can a skier be sped up-mountain in such a variety of space-age lifts? Where else do you ski a world of 40 slopes and trails...where men and machines pamper even deep powder to perfection?

The Mt. Snow Spirit is contagious. Come up soon.

Mt. Snow Ski Weeks *from* **$59.95**
■ 5 full days and nights with lodging and meals ■ all lifts
■ 5 two-hour lessons (famed Harvey Clifford Ski School)
■ heated outdoor swimming pool ■ Après-ski entertainment.
Monday thru Friday except Christmas & Washington's Birthday Weeks.

MT. SNOW
In the Heart of the Ski Country. **Write for color brochure: Box A-4, Mt. Snow, Vt. Phone 802-464-3333**

FIGURE 5.4. *The Mt. Snow Spirit,* 1966. By the 1960s, "space-age lifts" and a landscape where "men and machines pamper even deep powder to perfection" had become central to how many people defined and experienced Vermont. Modern, fast-paced, and technological, such ski resorts added new meaning and new depth to the state's traditional identity. New England Ski Museum.

States also turned to pseudo-Alpine designs, reproducing in the resort land-scape a racialized identity for skiing defined predominantly by its associa-tions with white, European culture. According to the historian Annie Gilbert Coleman, that identity made skiing all the more attractive to white middle- and upper-class Americans, reinforcing a dominant and persistent link be-tween skiing and whiteness.[44] Similar patterns unfolded in Vermont, where the ski clientele has remained almost entirely white to this day, thereby re-producing Vermont's established reputation as an Anglo-American strong-hold.[45] Well-to-do white skiers in Vermont, like their counterparts in other areas of the country, could connect with a sense of European heritage by staying at places like the Hotel Tyrol in Stratton (a "European Alpine Hotel") or the Alpenhorn Lodge in Chester (a "traditional Swiss chalet"). Or, they could buy a new A-frame ski house with gingerbread trim or even a prefabricated European chalet ready for assembly.[46] Such architectural styles had little precedent in Vermont, but then again, neither did chair-lifts, snowmaking, or outdoor swimming pools.

Finally, new patterns of social behavior among tourists also contributed to Vermont's growing associations with modernity and change. Class-based ideas about consumption, for instance, played an important role in defin-ing postwar skiing, making the sport, in Coleman's words, "a means of ex-pressing oneself and impressing others."[47] Flashy, fast-paced, and expen-sive resorts from Vermont to Colorado gave well-to-do skiers an outlet for performing and reproducing their class identity through consumption and social display, or what one journalist referred to as "social climbing on the slopes."[48] Vermont's Sugarbush resort became the state's best expression of this. Sugarbush opened in 1958 and quickly assumed a glamorous reputa-tion, becoming known in some circles as "Mascara Mountain." With a modern gondola, four chairlifts, excellent ski terrain, and a vibrant night-life, Sugarbush became the premier winter vacation spot for a young New York crowd, some of whom chartered Friday evening bus services from Manhattan, complete with cocktails. In addition to these young, aspiring socialites, Sugarbush was New England's home base for some of skiing's most elite national and international celebrities — the nobility of the sport, who mingled at the resort's private "Skiclub 10" and set a high social bar to which others aspired. Whether one was a budding or an established so-cialite, one simply had to be seen socializing at Sugarbush's noted bars.[49]

Gender and sexuality became additional social contexts through which Vermont's modern ski landscapes and modern ski identity evolved.[50] As other geographers have argued, tourist promotions often privilege white, male, heterosexual perspectives, leading women to be "represented as exoticized commodities which are there [in tourist destinations] to be experienced."[51] Such was the case at times in Vermont, where women were expected to contribute to an attractive identity for the state's ski landscape. Ski fashion had a significant role in this. Designers and advertisers worked hard to sell fashionable ski clothing to women, suggesting repeatedly that women should always remain one step ahead when it came to fashion on the slopes. "The world of high fashion has conquered the hills of Sugarbush," one journalist observed, "and so has the skier who wouldn't think of wearing the same fur parka two days in a row."[52] "Stretch pants" became the most popular example of women's ski fashion during the 1950s and 1960s. These tight-fitting elastic pants allowed for athletic movement while at the same time enhancing women's sexual appeal. An astounding number of magazine articles, advertisements, and ski-area promotions featured attractive women wearing stretch pants, thereby setting restrictive standards for beauty and dress on the slopes.

Discussions about dress and appearance also suggested links between sex and the ski landscape. Romance and flirtatiousness had long been equated with skiing, often in quite innocent ways. A 1947 *Life* magazine article, for instance, portrayed Vermont's ski resorts as places where male and female college students could mix innocently over weekends of "strenuous exercise and quiet evenings." Such encounters, *Life* was quick to add, seldom included smoking and drinking, or, presumably, any other questionable behavior.[53] Suggestions of romance grew more common in the coming years, although they often remained obtuse. An early 1960s promotional film produced by the Vermont Development Department, for instance, describes a fictionalized, slope-side meeting between an expert male skier from Ohio and a struggling "pretty little gal" from New York City. After a playful collision on the slope, the two get to know one another over coffee and make plans to "have a date to meet again." Nothing about the couple's meeting and courtship was out of the ordinary, the film suggested, and viewers were left to fill in the gaps about their future together.[54] Handsome ski instructors (often recruited directly from Europe for added flair and authenticity) also lent a romantic appeal to ski resorts, many of which competed to see who could attract the most prestigious and attractive instructors. Resort promotions frequently presented an air of flirtation associated with ski instructors and ski classes: Strong, handsome instructors, they suggested, promised to guide pretty girls safely down the slopes, and maybe meet them later for a drink or a meal in the ski lodge.[55]

By the late 1960s links between sex and skiing in Vermont were becoming more direct, particularly as attitudes towards sex loosened among young

Stretch clothes unveiled
the sex in skiing.

FIGURE 5.5. Stretch clothes, 1967. This illustration appeared in an article entitled "Sex and the Single Skier" in the winter 1967–1968 issue of *Vermont Skiing* magazine. The article, the image, and its original caption, all contributed to an emerging association between Vermont and a fast-paced world of skiing, modern fashion, and sexuality—a world where women were expected to shed a layer of clothing to add a layer of attractiveness to the ski landscape. Illustration by Mike Ramus, courtesy of Anne Gray and Grace Ramus, and courtesy Special Collections, University of Vermont Libraries.

Americans. In a 1967 article entitled "Sex and the Single Skier," *Vermont Skiing* magazine noted that the state had become a pleasuring ground for young, single, sexually inclined skiers, and it offered advice to single men on how to meet and impress women skiers. The sexual revolution had overtaken skiing, the article suggested; even in Vermont, a state traditionalized as a Puritanical stronghold, skiers could find more than sun and snow. "Take a roomful of tanned, active, healthy *glandular* people, both sexes, lounging unchaperoned like young gods around the lodge hearth," the magazine's author noted. "Winch them into stretch clothes and pour a couple of drinks into them. Isolate them. Stick them in a remote mountain fastness, miles from the nearest dunking stool and with nothing to read but ski publications. Barring extreme cases of frostbite or anoxia, what do you think they're thinking about or liable to do?"[56]

Connections between sex and skiing were often drawn in ways that made women appear like an attractive supporting cast in what remained a decidedly rigorous, masculine sport. Sexual references in ski literature frequently treated women as inherently weak skiers or, worse, as self-absorbed opportunists who came in search of men rather than sport. As one journalist wrote, "Gentle trails and broad slopes [are provided at Vermont resorts] for the weaker sex trying to keep up with their prey."[57] As another suggested in an early *National Geographic* article on skiing, most women only "became interested [in skiing] when attractive ski costumes were made available." More concerned with looking good than anything else, he argued, women's winter footwear was more fashionable than sensible: "When she falls and twists her ankle, as she well may," the author added, "her suffering is a just reward for her stupidity."[58]

Less than generous commentators like these seemed designed to preserve skiing as a bastion of masculinity. Yet even so, many women were equally, if not more, accomplished as skiers than their male counterparts. Their participation in the sport at mid-century, however, often had its limits. As Coleman notes, for example, women who were accomplished skiers were often expected to maintain a feminine appeal, lest they appear "too masculine for society's taste."[59] As quickly as the *National Geographic* journalist quoted above judged women for dressing inappropriately, then, he and others like him would have wanted women to remain a part of their ski experience, even (or perhaps particularly) if their attractive clothing put them at a competitive disadvantage on the slopes.

New social norms, new architectural styles, and new technological encounters with recreation all helped to redefine landscape and identity in Vermont according to the evolving dictates of a modern leisure economy. An authentic vacation experience for Vermont's mid-century skiers now included encounters with space-age technologies, modern vacation homes, and glamorous ski resorts as much as with white-steepled churches, farm scenery, and covered bridges. Indeed, if resorts such as Mount Snow seemed "strikingly modern" to some, or like something "out of the next century" to others, that was largely because they stood in such sharp contrast to traditionalized images of rural timelessness. The ski industry had become "new sugar in an old state," *Sports Illustrated* proclaimed, and for many that was cause to celebrate.[60]

Where, then, did that leave "rural" Vermont — or at least conceptions of rural Vermont that hinged on tradition and anti-modern imagery? Skiing may have been "new sugar in an old state," but that did not mean that the spaces and meanings associated with "old" Vermont gave way entirely to those associated with "new" Vermont. Rural Vermont did not stop being "rural" because of skiing, but it did undergo changes that, more often than not, amounted to a pronounced overlap between vestiges of the old and vestiges of the new. Rather than forego the old entirely in favor of the new, then, ski promoters advanced a new conception of the rural in which Vermont's traditionalized heritage and modern ski development mixed and mingled with one another to create a new rural blend.[61]

One can find examples of this blending in both pictorial and written form. Traditional images of a distant mountain wilderness framing a foreground of pastoral valleys, for instance, now gave way to images of pastoral valleys framed by mountains carved with ski runs. As one ski journalist described the view from below: "As if it had been planned that way from the start, the pastoral valleys and the tiny settlements give way at intervals to the white squiggle of ski runs dropping down from the peaks and to the straight, white lines of 67 ski lifts rising up to the rime frost and the dazzling sparkle of trees on the summits."[62] Advertisements for resort services also promised visitors simultaneous access to the nation's best and most technologically advanced skiing as well as all of the trappings of Vermont's traditionalized rural charm. Skiers could expect to find services in Vermont that were rustic, charming, authentic, traditional, and historic on the one hand and modern, dramatic, unique, and exotic on the other. Some promoters told of people who converted old barns or abandoned farmhouses to ski retreats, altering them to accommodate new uses while retaining the spirit of the original design. Others described new ski lodges that were entirely modern while at the same time retaining a connection to the "keen sense of proportion found in centuries-old Vermont structures."[63] Even in Stowe, where postwar ski development touched all aspects of the town's life, skiers were told that they could "still get the feeling of finding yourself in a clean-scrubbed, church-spired New England village."[64]

Imagery such as this continued to play on old themes while at the same time reworking landscape and identity in the state according to skiing's modern associations. By blending a "church-spired" image with a gleaming technological one, a new generation of visitors could continue to think about the rural as being traditional while at the same time accommodating the pace for leisure set by their brand of tourism. The Vermont landscape had not stopped being "rural" in the popular mind, but neither had tourists or tourist developers allowed it to stop changing in favor of a tourist-directed image of oldness and tradition. After all, it had never paid to leave Vermont as it was.

A New Face for Work

Vermont's new pace for leisure also brought a new face for work to those who lived in and around the state's ski towns. Although many rural Vermonters took to skiing as eagerly as vacationers did, many more encountered the

FIGURE 5.6. Steeple and ski trails in Stowe, Vermont, no date. New kinds of promotional imagery associated with skiing often merged traditional icons of rural Vermont (the church steeple) with newer icons of the ski industry (the white ski trails in the photograph's upper-left-hand corner). Rather than supplementing one another, each became mutually reinforcing elements in the state's evolving rural tourist landscape. Photograph by Stowe Photo for the Vermont Travel Division. Courtesy Vermont Historical Society.

state's ski landscape through the context of work rather than leisure. And as growing numbers of residents took jobs in the state's ski economy, they often found themselves struggling to make sense of their place in this latest chapter in the reworking of rural Vermont.

To understand skiing's impacts on rural work in mid-century Vermont, we need to first consider simultaneous changes occurring in the state's agricultural economy. As we saw in chapter 4, Vermont's mid-century dairying industry was undergoing a period of consolidation and retrenchment marked by falling numbers of farms and rising sizes of both landholdings and herds. That trend toward consolidation was not unique to Vermont but was unfolding on a national scale. Following World War II the number of dairy farms in the United States began declining steadily while the percentage of remaining farms with over five hundred cows rose.[65] Vermont's farm economy lost roughly ten thousand workers between 1950 and 1959, as the number of farms in the state dropped from 15,444 to 11,022.[66] Behind such changes lay a general modernization and mechanization of farming enhanced by rural electrification and the increased use of tractors, automobiles, milking machines, and refrigerators. The downside to this, of course, was that production costs rose, challenging the ability of many farmers to remain competitive in regional markets. Consequently, many mid-century dairy farmers either went under or fell deeply into debt.

Some Vermont dairy farmers sold out completely and moved on, while others kept a few cows around out of habit but looked elsewhere for their primary source of income. In either case, Vermont's expanding mid-century ski resorts often became one of the Vermonter's best new sources for work. Although it is difficult to know the exact numbers of mid-century farmers who found work in the ski economy, it is clear that many thousands of rural residents did, whether as full-time, part-time, or seasonal employees. From the start, actually, Vermont's ski economy had offered enterprising farmers with the right kind of property and a good location an opportunity to make a few extra dollars in the winter. Some started single-family ski operations or teamed up with neighbors to build homemade rope tows on otherwise idle, unproductive winter pastures, merging work and leisure on the farm in ways that were not entirely dissimilar to what generations of Vermonters had done before. Publications like the *Farm Journal* and *Popular Mechanics* even offered advice to farmers on how to construct modest ski tows from simple equipment, some of which New England's farmers would have likely had lying around anyway. There was no need to build a large resort, some advised, and no need to invest in expensive new machinery; if a farmer wanted a little extra income and a little fun for his own family and friends, he only needed to build a small rope tow, open a warming hut, and hang a sign out front.[67]

But as the scales of investment and infrastructure at Vermont's postwar resorts grew, rural residents, whether they were farming or not, were more likely to find work in the ski industry by becoming wage laborers in or around the state's new resorts. In one sense, skiing became another example of farmers using winter work to supplement their families' incomes: where grandfathers had gone to the woods to log for the winter, grandchildren now went to the ski resorts, where they helped people onto ski lifts, sold tickets, repaired machinery, cooked, and cleaned. Even more people found work in the towns around the resorts, where most of the ski-related growth was occurring.[68] By 1948 skiers were spending eighty-three cents on every ski dollar not on lift tickets at the resorts themselves but on related goods and services such as transportation, equipment, entertainment, food, and lodging.[69] Men under the age of thirty made up the majority of skiing's mid-century workers, although women accounted for 30 percent of the seasonal employees at Vermont resorts by the late 1950s.[70]

Some enterprising Vermonters looked beyond the resorts themselves

FIGURE 5.7. Landscape of farming and skiing, 1940. With his ski hill rising on a pasture in the background, Clinton Gilbert and a helper haul wood on Gilbert's Woodstock farm. Despite his modest success as a ski entrepreneur and his role in creating one of North America's first rope tows, Gilbert remained a farmer and his property remained a space where rural work and rural leisure existed side by side. Photograph by Marion Post Wolcott, Library of Congress, Prints & Photographs Division, FSA-OWI Collection.

and opened their own businesses. One Waitsfield farmer, for instance, sold his cows in 1957 to open a gas station after realizing just how difficult it was for skiers in central Vermont to get gasoline on weekends or during evening hours. He was soon making money not only by pumping gas but by towing cars, jump-starting dead batteries, and giving rides to stranded skiers.[71] But most worked in a dizzying array of part-time or limited-term employment as carpenters, retail clerks, cooks, wait staff, hotel employees, ambulance drivers, trail builders, road workers, and even woodcutters for the hundreds of new fireplaces burning each winter in ski lodges and vacation homes. Then as now rural residents living in areas with little economic opportunity had to combine hard work, skill, creativity and flexibility in order to get by.

Many Vermonters welcomed the kinds of changes that skiing brought to the economic geography of their state. Many were pleased, for instance, to see high-elevation terrain being put to use — terrain that had once been logged or perhaps even farmed but that had often sat idle for a generation or more. That kind of change in land use, for instance, was often considered good news for struggling towns like Waitsfield, Warren, and Fayston in Vermont's Mad River Valley. By the early 1960s the Mad River Valley was home to three major ski areas, and all of its towns seemed to be benefiting as a result. According to optimistic reports in the *Burlington Free Press*, a call to growth was "being hammered out by carpenters, bricklayers and heavy construction equipment" across the Mad River Valley. Hundreds of new jobs had been created in these towns alone by 1963, and with new trails, new lifts, and new vacation housing under construction, the area's economic outlook was looking up for the first time in decades.[72]

The reality for most mid-century residents was undoubtedly more complex than the glowing reports suggested, however. For one thing, ski industry growth did not touch all of the towns in Vermont, and over time there developed grave differences in economic opportunities (however those opportunities were defined) between ski towns and non-ski towns. Just as some Vermonters talked of new opportunities and a promising future, others pointed to a starker reality of persistent poverty, alcoholism, domestic violence, and racism in rural communities across the state. Vermont's national image was tarnished in 1968, for instance, when a vigilante in Irasburg fired into the home of a transplanted African American preacher, David Lee Johnson. This notorious "Irasburg affair," as it became known, turned even more scandalous after state troopers brought Johnson up on what appeared to be trumped-up adultery charges. Local thugs continued to harass and threaten the family, and many people in Vermont expressed their outright support for the convicted perpetrator, who was ultimately fined a mere $500 for a "breach of the peace."[73] For all of Vermont's glittering ski development, publicity like this suggested a more troubled situation, again drawing attention to the kinds of problems that tourism promised to solve but never seemed quite able to do so.

Problems existed in booming ski towns as well, where public officials and citizens complained of rising rates of crime, violence, and alcohol abuse in towns that in many cases had only recently been dry. Skiers brought as many problems as blessings, they complained: they drove recklessly, they partied too much, and they placed undue burdens on local services.[74] Perhaps none of this would have seemed quite so bad (Vermonters drove recklessly and partied too much, too) had local residents not felt like they were getting the short end of the stick. Despite all of the excitement about new economic opportunities in ski towns, ski industry workers were often not as well off as they had hoped they would be. Without question, some found successful, stable careers in the ski industry, or at the very least they earned a little extra money to help the family along. But others fell into a hopeless spiral of seasonal unemployment and low wages. Roughly one third of the mid-century labor force at Vermont's resorts was part-time, unskilled, and poorly paid. Work for about 80 percent was seasonal, and employment (even during the winter months) was often unsteady due to the vagaries of weather.[75] Quite simply, the promises of ski industry employment were not always what they were cracked up to be.

Compounding this were the effects that such employment was having on the self-identity of rural residents. In some cases, ski-related jobs meshed well with Vermonters' traditional farming experiences. Men and women who had spent their lives on farms, for instance, were often well suited to tasks required of them as ski industry employees. Not only were they familiar with heavy labor and working outside in all types of weather, many had practical carpentry or mechanical skills, both of which came in handy at expanding and technologically sophisticated resorts. Therefore, some of those who made the transition from farm to ski area would have been on somewhat familiar ground, or at the very least could feel proud that they were putting some of their traditional skills to use.

But not all ski jobs were so skilled or so familiar. Nor, for that matter, did employment in the ski industry offer the same sense of self-reliant independence that proud Vermont farmers so often mapped onto their and their ancestors' lives. Rather than embodying a Jeffersonian heritage, thousands of mid-century rural Vermonters were now forced to seek work in a wage-based, corporate economy — one where the practices and spaces of work were geared not toward the production of a tangible commodity but toward the production of fun for people from outside the local community. Rural Vermonters remained proud and independent in spirit, and they continued to portray themselves and to be portrayed by others as both dignified and bound to tradition. But most also knew that times had changed, and that the celebrated working heritage of their ancestors was becoming a thing of the past. According to one 1972 commentary on the Mad River Valley, "You've gone from an independent, self-reliant community to a group of people who provide services to the out-of-state wealthy . . . The developments bring more jobs for slinging hash and plowing snow. But the identity of our towns is changing significantly in the process and it's not changing for the better."[76]

That impression would have felt even starker if viewed in contrast to the growing numbers of young non-residents — so-called "ski bums" — now joining Vermont workers on the slopes. The mid-century ski bum was a new kind of social figure in rural Vermont — a hybrid of sorts, for whom work in the ski industry was tantamount to leisure. Ski bums were mostly young men born in other states, many of whom were from relatively privileged back-

FIGURE 5.8. Mechanics at work on a chairlift tower, 1965. Many rural Vermonters who grew up on farms possessed basic skills in mechanics and carpentry — skills often in demand at the state's new resorts. New opportunities for work like this allowed some to leave farming for what they hoped would be a better future in the state's leisure economy. Photograph by Sherman Howe, Jr.

grounds and many of whom worked just enough as dishwashers or chairlift operators to finance a winter or two of skiing. For some there was a certain romance — a "Hemingway-like" quality, as one called it — associated with setting college or better-paying jobs aside in pursuit of the sport they loved.[77] In one sense, then, the romance of the ski bum echoed the kind of urban escapism that for decades had characterized visitors' sentiments about Vermont.

But ski bums were not quite like these other visitors, for they merged work with leisure in ways that provided a telling contrast to many residents' experiences with the ski economy. Ski bums worked in kitchens until late at night in order to wake up early and ski the next day. Or they stood outside in the cold, loading skiers onto chairlifts in order to take a quick run or two during their lunch breaks or to ski on their days off. But so did local residents, for whom Vermont was an ancestral home as much as a temporary stop on the road of life, and for whom a future of loading skiers onto chairlifts was perhaps the best foreseeable economic opportunity they had.

Many ski bums stayed on permanently in Vermont, raising families, buying property, and investing themselves entirely in their new communities.
But most moved on from what was a temporary and conscious choice to
put off middle-class responsibilities and expectations—to have a little youth-
ful fun while they could still get away with it. And whether they realized it
or not, their rural playground was, for some, a rural disaster.

The Politics of Identity and Landscape
Control in Mid-Century Vermont

Skiing's dramatic rise in popularity, along with the new brands of work and
leisure it produced, turned the sport's resort landscape into an emerging
focal point for social conflict. At times, diverse groups of visitors and residents worked together to negotiate the changes skiing brought to the state,
but more often than not skiing sparked a more contentious politics of identity and landscape control—a new, modern middle landscape where social
groups collided over skiing's place in the future of rural Vermont. In contrast to the Green Mountain Parkway debate of the 1930s, which had caused
statewide interest but had remained a singular event, debates about skiing
and postwar tourist development more generally were both statewide and
ongoing. As the remainder of this chapter shows, and as we shall see in more
detail in the next chapter, such debates were becoming a standard feature
of the reworking of rural Vermont.

Vermont's postwar ski economy promised great things, yet, as industry
observers recognized early on, Vermonters were often less than enthusiastic about the sport. As the magazine *Ski News* noted in 1948, "The lack of
cooperation on the part of certain native residents has in many cases served
to overcome the natural advantages of several of the areas, through short-
sightedness, petty jealousies, and a resentment for 'foreigners.'"[78] Part of
this resistance stemmed from residents' concerns about access to property.
Privately owned land that had once been open to community members for
hunting or fishing was, in many cases, now being sold to ski area developers and non-resident vacation home owners, many of whom posted their
land against trespass. Vermont's total acreage posted against trespassing increased from 167,224 acres in 1955 to 395,492 acres in 1963, with the greatest change occurring in Windham County, where some of the state's most
extensive ski area development was underway.[79] Residents also expressed
frustration about the legal status of ski areas constructed on land leased from
the Green Mountain National Forest, arguing that powerful ski companies

were being given special land-use rights by the federal government that local residents were unfairly denied.[80] Fears such as these indicate a growing concern that ski area expansion came at the expense of locals, whose control over land use in the state was diminishing. That concern, we have already seen, had roots going back to the 1890s, but its potency was intensifying now as never before.

Other residents responded negatively to the aesthetics of ski area development, particularly with reference to what one journalist called Vermont's "galaxy of lifts, slopes, trails, and hotels."[81] As early as the 1930s residents in Stowe had objected to the idea of a chairlift being constructed on Mount Mansfield and had forced developers to consider locating the lift in such a way as to hide it from sight in the valley below. "Those interested in the construction and others," one observer noted, "have been insistent that nothing be done which will in any way mar the beauty of the mountain."[82] Charles Edward Crane, author of *Vermont in Winter*, viewed such opposition to ski development as perhaps understandable but not entirely justified. From his perspective there was virtue in the fact that "a few aesthetes find spiritual uplift in nature wholly untouched." However, he added, "they seldom if ever went near the mountain in winter." Instead, Crane argued, ski trails and ski lifts should be applauded for making the mountain "so accessible that thousands of human beings can enjoy it."[83] Other residents never bought into arguments like these. As one folksy writer recounted of a new ski area in Vermont, "About the first thing them fools did was cut down one of the best sugarbushes in town, and then that wa'n't enough, they goes right on, cutting ugly slashes right up the mountain. That's all we can see, now, when we look out the west windows is them scars in the hill."[84] Those "scars" may have opened the mountain up to more people and may have even provided jobs, but for many rural residents they were troubling, unwanted eyesores.

Residents also reacted in a more general sense to what many perceived as the excessiveness and outlandishness of skiing. As one Vermont farmer later recalled, "When skiing first started I think we were very skeptical . . . It was a little hard to really envision that skiing could be a worthwhile operation other than just plain foolishness."[85] For many that "foolishness" was harmless enough, but if kept unchecked it had the potential to become a larger threat. Vermonters and visitors alike were alarmed, for example, when in 1963 Walt Schoenknecht announced that he had initiated negotiations with the Atomic Energy Commission to use an atomic blast to enlarge and improve Mount Snow's skiable terrain. Although not everyone was entirely sure whether or not to take him seriously, and although he never received permission to use atomic power to enlarge Mount Snow, Schoenknecht's reputation for outlandishness convinced some that he was willing to do whatever it took to make his mountain a success, regardless of the potential ramifications of his actions. Sure, a few windows may break because of the bomb, he nonchalantly told the press, but the blast would occur on the less inhabited side of the mountain, and fallout from the bomb would reach thirty thousand feet before descending back to earth thousands of miles from Vermont, most likely in the Soviet Union.[86] Visitors and residents expressed concerns not just about the proposal itself but about what it suggested about power dynamics in the state. As one summer home owner argued in a complaint to local officials, Vermonters typically "do not permit any one person to gain such influence or self-importance that he could even consider doing such a monstrous thing."[87] Too much power in the hands of those who ran a "foolish" ski industry, such criticisms suggested, would prove disastrous for all who had a stake in Vermont's landscape.

Frustrations spilled over into everyday encounters between residents and skiers, particularly as residents tried to shore up their control over the state's rural identity. Some residents tried to put skiers in their place by deriding them as ignorant "flatlanders" who lacked both common sense and the ability to survive the most basic challenges of rural life. According to one resident, skiers were enigmatic fools who, astoundingly, "risk their neck[s]" riding up ski lifts "just so's they can risk life and limb on the way back down." He added, "Until that ski tow come to town, I didn't believe idiots could be collected by the gross." Skiers were foolishly willing to pay any price for goods and services, he jeered. They wore outlandish clothes, and what was more, some did not even speak English.[88]

Running throughout commentaries such as this was a clear sense of local superiority and a related effort to use traditional rural identity as a privileged social tool to put urban visitors in their place. A familiarity with rural life gave some residents a sense of self-worth in much the same way that an urban sophisticate might scoff at a rural "rube" in the city. Residents often articulated this sense of superiority by commenting on skiers' notoriously bad driving habits. Not accustomed to driving on mountain roads in winter, skiers often became stuck in the snow or stranded on the side of the road, making the winter roadway a place where residents could either express kindness, as many did, or assert at least some degree of power over helpless skiers. One employee at Mad River Glen, for example, recalled pulling a stranded visitor's car out of a ditch and then pushing the car right back in again when the visitor refused to give him a tip for his service.[89]

Aside from social tensions between visitors and residents, visitors disagreed among themselves about how best to use and define the state's rural tourist landscape. By the 1950s and 1960s many visitors had been traveling to Vermont for long enough to claim a privileged, established sense of legitimacy within the state. After generations of owning a summer home or visiting the same lake, for instance, many visitors felt almost like honorary residents, particularly by comparison to the droves of skiers who had only recently discovered Vermont. At the very least, long-standing summer visitors could make comparisons from personal experience between Vermont's post-ski and pre-ski leisure landscape. What they found in those comparisons often did not sit well, for although many summer visitors were also skiers, many others were not.

Visitors who owned summer property adjacent to ski resorts often felt particularly squeezed. By the early 1960s some were struggling to maintain the traditional feel of what in some cases had been long-standing summer colonies as new hotels, vacation homes, and ski trails began encroaching on the borders of their land. At times, the fact that they owned property (often sizable amounts of property) could work to their advantage. One recalled trading property with an encroaching ski area developer, for instance, in order to maintain a more effective buffer around his summer home and, as he put it, avoid "having a ski area right down around my ears."[90] Other summer home owners were less lucky, however, as new ski lodges sprouted in meadows across the road or formerly wooded mountain views became striped with ski lifts and trails. The more such patterns developed, the more antagonistic some summer residents felt toward their new ski industry neighbors.

Differences of opinion also developed among skiers themselves, particularly over the modernization of their sport and the technological development of the ski landscape. By the 1940s and 1950s some skiers were worrying about the effect that a national lift-building bonanza would have on the industry's safety record or on the financial integrity of overextended resorts. Lifts cost money to build, which meant that resorts charged higher prices for lift tickets. That challenged the ability of some skiers to afford the sport. In response to that challenge, one commentator went so far as to suggest that skiers boycott lifts and return to walking uphill instead.[91] Concerns like this merely reinforced the fact that postwar skiing had become big business. Like it or not, industry observers argued, modern ski areas cost a great deal to run. As the popular ski writer A. W. Coleman suggested, skiing had become a truly "commercial enterprise," adding, "To those of us who knew the old days . . . something undoubtedly has been lost in the transition — something difficult to put your finger on. In a way it's too bad the old atmosphere could not have been maintained. But that's progress, as the phrase has it."[92] As lamentable as that shift was, though, a growing generation of younger skiers never knew the sport as anything other than a highly technological and commercial activity. By the 1950s fewer and fewer skiers and industry observers stopped to question the sport's trajectory.

Finally, purists scorned the gimmicky promotions and haughty atmosphere of large resorts like Mount Snow or Sugarbush as a cheapening of the sport. Although some defended Sugarbush as a great place to ski, its reputation as "Mascara Mountain" often made it feel like a shallow place where an obsession with image left little room for the actual sport of skiing.[93] Such purists were joined in their critiques by others who criticized ski lifts as a perversion of the "noble challenge" of the mountains and the destruction of "the skier's only chance to fathom by his striving the greatness of the mountains and to become aware of the minuteness of himself." By becoming as technological as it had, the new ski landscape undermined the soul of the sport.[94]

Some purists rejected skiing's technological turn entirely, embracing what many by the late 1960s were referring to as a "new" sport on the ski scene — cross-country skiing. With millions riding up Vermont's chairlifts each year and skiing back down on artificial snow, and with thousands more now taking to recreational snowmobiling, cross-country skiing amidst Vermont's fields, forested hills, and country lanes became something of a statement about landscape and identity in postwar Vermont. Of course, cross-country skiing had been around a lot longer than modern, resort-based downhill skiing — if not as a recreational sport than as a means of practical transportation — but its rediscovery in the late 1960s and early 1970s provides a glimpse into some skiers' feelings about the status of recreational leisure in rural Vermont. For those to whom downhill skiing had become too technical, too expensive, or simply too big for its own good, and for those to whom snowmobiling was simply too offensive, cross-country skiing offered an inexpensive escape to an entirely different geography — one of quiet solitude, of communion with natural or pastoral beauty, and of the personal challenge of using one's own energy to move through the snow. By turning their backs on what some called the "giant lift complexes," the "subway-rush-hour lift lines," and the "machine-dominated life here in America," cross-country skiers redefined the ski landscape, not according to a modernist discourse but according to something more akin to older images of anti-modern rural simplicity.[95] Gliding up and down snow-covered pastures, moreover, they even managed to reforge connections between skiing and the landscape of work in which it had evolved nearly a half century earlier.

By the end of the 1950s many were taking a more critical look at the complexities and contradictions at work in the making of Vermont's tourist landscape. As argued by Miriam Chapin, a sixth-generation Vermonter writing for *Harper's* magazine in 1957, Vermont had become "more full of assorted baloney, hokum about unspoiled Vermont, snobbery about ancestry, guff about noble Vermonters, maple syrup, and Calvin Coolidge . . . than any other state with the possible exception of Virginia." Instead, Chapin wrote, the state's farms had been given over to professors and retirees or were inhabited by Vermonters with work-a-day jobs who only dabbled in farming. The new Vermonters were individuals who trafficked in nostalgia, sold their land, or opened tourist-related businesses such as hotels or retail shops. There were a few "pristine," "old-time" Vermonters left, she conceded, but they were nothing like the popular image suggested. Instead, "They get drunk on Saturday nights, pillage the empty summer camps in the fall, go to jail for driving their old cars without licenses, go fishing when they feel like it, pick berries, shoot deer and rabbits without too much care for the seasons set by the game laws."[96] Whether Vermont's rural tourist ideal had ever actually been a reality was irrelevant in the face of such critiques. What mattered most, at least from Chapin's perspective, was that modern Vermont was something other than ideal at all.

The tremendous effects that skiing had on Vermont's rural communities made it instrumental to the kinds of self-reflections that lay behind critiques like these. The dazzling technological scenes, new architectural styles, and new social codes that defined the mid-century ski landscape had spawned a new middle landscape where visitors and residents were forced to negotiate the tensions that modern and extensive development was bringing to Vermont. New expectations about leisure and new kinds of work were transforming the ways in which social groups defined and interacted with the state's rural landscape. For many, skiing offered great promise, both in terms of the new outlets it generated for leisured visitors and the new opportunities it offered for working Vermonters. For others, skiing was becoming a curse and a blight on the state. Whether one approved of skiing or not, the ski landscape represented a new pace of change in a state long popularized for its oldness, tradition, and stability. It would be left to Vermonters in the 1960s and 1970s to deal with the consequences of that shift as best they could.

CHAPTER 6

Balancing the Rural

Planning, Legislation, and the Search for Control

Tourism's twentieth-century expansion into all corners of Vermont and into all four seasons of the year, coupled with skiing's dramatic effects on the state, placed enormous pressures on Vermont's rural communities. During the 1960s and 1970s concerned residents as well as concerned visitors struggled to manage problems associated with rising numbers of visitors and a widening range of recreational land uses. This chapter explores the planning initiatives and the land use and environmental legislation to which they turned in their efforts to negotiate present and future challenges.

The kinds of challenges Vermonters faced at this time were not entirely unique; residents in thousands of postwar rural communities from New England to the American West were faced with growing numbers of newcomers and the problems they posed. Millions of Americans embraced nostalgic conceptions of rural life with renewed vigor following World War II, and many acted on that embrace by purchasing vacation homes in the country or at the very least pursuing recreational activities that took them to and through rural communities. They were joined by millions more who moved permanently to rural communities and small-town suburban developments in search of a lifestyle unattainable in urban America. The escalating pressures that visitors (as well as new residents) exerted on the American rural landscape intensified tourism's "devil's bargain" by threatening the very qualities that made rural places attractive to outsiders.[1] "Today, more and more of the 60-odd million people who are clustered in the northeastern cities are taking advantage of Vermont's fresh air and striking natural beauty," one Vermont commentator argued in 1966. "If not controlled, this use of Vermont's resources by so many visitors and new residents could easily create the very things the city-dweller seeks to escape."[2]

State officials across the United States turned to planning and legislative initiatives to cope with this kind of concern. In the process, the historians Richard Judd and Christopher Beach have argued, they produced innovative and powerful new scenic and environmental controls, helping to seed the American environmental movement at the state level.[3] The state of Vermont emerged as a leader in that trend, as residents at all levels of society and government increasingly directed the course of landscape and identity formally through public policy. They were motivated in the broadest sense by a search for balance between the overlapping demands of *development and restraint*. Questions about balancing the demands of those who advocate development and those who advocate its restraint are, of course, at the very center of what land use planning is about. Their ties to rural tourism in Vermont, however, gave such questions a specific focus and a set of place-based associations that made the state an important historical site for their negotiation. The fundamental question facing Vermonters as the 1960s passed into the 1970s was how to balance demands for economic growth through tourist development with demands for restraints on that growth designed to protect the state's rural scenery and rural "way of life." This search for balance echoes Sidney Plotkin's arguments linking the history of American land use conflicts to capitalism's inherent tension between the forces of expansion and those of exclusionary property rights. While capitalism is predicated on the idea of "perpetual growth of production for profit," Plotkin notes, it is simultaneously dependent on the exclusion of the working class from "direct access to the means of production." For Plotkin, the nation's history of land-use conflicts embodies this tension: "Every major property owner that seeks to benefit from the wider circulation of wealth," he writes, "has simultaneous interests in both exclusion and expansion."[4]

Expansion and exclusion — development and restraint — became a recurring framework through which Vermonters debated the relationship between tourism on the one hand and land use planning and legislation on the other. No one doubted tourism's importance to the state's postwar tourist economy. Between 1948 and 1958 alone, annual expenditures for tourism and recreation had risen 81.3 percent, from $42.4 million to $77.7 million. Tourism ranked a close third in its value to the state, behind agriculture and manufacturing, and it had become a major source of employment for Vermonters. Clearly, one group of planners noted, tourism was "a decisive element in the viability of Vermont's economy."[5] But just as no one doubted tourism's importance, no one could agree on how to manage its consequences, particularly when that management involved public policy. There was a simple reason for this: the fates of work and leisure had never been more interconnected in Vermont, nor had their relationship ever been more complex. By the 1960s Vermonters agreed as never before that threats to the scenic landscape — itself a product of rural work — were also threats to tourism, and thus to the economy of the state as a whole. That equation reminded everyone that efforts to balance the demands of development and restraint were linked inextricably to efforts to balance the demands of work and leisure. Of course, that link was by no means straightforward; there was no simple equation between scenic protection and the protection of leisure for the sake of visitors, nor was there a simple equation between unfettered development rights and the protection of working opportunities for Vermonters. The protection of rural scenery also had everything to do with maintaining those opportunities (for farmers as well as for those who worked in tourism), just as the protection of development rights had everything to do with maintaining a robust leisure economy.

No one could define this relationship between development/restraint and work/leisure with precision or to the satisfaction of all. But what they could try to do was manage and control it in ways that seemed, on the surface at least, to benefit as many social groups as possible. This chapter explores the planning and legislative initiatives used in that effort across two different yet related periods. The first of these periods was the early and middle years of the 1960s. During this time state officials focused their attention on "scenic preservation," particularly along the state's roadsides. Their efforts were guided throughout by the belief that scenic protections were crucial to the continued expansion of the state's tourist trade. The second of these periods was the late 1960s and early 1970s. During these years two factors converged to shift policy discussions about development and restraint. One was a growing sense of public urgency about tourism, which as we shall see here had a great deal to do with concerns about vacation home developments in Vermont' ski towns. Echoing debates from the 1930s about accessibility and the Green Mountain Parkway, Vermonters and visitors alike were now defining tourism as a *threat* to scenery rather than a mere beneficiary of it. A second factor was the emergence of the national environmental movement. Environmentalism gave Vermonters a new language through which to articulate their concerns — a language that moved beyond "scenery" and "scenic preservation" and into the realm of "the environment" and "environmental controls." That new language reflected new kinds of concerns and it spawned a new kind of identity for Vermont.

Whether concerned social groups directed their attention toward "scenery" or the "environment," their goal typically was to protect a rural landscape and a rural identity that echoed familiar nostalgic discourses. But while the image may have often been nostalgic, the means of protecting it were not. There was nothing conservative or reactionary about Vermont's use of land-use planning and environmental legislation. In fact, Vermont gained a national reputation during the 1960s as a leader in the environmental movement, and the state's rural landscape became equated in the popular media with a progressive discourse of planning and environmental control. Rural Vermont now became even more attractive in the public mind, not because it was traditional but because it seemed to reconcile tradition with cutting-edge policy.

Concerned social groups shared another common goal as well. What they were after was a greater sense of control over the future of landscape and identity in a state where tourism left little untouched. But questions about who got to maintain that control and to define its contours were obviously another matter. Planning and legislation were always divisive issues. They were opposed by some who felt that the language of balance and fairness used by their backers slighted the interests of working Vermonters, or that they favored the interests of leisured visitors. They were opposed by some who said they did not go far enough to stop things like vacation home development. They were opposed by others who saw them as "socialist" or "communist." In other words, formal efforts to plan for and legislate the relationship between development and restraint yielded a middle landscape where debates about public policy became debates about the right to control the reworking of rural Vermont. A simply stated but exceedingly difficult question now filtered its way into every corner of Vermont: How does one use planning and legislation to guide the future of work-leisure relations?

Planning, Scenery, and Economic Growth, 1962–1968

Vermonters have a history of resisting government-sponsored planning and land use controls — a history that, as we saw in chapter 3, impeded the use of tactics such as zoning. That resistance began to break down during the 1960s, however, both among politicians in Montpelier and among some in the general public. This was a time of dramatic change in Vermont, when residents faced a range of challenges for which government-sponsored action seemed more appropriate than ever. For some, Vermont's interstate highway system was perhaps the best symbol of this change. Vermont's interstates opened piece-by-piece throughout the decade, connecting Vermont to the rest of New England as never before. That connection promised to bring new businesses, new job opportunities, and new permanent residents to the state. And in many respects, such promises came true. After decades of out-migration and marginal population growth, the state's postwar population grew at what was for Vermont an astonishing rate — a fact many in the general public took as the best marker of economic growth. For decades, annual rates of natural increase in Vermont's population had only narrowly matched annual rates of out-migration. The state's total population had declined by 3,500 between 1910 and 1920, and again by 380 between 1930 and 1940. That trend turned around dramatically in the coming decades, however, as reduced rates of out-migration and increased rates of in-migration from other states raised Vermont's population between 1940 and 1970 by 85,501 to a total of 444,732. The most dramatic growth during that period came between 1960 and 1970, when the state's population rose by 54,851. Nearly 30 percent of that growth was attributed to people from outside Vermont. Trends like these were particularly notable in Chittenden County, home to the state's largest city, Burlington, and home now to a handful of high-tech manufacturing companies, including IBM.[6]

Vermonters were well aware that their state was changing, and that with those changes came new threats to its rural beauty. How, many now wondered, would they protect the state's traditionalized rural scenery — its farms, villages, and mountain views — from the unsightly shopping centers, billboards, suburban developments, and new roads that seemed to be creeping into the state? Part of their concern stemmed from a desire to protect a landscape that many residents saw as essential to their quality of life. Rural beauty made Vermont a special place to live, and Vermonters wanted to see that continue. But concerns about rural scenery — particularly among politicians and entrepreneurs — also grew out of a desire to protect tourism and its privileged place in the state's economy. State-sponsored programs in land-use planning became crucial to that project, wedding the future of work-leisure relations to the realm of public policy in ways it had never been before.

State-sponsored planning got its first formal boost in 1962 with the election of Governor Philip Hoff, Vermont's first Democratic governor in over a century. Hoff capitalized on the availability of new federal funding for planning and social programs under Lyndon B. Johnson to initiate an ambitious planning agenda. He maintained a focus on that agenda over the next six years, normalizing planning as part of the state's political process and centralizing political power in Montpelier to an unprecedented degree. In 1963, Hoff established the Central Planning Office (CPO) to fund and manage a "comprehensive state planning program" focusing on issues of economic development, education, transportation, and land use. He established new planning task forces and advisory groups, the most notable of which was the Vermont Planning Council, whose 1968 report, *Vision and Choice*, became a key text in the annals of Vermont planning. Hoff oversaw the creation of regional planning commissions to coordinate planning initiatives throughout the state and to facilitate interactions between the CPO and the local planning boards that many towns were now establishing as well.[7] Vermont was changing across a host of social, economic, and geographic fronts, planners argued; meeting these challenges demanded coordinated action. "Through state planning," the CPO announced in a 1964 report, "these changes can be directed and controlled for the State as a whole or the State can remain indifferent until it is too late and the change controls it. The State must plan in order to avert undesirable change, meet impending change, and produce desirable change."[8]

"Scenery" and "scenic preservation" became keywords in Vermont's planning circles and in the popular press, particularly after 1966, a year in which the CPO released two landmark studies, *Vermont Scenery Preservation* and *The Preservation of Roadside Scenery through the Police Power*. Taken together, these reports highlighted two important trends. First, they reflected a growing sense of urgency about the future of rural scenery in Vermont. As the CPO emphasized in *Vermont Scenery Preservation*, "It is imperative to realize that [Vermont's] scenic assets are now seriously threatened, for the first time, and in several different ways. From time immemorial Vermont has been green, peaceful, unspoiled; now large parts of the state may quite suddenly be transformed, possibly almost beyond recognition. (Some already have been.)"[9] Second, the reports revealed a widespread belief that the protection of scenery and the future of economic development in Vermont went hand in hand. As the CPO argued,

The general welfare of every city, village, and town in the State will be improved by the preservation and protection of the scenery in Vermont. The logic is quite simple:
   a. Vermont's scenery is exceptionally beautiful;
   b. exceptional beauty attracts people;
   c. people impressed by the aesthetics of Vermont, spend money in the State, and some open new businesses and industries;

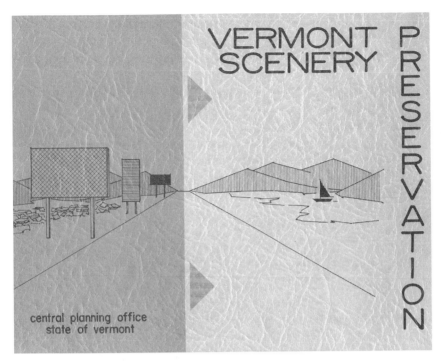

FIGURE 6.1. *Vermont Scenery Preservation*, 1966. This planning report was the culmination of years of study by the Central Planning Office. Reports like these became increasingly common in Vermont during the 1960s and early 1970s as state officials, research consultants, academics, and environmentalists sought to bolster their arguments about the future of land use in rural Vermont. 1966 Vermont Central Planning Office, *Vermont Scenery Preservation Report*. Courtesy of Special Collections, University of Vermont Libraries.

d. therefore the preservation of the State's scenic resources and beauty will help insure its economic well-being.

Vermont cannot long survive without the economic value of its scenic assets. These scenic assets have been responsible for attracting new businesses and industries, permanent residents, part-time summer and winter residents, and tourists.[10]

The "quite simple" logic outlined above pointed to an inescapable conclusion: scenery was far too important to Vermont's economy to leave it unprotected from threats such as commercial and residential development.

The question Hoff and his planners faced in the wake of findings like these was not how to *stop* development before it consumed Vermont but how to *guide* it in ways that maintained the state's classic identity, thereby attracting more visitors to Vermont. The equation between development and restraint, after all, was never clear cut. Vermonters may have valued rural scenery as a crucial component of their identity and quality of life while at the same time celebrating a new local shopping center, a new gas station closer to their house, or an influx of new permanent residents to their town. Likewise, tourists may have wanted to see timeless rural scenery while at the same time expecting gas stations, shopping centers, and motels as well. Vermonters and their visitors were entirely capable of carrying competing demands around in their heads — demands that challenged the neat dichotomies we might be tempted to draw between development and restraint.

Roadside beauty moved to the forefront of public discussions about scenic preservation in the 1960s, as Vermonters once again linked the success of tourism to the view from the road. The CPO's *Vermont Scenery Preservation* devoted the bulk of its attention to that view, for instance, arguing that new commercial and residential development posed some of the most potent threats to roadside scenery. In response to that threat, the CPO advocated passage of legislation designed to protect the "typical Vermont" scene of farms, forests, and villages from advertising, junk cars, and the appearance of commercial buildings visible from the road. Hoff publicly backed the idea of such legislation, in addition to supporting civic groups active in beautification and sponsoring contests and conferences on roadside scenery statewide.[11]

Hoff's efforts on behalf of roadside scenery and the attitudes of those who supported him were not unique to Vermont during the 1960s. Spurred on by influential publications such as Peter Blake's pictorial lament for American roadsides, *God's Own Junkyard* (1964), thousands of citizens began taking part in tree-planting and anti-litter campaigns designed to improve roadside beauty.[12] Thousands more began expressing support for legally regulating junkyards, advertising, and commercial developments visible along the nation's roadsides. Support for roadside beautification came directly from Washington as well, where Lady Bird Johnson, wife of President Lyndon B. Johnson, helped win passage of the Highway Beautification Act of 1965. A foundational piece of legislation in the modern environmental movement, the so-called "Lady Bird's Bill" imposed tighter restrictions on billboards and junkyards visible from federally funded highways, raising public sensitivity toward roadside beautification nationwide.[13]

Vermont's policymakers recognized the need to act on the heels of this newfound sensitivity toward roadside scenery. For some, like the CPO's director of planning, Richard G. RuBino, Vermonters faced a clear choice by the mid-1960s: "Unless we move rapidly," he argued, "our scenic roads could soon become nothing more than aisles bordered by cash registers (unguided development) obscuring the merchandise (scenery)."[14] The question now was whether or not Vermonters would follow through and pass new state-level legislation to protect roadside scenery rather than drawing up planning studies and making recommendations alone. Vermont lawmakers took a widely publicized step in that direction in early 1966 by passing a law regulating the visibility of junkyards along federally funded highways. The law brought Vermont into minimal compliance with the 1965 Highway Beautification Act, and it wrote into the books core ideas about the relationship between scenery, tourism, and economic growth. "The general assembly finds and declares," the final version of the bill stated, "that the recreation and tourist industries have become a significant part of the economy and consequently of the general welfare and public good of the people of this state."[15] The needs of working Vermonters depended on the integrity of roadside scenery, such arguments suggested. It followed, then, that the legislative protection of that scenery was justified by its protection of economic growth. That kind of logic would creep into public discussions about scenery preservation repeatedly in the years to come. The intent operating behind it would never be to prevent roadside development outright, but neither would it be to let those roadsides develop in a haphazard, unregulated way. Rather, some kind of balance was needed between regulation and development, between scenic protections and the protection of the free market, if tourism were to keep its promise to benefit all Vermonters.

Renewed calls for billboard regulations in the mid-1960s would become a proving ground for those who argued on behalf of legislative protections for scenery. Billboard debates had not disappeared entirely from Vermont's political scene after the 1930s, but they returned with renewed intensity during the 1960s as public opinion once again fell into step with the cause of roadside beauty. The Highway Beautification Act had a lot to do with raising the profile of billboards, both in Vermont and beyond. The law required that states remove billboards located closer than 660 feet from federally funded roads except in areas zoned as commercial or industrial. Although loopholes in the act made it possible for states to subvert billboard restrictions, some politicians saw the law as an opportunity to step up their own state-level controls.[16]

For their part, Vermont's general assembly enacted new billboard restrictions in 1966, forcing the removal of hundreds of signs across the state. That set off waves of protest from hotel owners, billboard owners, and tourist industry officials, all of whom argued that billboards were essential to Vermont's tourist economy and hence to the economic wellbeing of all working Vermonters.[17] As they had a generation before, business owners wondered what kind of protections they would get in the face of growing support for scenery preservation. Who would control the future of commercial development along the state's roadsides, they asked, and for what purpose? Would Vermont remain a place where entrepreneurs were allowed to pursue economic gain by developing its landscape, or would the state become a caricatured, static image of itself designed to please the tourist?

Those kinds of questions took on a new urgency in the spring and summer of 1967. In May the state established the Committee to Study Outdoor Advertising, chaired by State Representative Theodore Riehle of South Burlington. After three months of research and inquiry by the committee, Riehle announced that he would soon be introducing tough new billboard legislation. On the one hand, Riehle's bill would establish strict new regulations for on-premises signs, while on the other it would establish a *total ban* on all off-premises billboards. It would give Vermonters a chance to settle the billboard question once and for all, he argued, and it would give them a chance to make a powerful name for the state as a national leader in scenic preservation. Vermont's tourist economy was dependent on scenic roadsides, Riehle's committee argued, and billboards were a threat to the future of that economy. What better way to protect the state's scenic beauty while at the same time appealing to the growing pool of American travelers for whom a billboard ban would have been an attractive concept?[18] A total ban on billboards thus promised to increase Vermont's tourist appeal by improving its scenery and forging a progressive identity for the state. Free from billboards, Vermont's rural landscape would embody and reproduce a discourse of scenery preservation — one that made the state an innovative leader in the national fight to protect roadside scenery.

Not surprisingly, industry lobbyists opposed Riehle's bill, as did many business owners across the state, some of whom branded it as "communist" and "dictatorial," and all of whom demanded to know how they would be compensated if their billboards had to come down. Tellingly, some of the most vocal opposition to the bill came from business owners who were directly involved in the state's tourist economy. For those who owned or managed businesses such as hotels, ski areas, and retail shops, the nexus between scenery, tourism, and economic growth looked different than it did for many who supported the billboard ban. Independent hotel owners, for example, saw signs (both on and off their property) as a cost-effective way to advertise services that were vital to Vermont's tourist economy. If the state denied them the right to advertise along roadsides, how would visitors find the services they needed? Indeed, what would prevent uninformed tourists from leaving Vermont for neighboring New York and New Hampshire?[19]

Riehle and the Committee to Study Outdoor Advertising had anticipated arguments like these, and they knew that their law's success depended on the legislature's willingness to provide funding for non-landscape-based advertising for those in the tourist trade. The fact that the state was willing to do just that helped win passage of the billboard law in March 1968, making Vermont (along with Hawaii) one of only two states to ban off-premises billboards entirely. Like the junkyard law of a year before, the billboard law reiterated the importance of protecting scenic beauty for the sake of protecting economic growth. But in recognition of tourism's dependence on advertising, the bill also provided for the creation of the Travel Information Council, a state-funded council designed to create new advertising outlets and new tourist information centers at key entry points into the state. Most importantly, the law mandated the creation of a new system of state-sponsored roadside advertising. Uniform and unobtrusive, these "business directional signs" would provide motorists with the information they needed to find businesses located off the main highway. These signs are still in use today.[20]

Support for land-use planning and scenic legislation in mid-1960s Vermont was rooted in the belief that scenery, tourism, and economic growth were closely interconnected. Hoff, Riehle, the CPO, and others pursued an aggressive program of planning and legislation designed to protect roadside scenery for the sake of the state's tourist economy. Their goal was not to stop commercial and residential development entirely in favor of a timeless, tourist-directed image of rural life. Rather, their goal was to plan for and control that development in ways that protected scenic beauty, tourism, and working opportunities for all Vermonters. As the CPO argued in 1964, "Everyone likes some things about Vermont the way they are. They like its celebrated untouched quality. At the same time, everyone wants the Green Mountain State to progress along with New England and the rest of the Nation. This means change, and unfortunately change can sometimes be damaging and ruinous; but if handled properly, it can be beneficial to our State, our economy and our people."[21]

Vacation Home Development

Much of the impetus for scenery preservation during the Hoff years stemmed from an effort to *promote* tourism. There was never an overwhelming sense during the middle years of the 1960s that tourism was itself a threat to the scenery on which it depended. That would change very quickly and very dramatically in the final years of the 1960s, as concerns about tourism's effects on the state's landscape escalated. The primary focus of those concerns were Vermont's vacation home developments. Here we look at the evolution and nature of those developments, before turning to public reactions to them.

Vermont's expanding postwar tourist economy created new development pressures in communities across the state. Nothing reflected this better than lodging. Vermont's annual increases in tourist lodging were some of the highest in the nation, particularly as the state's numbers of skiers grew and as greater numbers of tourists began choosing motels over the tourist homes of the 1930s. The number of motels in Vermont increased by 200 percent during the 1950s alone, and expenditures for tourist-related construction of all kinds were reaching as high as $15 million annually by the early years of the 1960s.[22] Ski towns saw the bulk of this development, as entrepreneurs constructed a new ski-landscape ensemble of roads, restaurants, shopping facilities, motels, and vacation homes that stretched far beyond the individual resorts themselves.[23] More so than other tourist activities, skiing concentrated large numbers of people in single towns during weekends and holidays, generating tremendous pressures for goods and services. That pattern accelerated during the middle years of the 1960s, reaching a peak by the end of the decade.

Vacation homes became the most dramatic and contentious part of this ski-landscape ensemble. Visitors to Vermont had been constructing summer cottages on lakes and in mountain communities for decades, and they would continue to do so throughout the postwar decades. But the real growth in vacation home construction by the 1960s was occurring in ski towns, where a highly profitable market in the sale of building lots and finished homes was now underway. Property owners in communities across Vermont responded to growing demands for second homes by dividing land into development projects ranging in size from a handful of houses to over a thousand half-acre or one-acre lots. Those who participated in this real estate trade were drawn from a variety of different groups. Some were local farmers interested in selling all or some of their land to a developer and moving on. Some were handy rural residents who built a couple of houses to sell on the side. Some were large investment, construction, and real estate firms with prior experience developing resorts in places like Florida and California. Others were timber companies like International Paper who saw an opportunity to make money in new ways from Vermont property whose value for timber production had declined. Some constructed ski chalets and A-frames and provided for basic services such as water and electricity. Others simply built a road or two, surveyed and divided their

FIGURE 6.2. Vacation home development sign, 1971. Signs for second-home developments like this one became increasingly common in Vermont during the 1960s. For some, such developments were healthy indicators of economic growth and an unfettered market in land. For others, they raised troubling questions about issues ranging from taxes to sewage. Photo by John Karol.

land, and began selling vacant building lots to families from out of state. But as different as they were, all of these groups shared one thing in common: they all enjoyed the freedom of developing their property in the absence of strict zoning ordinances or other land-use regulations, and so they all saw dollar signs when they gazed out across the Vermont landscape.

Vermont had 22,548 vacation homes by 1968, with thousands more on the way.[24] Some of the state's largest concentrations of existing and planned vacation homes were located in its southern ski towns, whose proximity to major urban centers outside the state had always made them attractive to visitors who did not want to drive an additional couple of hours in winter conditions to northern ski towns like Stowe. Speculative land sales and vacation home development accelerated in towns such as Stratton, Londonderry, Winhall, Dover, and Wilmington, where the Stratton, Magic Mountain, Bromley, Haystack, Mount Snow, and Carinthia ski resorts were attracting tens of thousands of skiers each weekend. By the late 1960s roughly one hundred developers of various sizes were at work in southern Vermont's Windham County, over half of whom were operating in the towns of Dover and Wilmington alone. In fact, although Dover had about five hundred permanent residents in 1968 and Wilmington about twelve hundred, the two towns had over six hundred vacation homes combined.[25] Five developments in Windham Country — "Stratton" in the town of Stratton; "Chimney Hill" in Wilmington; "Dover Hills" in Dover; "Haystack" in Wilmington; and "Whitingham Farms" in Whitingham — had a combined size of 30,000 acres, at least 4,500 of which were being actively subdivided by 1969. Over 3,700 lots were on the market at these developments already (ranging in price from $4,000 to $6,000), and nearly 1,200 had been sold. Aside from houses, some of these projects also included amenities such as hiking and snowmobile trails, community centers, golf courses, artificial lakes, and even shopping areas.[26]

Developers and real estate speculators often had considerable amounts of money tied up in projects like these, which meant that they were eager to sell lots (with or without houses) as quickly as possible. That required some fancy salesmanship and, in some cases, some outright over-embellishment of their property's attributes. For at least one large developer, the sales process began by mailing glossy brochures to potential buyers who were identified based on "occupational listings and referrals." If prospective buyers responded to these brochures, sales representatives would make a personal presentation in the comfort of the customer's home. Next, promising candidates would be offered a free trip to the vacation home development, complete with lodging in a nearby motel. There they would be treated to tours of the development and the surrounding towns.[27]

Vacation home developers attracted buyers by crafting an identity for their projects that combined the traditional and the modern. According to a promotional film for the Quechee Lakes resort near Woodstock, the resort offered all the charm and scenic qualities one expected from rural Vermont. "But despite all its old-fashioned charm and tranquility," the film added, "Quechee has another side as well: a feeling of excitement and action, knowing that you're in on the ground floor of a multi-million dollar creation, that you're taking part in the growth of a most extraordinary leisure-living community."[28] As had been the case for generations, then, landscape and identity in rural Vermont were based on more than "old-fashioned charm" alone. Visitors who wished to purchase into the "excitement and action" of resort and vacation home complexes now had an unprecedented array of options.

Vacation home developments generated new job opportunities for rural residents, both in construction and in service jobs such as housecleaning, gardening, and snowplowing, once the new houses were in place. In this sense they represented a new avenue of economic growth and another expression of the reworking of rural Vermont. Vacation homes redefined landscape and identity in rural Vermont, just as other kinds of leisure land uses had for decades, in part by generating new sets of relationships between work and leisure. Much of that process stemmed from their associations with land speculation and a high-priced real estate market. During the second half of the 1960s, real estate agents were enthusiastically reporting average prices of $1,000 to $1,500 an acre, with properties selling for $4,000 to $6,000 an acre around ski areas such as Sugarbush, Stowe, and Stratton. In some cases, these figures were up from farm-based prices of only $200 an acre ten years before. Of course, prices for properties with houses already on them were even higher: modest ski chalets on half-acre lots could run as much as $60,000.[29] Numbers like these reinforced the sense among many rural landowners that property was far more valuable for leisure than for farming, and that now was the right time either to sell the family farm or get into the real estate businesses on the side. That is precisely what many Vermont dairy farmers did, both by choice and by necessity. Here was a once-in-a-lifetime chance to get out of farming and get comparatively wealthy, many reasoned. Now was the time to act.

Inflated claims about the profitability of property in Vermont only added to the sense that land was more valuable for leisure than for work. "Land development prices in and around Haystack and Mount Snow ski areas [sic] have risen to heights that few of us ever dreamed possible," one real estate broker wrote in 1965. "It seems as though everyone is inquiring for undeveloped land in hopes of subdividing it into one-half acre or one-acre lots for chalets."[30] Representatives from development companies also fueled a get-rich-quick mentality among rural Vermonters by offering unheard-of prices for land that in some cases had *never* been worth very much, and by hosting public-interest meetings where they praised vacation home development as a tremendous new opportunity for Vermonters who worked in the building trades.[31] Others put a happy face on their development plans, cultivating support among local residents by making donations to local churches and promising to make public recreational land in their developments available for local use.[32] Even agricultural economists and professional planners saw the sale of farmland for vacation homes as a welcome opportunity in some rural towns. As dairy farming became more competitive, costly, and consolidated, it seemed to make sense now more than ever for struggling farmers to sell or develop their property. Indeed, some of the most valuable vacation properties — hilltop land with views and access to ski areas, for instance — had always been the least valuable for farming, due to their steep terrain, shallow soils, and isolation.[33]

None of this means that all Vermonters were necessarily eager to sell their

FIGURE 6.3. Interest groups vie for landowner's attention, 1967. This illustration, which appeared in *Yankee Magazine*, shows a Vermonter with gold at his feet resisting pressures to sell his land from groups such as lawyers, college professors, developers, skiers, and attractive women. By the late 1960s, many Vermonters had seized the opportunity to make a quick profit on their land, while others, like this illustrated character, had refused. Illustration by Austin Stevens.

land for a one-time profit. Many Vermont landowners resisted the temptation to sell out, using the power that land ownership gave them to thwart development for as long as they could afford to do so. Nor does it mean that Vermonters advocated the complete abandonment of farming in favor of tourist development. Rather, most policy makers hoped that farms and vacation homes would be mutually reinforcing — even in the state's rapidly developing ski towns. In the opinion of the noted agricultural economist Malcolm Bevins, for instance, farming and tourism should be thought of as "compatible and complementary" rather than conflicting. Echoing nearly a century of summer home promoters, Bevins argued in 1968 that recreational development, such as the construction of vacation homes, put less-valuable farmland to good use. Moreover, he added, vacation homes lowered the property tax burden carried by farmers by distributing that burden across a greater number of landowners. In the best of cases, that reduction in taxes would actually make it *easier* for farmers to stay in business. And, of course, the more farmland Vermont preserved, the more attractive its landscape became to visitors, thereby encouraging even more investment in the state's tourist economy.[34] The sale of land for leisure, then, would actually maintain rather than undermine a landscape and identity defined by agricultural work. That logic harkened back to the turn of the century, when many hoped the sale of farms as summer homes would achieve similar results. The particulars of how that process was now supposed to play out were new, but the close interconnections and assumed compatibility between work and leisure that guided it were not.

The Emergence of "Environmental Controls"

Optimist arguments like these had been around for generations. But their viability by the end of the 1960s was getting far more difficult to sustain, particularly as commentators in both the regional and the national press gestured with growing concern to the development crisis bearing down on many of Vermont's rural communities.[35] Tourism, it seemed, was becoming as much a threat to the entire scope of rural land and life as a beneficiary of it, and vacation home developments, more specifically, were becoming potent symbols of what happens when work-leisure relations go terribly awry. Concerned residents and concerned visitors drew on the emerging national discourse of environmentalism to express their anxieties about vacation homes, shifting public discussions about scenery preservation to new concepts like "ecology" and to new methods of "environmental control" designed to rein in unregulated vacation home developments. Tourist development had now become something to protect the state's rural landscape *from* as much as *for*, and vacation home developments lay squarely in the middle of that shift.

Public criticisms of vacation home developments typically focused on three main categories. First, many residents worried about the social consequences that these developments seemed likely to have on their communities — a concern that drew strength from broader fears about allowing unknown outsiders to control the future of so much land in the state. Vermont's largest developers were often based in faraway places, and if for no other reason than this, some considered them unfamiliar and threatening. The fact that the giant International Paper Company owned Stratton's largest vacation home development was reason enough for concern. The fact that they passed responsibility for its construction and management to American Central, a subsidiary based in Lansing, Michigan, seemed even more dubious.[36] Dover Hills's new developer, the Cavanaugh Leasing Corporation from Palm Beach, Florida, seemed no better than its previous one, a New York City firm called the Vermont Land Corporation, whose subsidiary, the Vermont Lumber Company (a company with no actual ties to the timber trade), had been in charge of developing the land.[37]

Many Vermonters looked at confusing corporate arrangements like these and wondered just how committed such developers were to the long-term quality of life in their communities. Indeed, some developers did little to quiet their fears. Speaking to *Time* magazine about his post-development plans, Whitingham Farm's developer, Clifford Jarvis, stated bluntly, "I personally have no intention of staying in Vermont."[38] And once they were gone, once their work was done, what kind of social order would someone like Jarvis leave behind? What effect would thousands of new visitors have on day-to-day life in rural communities? How would one ever get to know these newcomers, particularly when they were only in town a few weeks a year, when often they did not attend church, and when their kids were not enrolled in local schools? As one resident of Waitsfield (near Sugarbush) later recalled, "It used to be you went to the store you knew everyone. And there were houses on the street where people lived and now they're all shops . . . But you go up and down the street now and there might be three houses where I know who's there. And I don't like that."[39]

Second, many Vermonters objected to vacation home developments on economic grounds, fearing that vast new tracts of homes would reduce the ability of local communities to pay for and manage town services such as road maintenance, snow plowing, police, and fire. Some worried that vacation homes would eventually turn into *permanent* homes for families seeking to escape from the city. The influx of new children that this change would bring, they warned, would place an impossible burden on small rural schools.[40] Supporters of vacation home development typically countered such fears with the well-worn argument that vacation home owners added a new source of tax revenues to pay for services such as roads and schools — services that part-time residents did not even use as often or with as much regularity as permanent residents. In fact, by 1973 22 percent of Vermont's roughly 250 towns received more than 25 percent of their annual property tax from vacation homes. Vacation homes accounted for over $9.5 million in property taxes statewide — over 9 percent of Vermont's total property tax collected.[41] To many these seemed like encouraging numbers indeed.

By the end of the 1960s, however, many Vermonters were no longer buying into the argument that vacation property was necessarily so good for taxes. While it was true that tax revenues increased in some towns, thereby offsetting some of the costs of new public services, the sale of vacation property also prompted an overall increase in property values, thereby raising the assessment figures on which towns developed their tax codes. As property values went up, then, so did everyone's taxes, making it harder for some to afford to keep their property at all. This produced a dangerous cycle, described by one observer as such: "The demand for land, which raised the price of land, which pushed up taxes, which forced many more parcels on the market, which displaced natives and encouraged massive, gigantic developments, will threaten the very landscape that caused the demand in the first place."[42] Rising property taxes, some warned, were threatening the ability of farmers to stay in business, thereby undermining the state's traditional way of life as well as its scenic rural landscape.

Third, Vermonters increasingly framed their critiques of tourist development through an emerging national discourse of environmentalism. Unlike its antecedent, the conservation movement, the nascent environmental movement of the late 1960s and early 1970s was concerned less with insuring the productive viability of resources such as timber than with broader quality-of-life issues relating to problems such as scenic blight, air and water pollution, and the protection of recreational space. As environmental concerns broadened, so too did the environmental movement itself, spreading from conservation's professional and policy circles alone into community groups, schools, and individual households.[43] Environmentalists spoke in a language of natural systems and the interconnectedness of all life — a language shaped by its association with the developing science of ecology. If human life was dependent on a larger natural system, many reasoned, then anything that threatened the integrity of that system was a threat to all humanity.[44]

By 1968 keywords such as "environmental" and "ecological" rather than "scenery" and "beauty" alone were beginning to inform popular discussions about tourist development in Vermont. This applied quite often to vacation home developments, which many now considered Vermont's most pressing environmental crisis. For critics, Vermont's vacation home landscape embodied many of the same kinds of problems being cited by environmental advocates across the country — problems of visual blight, forest loss, and soil erosion produced by construction on steep slopes. The most common problems to which they pointed, however, were improper sewage disposal and the groundwater pollution that resulted from it. As the environmental historian Adam Rome has shown, concerns about water pollution and overloaded septic systems stemming from rapid postwar suburban development became foundational cornerstones of the nation's environmental movement as well as contributing factors in the passage of state and federal clean water legislation.[45] Similar concerns extended to vacation home developments in Vermont, where sewage disposal and water pollution fueled the popularization of environmental discourse as well as the passage of new environmental legislation.

Town officials throughout Vermont began raising earnest questions about sewage and water in early 1968, and they used those questions in some cases to justify the passage of local zoning ordinances. Wilmington residents passed their town's first zoning regulations in March of that year. Among these were rules that required landowners to abide by a minimum residential lot size of one acre. That posed a direct challenge to developers like William Thomas Cullen, whose Chimney Hill development was then just getting off the ground. What Cullen had hoped to do was sell seven hundred to nine hundred building lots, each of which would be less than an acre in size, and each of which would come with an individual well and individual septic system. Many of Cullen's lots, however, were located on steep land, where a combination of slope and shallow soils made it impossible to support the hundreds of septic systems and wells he had in mind. Consequently, Wilmington's newly minted zoning regulations forced Cullen to either apply for a special variance to the one-acre zoning rule or redesign his development to include larger lots and a centralized sewage and water system. Because he already had sold 310 lots on 210 acres of land, Cullen reluctantly chose to apply for a variance, arguing with the town even as he did so. After initially losing his appeal, Cullen and the town eventually reached a compromise that summer, allowing him to continue selling lots.[46]

Cullen's case caught the attention of developers, planners, and civic leaders across Vermont, moving sewage to the center of popular concerns about vacation home development. Voters in Dover responded by decisively rejecting a 1969 appeal by the developer of Dover Hills for a variance from one-acre zoning. Town leaders cited sewage disposal among the subdivision's 2,855 planned dwellings as a primary source of concern.[47] Just two weeks later, the CPO released a highly critical report citing "acute problems" in Dover Hills relating to sewage and water. According to one CPO official, the development would "destroy a mountain community and devastate the land" unless it included centralized sewage and water systems — a move its developers resisted as financially unfeasible.[48] In addition to individual developers, town governments now faced questions about their public responsibility for sewage handling and their obligations to construct new municipal sewage-treatment facilities. In a highly publicized set of remarks, State Attorney General James Jeffords labeled Stowe (the so-called "Ski Capital of the East") the "Sewage Capital of the East" and threatened legal action if local officials did not take immediate action to ameliorate their town's sewage problems.[49] And what, really, could they do? Sewage was a political liability and a highly visible problem that demanded action, from towns in some cases and from individual resort developers in others.

The rising tide of environmental discourse in the United States gave concerned Vermonters a powerful new framework both for articulating their fears about tourist development and for crafting legislative responses aimed at controlling growth. This "environmental" agenda echoed the tenor of the Hoff administration's "scenic" agenda, and its momentum had much to do with Hoff's groundbreaking work. Indeed, Hoff's Vermont Planning Council had hinted at this conceptual shift from "scenery" to the "environment" in the waning days of his administration, arguing that Vermont must "adopt a pattern of development that provides for growth consistent with environmental quality."[50] But whether one talked in terms of scenery or the environment, public policy throughout the 1960s and 1970s was shaped by its continuing focus on the need for some kind of balance between the demands of development and restraint. Policy debates in the late 1960s and early 1970s targeted tourism as a problem more often than they had a few years earlier, but as in the case of scenery preservation, those who backed the cause of environmental legislation did not do so with the intention of stopping tourist development entirely. Rather, their goal remained the protection of tourism through the protection of landscape — except that by the late 1960s many policymakers now saw themselves as protecting tourism from itself. The only effective way to do so, many now believed, was to craft powerful new land-use regulations that went beyond the regulation of roadside scenery.

That feeling deepened during the tenure of Vermont's new governor, Deane C. Davis, who was elected in 1968 in the wake of Hoff's retirement. Davis was a Republican, a proponent of small government, and a strong ally of Richard Nixon. For all of these reasons, he took a different (and for Vermont, more traditional) approach to governing than his Democratic predecessor. Indeed, when Davis took office in early 1969, state planning and land-use legislation (let alone "environmentalism") were not at the top of his list of priorities. That changed very quickly, however, due to mounting public opinion and to the accelerating problem of unregulated vacation home development. In mid-May the governor hosted the Vermont Conference on Natural Resources, at which he and others examined a series of pressing environmental issues now facing the state. Davis reiterated the paradox of planning under the Hoff administration (substituting the word "scenery" for "environment") when he asked conference goers, "How can we have economic growth and help our people improve their eco-

FIGURE 6.4. Golfers at the Quechee Lakes resort, 1971. Playing golf in the shadow of a former farm-house and barn was representative of the overlap that had occurred by this time between Vermont's traditional landscapes of work and its newer landscapes of leisure. Yet even a former-farm-turned-golf-course was a place of work as well, offering employment for the Vermonters who built and maintained it. Photo by John Karol.

nomic situation without destroying the very secret of our success, our environment?" The answer, he argued, was by crafting a broad range of policies intended to control, rather than forgo, continued development. Pairing the new keywords of "environmental" and "control," Davis pledged to create an orderly, non-threatening climate of development in the state. To make good on his promise, he created a new seventeen-member, eleven-month "Commission on Environmental Control." Led by Arthur Gibb, the former chair of the House Committee on Natural Resources, this high-profile "Gibb Commission" was charged with assessing public opinion on environmental issues, meeting with business leaders and developers, and drawing up a series of recommendations on how best to protect the state's rural landscape both for and from continued development.[51]

If Davis had any lingering doubts about making "environmental control" a top priority of his administration, they were put firmly to rest during his visit to southern Vermont a little over a week after the Conference on Natural Resources. On May 27, 1969, Davis stopped in Brattleboro as part of a statewide fact-finding tour. On his agenda that day were meetings to discuss tourist development with members of the Windham Regional Planning and Development Commission (established during the Hoff years) and with officials from Wilmington and other area towns. All wanted to discuss spiraling property values and their impacts on local taxes, as well as the environmental impacts of vacation home development. Alarmed by the intensity of southern Vermont's problems, and by the apparent likelihood that those problems would continue to migrate northward in the state during the coming years, Davis cited "environmental control" as the most "urgent" issue facing the region. He would rearrange his schedule, he announced at a news conference later that night, and return to the area in a couple of weeks to find out more himself.[52]

Davis made a number of visits in the coming months to southern Vermont towns like Stratton, Wilmington, and Dover, where residents took him to visit proposed and existing vacation home developments. The first problem with these projects, he learned, was that civic leaders never knew exactly what developers had in mind. Without well-defined planning and review processes in place, the public could only speculate about how big these developments were going to become, when they would be completed, how they would be managed, and so on. The second problem was that there was an almost complete lack of legislative control over environmental issues like sewage disposal and water quality. Some towns had modest zoning ordinances in place, but most did not, making it easy for unscrupulous developers to cut corners and costs. Vermont's traditional laissez faire approach to land use, Davis and others readily concluded, had made the state fertile ground for irresponsible development. Davis's first course of action, then, was to meet with the largest developers at work in Vermont—even as bulldozers cut new roads and drillers dug new wells—and learn more about their plans. Between fact-finding and the anticipated Gibb Commission report, perhaps he and others might come up with reasonable legislative controls over unregulated vacation home development.

Davis soon had meetings scheduled with Vermont's largest developers, including an urgent meeting with the president and chief executive officer of the International Paper Company, Edward Hinman. "We welcome growth," Davis told Hinman in his request for a meeting, "but must preserve Vermont's scenic beauty and natural environment."[53] The Davis-Hinman meeting turned into a high-profile media event, and its outcome gave Davis a new sense of confidence in the emphasis his administration was placing on environmental control. Soon after the meeting, International Paper disclosed details about their plans in Stratton and then voluntarily halted those plans pending further deliberations with the governor and the public.[54] By September 1969 all of Vermont's largest developers had promised to submit their plans to the governor's office for a non-binding review. Whether they did so out of good will or out of fears about sinking too much money into projects that might need to be scaled back at a later date was almost beside the point. Davis and his advisors had scored a major public-relations victory.

Cautious voices warned, however, that this victory meant little without tough legal guidelines designed to hold developers to a higher standard. "Speed was essential," Davis later recalled, "if we were to prevent large-scale, improper, poorly planned development in Vermont."[55] Public discussions about environmental control and vacation home development continued throughout the summer and fall of 1969, growing more pressing and more anxious as the Gibb Commission prepared to release its report. In September the Vermont State Board of Health launched a shot across the bow by passing an emergency set of 120-day subdivision restrictions mandating tougher guidelines for sewage disposal on ten-acre parcels divided into three or more individual home sites.[56] The move threatened, worried, and angered some developers, who saw it as a harbinger of tougher laws yet to come. Tighter environmental controls would spell the end of their development objectives, they argued, robbing them of their right to profit from their land. They were certainly willing to cooperate, developers claimed, but they also had a great deal of money sunk into their projects already; it would be wrong to prevent them from selling their land, and it would be wrong to force them to install prohibitively expensive sewage and water systems.

The Gibb Commission released their long-awaited report in January 1970. It began with a commentary on "those factors affecting the environment in an esthetic sense," such as roadside blight, and "those affecting the ecological balance and thus posing a threat to the survival of mankind," such as erosion, water pollution, and anything else that tended to "disturb the balance of nature." But despite this diversity, the commission cited "ecological" problems rather than "esthetic" problems as the dominant issue facing Vermonters. Consequently, the commissioners argued, Vermont needed concise resource classification schemes, zoning for floodplains and high-elevation ecosystems, tighter pesticide regulations, and more effective control of recreational development. Two core themes guided these and other recommendations in the report. First was the belief that environmental regulations — both at the state and the local levels — were needed now more than ever. As the report flatly stated, "We must establish control over any act which has an undue adverse effect on the public health and safety, or the right of people to enjoy an unpolluted environment." Second was the continued belief that any new laws should balance environmental controls with continued economic growth, particularly through tourist development. As had been the case for nearly a decade, the objective here was not to deny Vermonters the right to use their property for economic gain. Rather, the commission argued,

our overall objective should be to insure optimum use of the resources of the State, including land, water, people, and space. We still have more unspoiled resources than do most parts of the Eastern United States. Once destroyed or lost, these resources may never be retrieved. Vermont is now enjoying the benefits of a substantial economic development. The function of its government must be to build upon that economic opportunity while making full use of what has been learned from the failures and consequences of unplanned development elsewhere.[57]

What the commission argued for, then, was greater environmental restraint as well as the protection of Vermonters' rights to develop their land, even if that development came in the form of vacation homes. Control, not outright prevention, was the goal to which they aspired.

Toward that objective, the Gibb Commission produced draft environmental legislation written by Vermont Attorney General James Jeffords and Assistant Attorney General John Hansen outlining the contours of a statewide system of land-use controls. That legislation — ultimately referred to as Vermont's Land Use and Development Law — was quickly introduced in the state legislature. After a few months of modest debate and revision, it passed almost unanimously on April 4, 1970, and was signed into law by Governor Davis the next day. The bill, which became known as "Act 250" for its sequential number in the legislative session, created a new review and advisory group, the State Environmental Board, to oversee the implementation of the law's core provisions. Those provisions included a mandatory permitting process and period of review for any development of ten acres or more, whether a vacation home development, ski area expansion, or new shopping complex. Developers were required to submit their plans to local review boards in the town of their development and to one of Vermont's eight new "district environmental commissions" established to help administer the law. Local and district commissioners would review each permit application based on *ten criteria*. The first three of these focused on protecting water and air quality. The next five focused on a range of issues including soil erosion, scenic degradation, historic preservation, traffic safety, education, and local government. The final two criteria required that developments conform to a series of follow-up state-level plans yet to be passed (criterion nine) and to local or regional plans (criterion ten).[58]

By the time Act 250 became law, it was clearly unrealistic to argue uncritically that what was good for tourism was good for the state as a whole. Vermonters believed more strongly than ever that tourist development was itself a threat to landscape and identity in their state, and they deployed social, economic, and, most importantly, environmental arguments to make a legislative case against uncontrolled tourist development. In doing so they implicated the state's landscape in the production of a new kind of rural tourist discourse — one predicated on using environmental control to initiate some level of balance between development and restraint. With state-level environmental controls now in place, Vermonters would have to figure out how to navigate the contours of that balance. That process would test the bonds of social relations in the state, both in the immediate wake of Act 250's passage and in the decades to come.

Balancing the Rural in the Wake of Act 250

Act 250's passage in April 1970 was really just the law's first step. The law mandated three additional steps, beginning with the passage of the Interim Land Capability Plan, a broad statement of development criteria and an inventory of land use statewide. Next, the law called for the passage of the permanent Land Capability and Development Plan. This plan would replace the interim plan, in addition to outlining a more detailed vision for the future of planning in Vermont and setting the stage for the third and final step: the passage of the comprehensive, statewide Land Use Plan. This plan would map the entire state according to four developmental categories: agriculture, recreation, forestry, and urban growth. It would determine where certain kinds of development should and could take place, and it would serve in place of local land-use controls in towns without zoning ordinances or development plans of their own. Taken together, these three steps were intended to make Act 250 a strong regulatory statute rather than a permitting process alone, and together they were intended to redefine landscape and identity along new, legislatively determined lines.

In the three years after the passage of Act 250 these mandates, coupled with the day-to-day challenges of learning how to implement the law's permitting and review provisions, sparked policy debates and on-the-ground controversies among developers, planners, politicians, journalists, rural residents, and vacationers. Strong opinions developed quickly about the law. Opponents criticized it as a concession to tourist-based perspectives on rural life, an unfair restriction on property rights, an example of overzealous government intervention into private affairs, and above all else a misguided impediment to economic growth and opportunity for working Vermonters. Supporters argued that Act 250 placed control over land use in the hands of individuals and local review boards as much as possible, and that the law was designed to balance the needs of economic growth with those of environmental restraint. Questions about the definition of rural identity in Vermont always lay at the heart of such debates — questions defined not merely through abstract, symbolic, or representational terms but, as we shall see, through struggles to control the use and management of the physical landscape itself.

During the late 1960s and early 1970s, the regional and national press picked up on Vermont's turn toward environmental control, redefining the state's rural identity according to its associations with environmental thought. That process of identity construction unfolded in a couple of ways. For starters, Vermont's environmental reputation was informed by a popular rediscovery and reinterpretation of the state's rural culture. Some environmental advocates now viewed Vermonters according to parameters established by a larger revival of interest in rural folkways and simple living in 1960s and 1970s American culture, in particular by romanticizing them as a people whose lives were guided by an innate sensitivity toward nature.[59] Indeed, it took no great stretch of the imagination for some to reconceptualize rural people as proto-environmentalists; rural Vermonters seemed to embody what it meant to live close to the land and in harmony with one's surroundings. That impression raised Vermont's appeal among the estimated 100,000 hippies and back-to-the-land enthusiasts who moved to the state between 1967 and 1973. Many of these transplants found their way onto one or another of Vermont's scores of communes before buying land for themselves or moving on from the state. For many of these young people, Vermonters seemed to embody a life of natural simplicity — one that had not been tainted by a suspect faith in modern materialism.[60] The reality was far more complicated, of course, and some newcomers were surprised to find that rural Vermonters (and rural life in general) did not always match their romanticized image. But the idea that rural Vermont was home to some sort of previously overlooked environmental wisdom provided a powerful new context for identifying the state and its people.

Vermont's new environmental identity was also shaped by the state's innovative and progressive legislative agendas, of which Act 250 was the centerpiece. Act 250 helped solidify Vermont's reputation as a leader in what by 1970 seemed to be the cultural and political mainstreaming of environmental thought. Enacted on the heels of the landmark National Environmental Policy Act and just weeks before the nation's first Earth Day celebrations, Act 250 recast Vermont not as a traditional and conservative place, but as a place on the cutting edge of environmental policy. According to an ambitious survey of environmental attitudes in Vermont conducted in 1971 by the Vermont Natural Resources Council, a "significant majority" of Vermonters supported Act 250 and the ideology of environmental restraint that it represented. "Vermonters show a high degree of concern about maintaining their state's quality environment," the report concluded, "and they support current efforts by their state and local governments to achieve this end."[61]

Additional legislation and new environmental programs enacted shortly after Act 250 seemed to confirm this leadership status. Vermont began a massive and well-publicized annual program in 1970 called "Green Up Day," in which volunteers removed litter and debris from local roadsides. The state's first annual Green Up Day drew support from an estimated 75,000 of its 450,000 residents, whom together collected thousands of cubic yards of trash.[62] Vermont gained national attention again in 1972 when it passed one of the nation's first "bottle bills," requiring a refundable deposit of five cents on all glass and metal beverage containers. The law went into effect in 1973, and studies soon showed that five cents was indeed enough of an incentive to discourage littering.[63]

New initiatives like these advanced an environmental identity for Vermont not just through the legal code but through the physical landscape itself. Carefully zoned towns, well-planned and closely regulated developments, and litter-free roadsides were meant to embody the state's commitment to environmental conservation and its identity as an environmental leader. Among some audiences, at least, the Vermont landscape became synonymous with environmental discourse, disseminating a favorable public image beyond the borders of the state and increasing its prestige both as a tourist destination and as a place to live. The more environmentally committed Vermont was, it seemed, the more appealing it would be to visitors.

There was some truth to this. In particular, many property-owning visitors and potential vacation home owners considered Act 250 a valuable means for protecting their investments and their vacation experiences, both of which depended on the preservation of rural scenery, not the creation of crowded and haphazard vacation communities. Many old-guard vacationing families who had purchased abandoned farms as summer homes long before scorned new vacation home developments as contrary to their ideas about what made Vermont a special place. But so did some of those who bought land and homes in the very same kinds of developments that had inspired Act 250. One woman from Connecticut who bought into the

FIGURE 6.5. Governor Davis (back row, in hat) and volunteers on Green Up Day, April 1972. The popularity of Vermont's Green Up Day program suggests the degree to which environmental sentiments were taking root in Vermont, in part on the heels of mounting concerns about tourist development. Green Up Day, which remains an annual tradition today, also helped to amplify Vermont's emerging reputation as a leader in progressive environmental initiatives. Photograph by the Vermont Travel Division, courtesy of the Vermont State Archives.

Quechee Lakes resort, for instance, sold her property back to the developer in disgust. "The area is being ruthlessly exploited," she complained, "and will soon look like a Vermont Levittown."[64] Uniform, suburban-style developments like New York's famous Levittown were not what she and many other visitors had in mind when they pictured rural Vermont.

State officials wanted people like this woman to know that Act 250 was intended to protect the state's rural landscape from the kind of mass-produced, homogenized, and environmentally unsound developments she feared. But not everyone was so optimistic about the law, nor were they so eager to embrace an identity for Vermont defined by its associations with environmentalism or with a traditionalized rural image that felt more like a concession to tourists than anything else. Surveys of hundreds of Vermonters following the passage of Act 250 revealed that a very narrow 51 percent were in favor of "state and local control of any private development." And while 65 percent felt that it would be worth giving up a "significant amount of personal freedom" to protect the environment, most also hoped that circumstances would never come to that.[65]

Numbers like these were often shaped by fears about the loss of property rights and the consolidation of political power among state-level bureaucrats.[66] "If I own a piece of land, I don't like to have the state tell me what I can do with it," one Vermonter argued. "There's too much state control already; that's the way I feel about it. It's just like Communists: they tell you over there what to do, and how much land you can use, and what you can do with it, and it's getting that way here."[67] Worries like this extended far beyond Vermont during the 1960s and 1970s, as landowners nationwide found themselves debating and adjusting to new federal, state, and local regulations. In addition to Vermont, the states of Hawaii, Oregon, and Maine all passed land-use initiatives that angered property-rights advocates.[68] Even those in Vermont who were willing to accept some degree of governmental control over private property tended (as they always had) to favor *local* rather than state-level regulations. Surveys showed that 67 percent of Vermonters favored local zoning ordinances over any sort of state plan, and only 49 percent favored the idea of requiring state permits (as in the case of Act 250) for developments in towns with local zoning regulations already on the books.[69] If individual landowners *had* to be regulated, many Vermonters argued, their respective towns knew best how to handle their needs.

Property-rights advocates often won support for their positions in ski towns, where large and small developers now found themselves faced with new zoning regulations and with the extra burden of shepherding their development plans through Act 250's permitting process. Among the most outspoken of these ski-town developers was Steve Chontos, a local businessman and small-scale vacation home developer in south-central Vermont. As Chontos flatly told the *Brattleboro Daily Reformer* in 1969, "A man is the best judge of what to do with his own land." Legal prohibitions on property owners, he argued, were an "overreaction" on the part of planners and politicians, most of whom seemed more interested in passing new regulations than enforcing those already in place.[70] Chontos ultimately laid out his displeasure with Act 250 (and the whole concept of environmental control) in a full-length, thinly disguised autobiography, *The Death of Dover, Vermont* (1974). The book tells the story of Mark Terrick, a Connecticut native who moves to Dover and opens a retail ski business before becoming a developer and member of the local selectboard. In Terrick's opinion, environmental legislation at both the state and the local level had undermined the rights of property owners like himself. That prompts him to rally against what he sees as the injustice of land use regulations, fighting against and alongside his neighbors to curtail the power of state, regional, and local officials.[71]

As Vermonters like the fictional Terrick struggled to understand the meanings and implications of environmental legislation, their state became something less than an idyllic pastoral refuge. During the early years of the 1970s, Vermonters fought continually amongst themselves over the fate of Act 250's next three steps — the Interim Land Capability Plan, the Land Capability and Development Plan, and the statewide Land Use Plan — embroiling the state in a powerful struggle for control over the form and meaning of the rural landscape itself. After the passage of Act 250, the Vermont State Planning Office (formerly the Central Planning Office) acted quickly to put these next three steps in motion. In June 1971 the office sent the completed interim plan on to the State Environmental Board and the governor, who, after a period of revisions and review, both approved the plan in March 1972. The interim plan offered a broad statement on development goals and environmental limits to growth, outlining in maps and in text those places in Vermont best suited for future development.[72]

The passage of the interim plan was simple by comparison to the Land Capability and Development Plan and the Land Use Plan, both of which required approval by the state legislature in addition to the Environmental Board and the governor. The State Planning Office completed drafts of both plans in the fall of 1972 and submitted over 150,000 copies to the public in mass mailings. But the draft's imprecise language and poor-quality maps led to confusion and concern, and widespread opposition grew quickly to the plan's rigid land-use categories and outright prohibition of development in ecologically sensitive areas of the state, including floodplains and high-elevation terrain. Did this all *really* mean, some residents demanded to know, that it would be illegal to sell such land to a developer or to develop it themselves? If so, who was going to compensate them for this potential loss in revenue? Environmental controls sounded good at first, some were now saying, but perhaps this was all going just a bit too far.[73]

Debates about Act 250's final two steps were fundamentally about the ability of one group or another to define the balance between development and restraint that the law had promised to deliver. They were debates about the entire legislative future of landscape and identity in rural Vermont. Many residents felt that these plans stripped them of the power to control their own futures, and they expressed outrage over the plans at public hearings held in towns across the state. For some, the statewide Land Use Plan (the more controversial of the two plans) amounted to nothing more than "statewide zoning" imposed on rural people by bureaucrats in Montpelier — a unthinkable act in a state where even *local* zoning was still suspect. One opponent from the village of South Royalton likened the plan to the work of Ghengis Khan and Hitler and promised to shoot or cut the throats of anyone who came on his property to enforce Act 250, including Governor Davis himself.[74] "We're 100 per cent in favor of controlling pollution," another argued. "This law controls people."[75]

Complaints like these are commonplace in the annals of land-use debates, whether in Vermont or elsewhere. But why did a law that "controls people" feel particularly offensive in the context of rural Vermont? The answer to that question has everything to do with the reworking of the rural landscape that followed in tourism's wake. One of Act 250's most objectionable qualities was that it seemed to forsake economic development in favor of a timeless, leisure-based conception of rural life. And in this sense Act 250 raised a host of difficult questions about the relationship between work-based and leisure-based conceptions of the rural landscape. Those questions were not new; for nearly a century now, tourism had been forcing Vermonters and visitors alike to debate the future of rural work, rural leisure, and the relationship between them. By the 1970s such debates were part of the fabric of rural land and life in communities statewide. What made them feel sharper and more painfully urgent now was their entry into the realm of state law itself. If Act 250's final two steps became law, some wondered, would that make quaintness and timelessness the law of the land? Would it lock Vermont into a static, sentimentalized image of timelessness and rural stability, precluding any kind of working activity or program of economic development that ran counter to that image? Or did it even make sense at all by this time to consider the needs and opinions of working Vermonters independently of those of visitors?

If you put questions like these to any of the new opposition groups forming against Act 250 or to someone like Steve Chontos's fictionalized Mark Terrick, you would have gotten an earful. For Terrick, "the right to own property and grow in business" is a crucial part of American culture, and while he agrees that people should act in environmentally responsible ways, he also maintains that they should never be forced to do so. Act 250, Terrick believes, had been passed through the support of self-centered summer visitors, idealistic newcomers, and misguided, even villainous planners and state officials, all of whom acted on "prejudice and emotion" rather than on sound democratic and business principles. Vermonters seemed to have forsaken their own heritage in favor of an unsustainable, leisure-based view of rural land use, and in so doing they had forsaken the capitalist system's most basic principles of supply and demand. As Terrick pondered the situation, a disturbing thought came to mind. "Is it possible," he wondered, "that a gigantic ball has been created in the minds of the people of Vermont; a ball that is rolling over the traditional desires to grow and expand; a ball that is crushing property and personal rights; a ball that is gaining momentum?" What was allegedly good for the environment and for the state's tourist image, he concluded, was not necessarily good for freedom and economic growth in Vermont as a whole. After Terrick reaches this conclusion, Chantos ends the book with Terrick and his family moving away in disgust, and then (apocalyptically) returning to Dover in 1984 to find the town impoverished by overregulation.[76]

Of course, most Vermonters knew that it was too simple to assume that the scenic preservation of the early 1960s or the environmental controls of the late 1960s and early 1970s precluded development, even when they seemed destined to protect a tourist-based conception of rural order. Pastoral charm and a sense of rural timelessness had always been important to the success of the state's tourist economy, but so, too, had development, growth, and a ready willingness to alter the rural landscape to create new tourist opportunities. Tourism had always involved reworking the rural landscape. It had always involved recreating and redefining work by recreating and redefining landscape and identity according to new leisure demands. It was no longer possible to talk about work, economic development, and landscape change in Vermont without talking about tourism.

Ski industry officials were particularly quick to remind their critics of that fact. But increasingly, their admonishments cut across the grain of public opinion, as Americans reassessed the environmental impacts of recreational activities like skiing, and as the ski industry became embroiled in a handful of high-profile land-use conflicts not only in New England but in California and Colorado as well.[77] Skiing was still extremely popular in Vermont, but its reputation had suffered through its controversial associations with vacation home developments and through its questionable impacts on the fragile, high-elevation ecosystems that Act 250 had specifically defined as being at risk. The comments of one state health commissioner demonstrated skiing's troubled reputation with particular clarity when he argued that thirty-five hippies living on sixty acres of Vermont land caused less environmental damage than thirty-five skiers spending the weekend in a two-room chalet. Because skiers came in such concentrated numbers, the health commissioner argued, they overloaded sewage facilities for short bursts of time, causing grave pollution problems.[78]

Needless to say, comments such as these won few friends among ski industry officials, many of whom already felt picked on by the public and disadvantaged by Act 250, which now required ski area expansion plans to go through a review process that had not existed before. They were not villains in this drama, industry leaders argued; they were victims. And if they were victims, then so were the thousands of Vermonters who depended on the ski industry in one way or another for work. By interfering with ski-related development of any kind, such logic suggested, environmental controls interfered with economic development and with the creation of new jobs. Industry observers predicted a sharp downturn in ski development in the wake of Act 250 — a trend that some gloomy analysts argued was already well under way by 1972. Supporters of skiing were too readily labeled as "black hats" in the modern climate of environmental opinion, they feared. Until the ski industry was treated as the responsible proponent of economic growth that it saw itself to be, until self-centered supporters of environmental control stopped denigrating the industry, the people of Vermont would lose out on valuable economic opportunities.[79]

Nearly everyone who had an opinion on the subject of Act 250 could agree on the importance of one word: balance. The goal of environmental legislation — if such legislation *had* to be a part of the state's future — should be the creation of a balanced rural landscape, one that accounted for the needs of visitors and residents, work and leisure, development and restraint. The trouble was, no one could agree on what that balance looked like.

Opponents of the Land Capability and Development Plan and the Land Use Plan saw Act 250 as a threat to any such balance rather than a creator of it, and it was with this attitude in mind that they stepped up their opposition during 1972 and 1973. One journalist described the attitude among members of an opposition group known as the "Green Mountain Boys" as such: "They argue that laws like Act 250 are driving away development and therefore jobs, that 'hyperenvironmentalists' who are rich or retired really want to stop all development rather than merely protect the streams and mountaintops, and that the Vermonter who has to make a living has therefore 'been treated like the weed in the garden.'"[80] Walton Elliot, chairman of the Vermont Ski Areas Association's environmental committee, echoed these sentiments, urging Governor Davis to define a set of "all-important guidelines of balance." Without them, Elliot argued, "Vermont will rapidly become an economic wasteland supported only by a few wealthy preservationists who, unhappy with their previous environment which yielded them their wealth at its own costs, have moved to Vermont and would like all of the state to be controlled as their own backyard."[81]

Environmental groups and defenders of Act 250 tried to drum up support by appealing to the concept of balance as well, arguing that it was balance that guided every decision they made about environmental control in Vermont. As one key architect behind Act 250 had warned in 1970, legislators needed to find a "balanced middle course to well-planned quality economic growth and environmental preservation."[82] Supporters of Act 250 were not placing tourist-inspired images of pastoral stability and quaint rural charm ahead of economic growth, they argued. They were protecting the environmental integrity and scenic beauty of the state's landscape for the benefit of all. Governor Davis himself urged audiences to "make peace with nature" but was careful to add, "By this I do not mean we should not be interested in development in Vermont." Instead, he argued, "The key word should be balance."[83]

Some remained optimistic that this balance was attainable, but the public was profoundly split on the relative importance of job creation versus environmental protection. Survey's revealed that 52 percent of Vermonters thought that jobs were ultimately more important than environmental protections. When asked if they supported industrial growth in the state, 62 percent of Vermonters said yes, even if new manufacturing jobs resulted in more pollution. Perhaps not surprisingly, statistics like these played out along class-based divisions. Among those with an income over $15,000, 58 percent supported state-level environmental laws, while only 43 percent of those with an income under $5,000 were in favor of greater environmental control. And when asked which was more important, job creation or environmental protections, 62 percent of those who earned under $5,000 said job creation, compared to 42 percent of those who earned over $15,000.[84]

Questions about balancing the rural may have been wrapped up in new terminology and a new legal apparatus, but they were actually a continuation of the century-long reworking of the state's rural landscape. Because jobs and economic growth were now so often tied to tourism, and because environmental protection and scenic beauty were so crucial to the state's tourist identity, the search for balance between economic growth and environmental restraint became another iteration of the ongoing negotiation of work-leisure relations in rural Vermont. By the early 1970s the stakes involved in that negotiation felt higher than ever. "Maintaining the economic vitality of the State is as critical as sustaining the quality of the environment," a 1973 report noted. "However, the two objectives must be regarded as complementary, and only development that is consistent with both permitted."[85] But was this even possible? Was legislation the best way to balance the rural landscape?

After months of negotiation, the Environmental Board and Governor Davis approved the Land Capability and Development Plan, handing it off to the legislature and to a new governor, Thomas P. Salmon, in January 1973. In the coming months the plan ran into countless committee roadblocks and underwent repeated revisions, all of which slowly weakened the measure as it worked its way through the legislative process. What remained of the original proposal passed in the spring of 1973, leaving only the statewide Land Use Plan unresolved. Amidst conflicts between the State Environmental Board and the State Planning Office, between developers and environmentalists, between property-rights advocates and zoning enthusiasts, support for the plan weakened as 1973 passed into 1974. By March 1974 there was little hope that the plan would ever make it through the state legislature. The Land Use Plan was unceremoniously labeled "dead" for 1974 and effectively thereafter. After languishing in House and Senate committees for years to come, the mandate to pass the Land Use Plan was stripped from Act 250 in 1984. What many had hoped would give Act 250 its sharpest teeth was no longer an option.[86]

During the 1960s Americans nationwide were confronted with the challenge of managing rural landscapes according to the demands of a growing number and diversity of users. That challenge plagued policy makers in Vermont throughout the decade, and it would continue to do so in the years to come. In a 1967 speech Elbert G. Moulton of the Vermont Development Commission summarized this challenge in the following way:

Gentlemen, I submit that our greatest threat to the future of Vermont does not come from outside, it comes from within . . . it comes from amongst us . . . it lies in the question of whether or not we are willing today to accept the fact that Vermont is going to change and rather than sit idle on our hands and let change occur on a haphazard basis, we use the tools that are available to us and plan for change. Decide the kind of Vermont we want in the year 2000. Set up some guidelines for growth and profit by the mistakes that other states have made. Can we do this? Can we have economic growth, population growth and not ruin the State we love? The answer to that question can be yes if you act now.[87]

Vermonters made many well-intended policy efforts to meet the core challenge to which Moulton pointed. They conducted research, wrote reports, established planning groups, and crafted new land-use legislation, all in the name of gaining and maintaining control over development, of creating a workable balance between development and restraint. What they were after was something akin to a legislated rural middle landscape; that is, a rural middle landscape in the classic, physical sense as a space embodying the best of urban and wild, nature and culture, development and protection. What they created was a rural middle landscape of another sort, where division and discord over formal policy were as much a part of life as any elusive balance.

That was not the outcome that Vermont's policy makers had in mind. They acted in ways that they thought would benefit all parties involved, one way or another. In the process, they declared their optimistic belief that Vermont's rural landscape could someday accommodate the needs of many different groups and many different demands simply because it — like *all* rural landscapes — was malleable enough to be transformed, to be reworked, without losing its larger cultural appeal. That was a powerful belief, for it revealed a simultaneous willingness to merge the traditional and the non-traditional in Vermont into a new, legislated rural landscape. It revealed a willingness, as in so many other cases, to construct a rural landscape that did more than reproduce a traditionalized rural ideal. No one had ever really wanted to leave Vermont as it was, or at least no one had ever acted consistently in ways that suggested they did.

And what of Act 250? The law's permitting process remains in force today as a legacy of Vermont's response to uncontrolled tourist development. And perhaps not surprisingly, it remains an ongoing source of debate. Its most ardent critics continue to complain that it discourages businesses from coming to Vermont, and that it poses obstacles to all avenues of economic growth, including tourism. Supporters have refuted such claims ever since the law went into effect. Within a year of its passage, supportive studies suggested that development projects in Vermont were not being undermined but were simply becoming more complex due to an expanded bureaucratic system. According to the Vermont Environmental Board, only 52 proposals out of 839 submitted under Act 250 by September 1972 had been either denied or withdrawn. The remainder had been approved.[88]

EPILOGUE

The View from Vermont

The mid-1970s was a time of deep uncertainty and concern for those with interests in Vermont tourism. Few who had an opinion about tourism would have failed to recognize its problems by this time, and few would have considered it to be anything other than a mixed blessing. After years of dramatic growth and political debate, the extent of tourism's impacts on landscape, identity, and social relations in rural communities was clearer than it had ever been before. What was less clear was tourism's future economic status. The national oil crisis was cutting into mid-1970s travel figures throughout New England, and regardless of what economists and policymakers said, many continued to believe that Act 250 would undermine development of all kinds, tourist and otherwise. Never had so many doubted the wisdom and future of tourism in Vermont.

Within a decade, however, record numbers of visitors were again coming to Vermont. Developers constructed massive new condominium complexes, golf and ski areas expanded their operations, and advertisers continued to promote Vermont as a four-season destination. Older activities were mixed with newer ones to create new kinds of vacation opportunities and new ways to interact with rural Vermont: ski areas capitalized on the emerging popularity of mountain biking to increase summer crowds; communities developed new trail systems to capture a share of New England's growing numbers of snowmobilers; and art fairs, music festivals, and flea markets gave vacationers new reasons to stop, look around, and spend a little money. As has always been the case, tourism in Vermont has continued to evolve, and its ability to do so has maintained its position as a leading sector in the state's economy.

Tourism's continued success in Vermont reminds us of the enduring appeal of rural land and life in American culture — an appeal that has been central to the stories in this book. Whether Americans make their homes in cities, suburbs, or small towns, they often envision the countryside as a scenic repository for core national values such as community, hard work, tradition, patriotism, and a connection to the land. Not all Americans feel this way, of course, just as not all Americans define concepts like community or patriotism in quite the same ways. But for millions, rural landscapes continue to embody some of their nation's best qualities, and for many of these people, Vermont ranks at the top of their list as an ideal rural landscape.

Tourism has been a beneficiary of this appeal as well as a force for perpetuating it. Since the nineteenth century, tourist travel and tourist-related literature have remained core mediating contexts through which Americans have defined, encountered, and shaped rural landscapes. This has been true for those who visit rural areas, as well as for those who live and work in them. When Vermont's turn-of-the-century host families redesigned their homes and farmyards, for instance, they did so to make them more attractive to visitors. When promoters advertised summer homes to visitors, they did so in ways that recast farm property according to their value on the vacation market. When summer home owners spoke out in opposition to billboards, they did so to protect the kinds of scenery that had brought them to Vermont in the first place. Or, when policymakers acted on behalf of scenery preservation, they did so because they understood the power that such scenery had to attract more visitors to the state. In these and many other cases, tourist travel and tourist-directed imagery have consistently informed and reinforced Vermont's rural reputation.

The contours of that reputation, however, have never been entirely fixed or agreed upon, and no universally accepted notion of a "properly" ordered rural landscape has ever emerged from the tourist experience. Rather, visitors and residents such as those we encountered in this book have always brought diverse and at times divergent perspectives about travel, nature, land use, outdoor recreation, scenery, and rural culture to bear on their encounters with rural people and place. Certain icons in Vermont — church steeples, town commons, and farms, for instance — have long been hallmarks of the state's identity, and many of the actors in this story have relied upon icons such as these to reproduce traditionalized notions about rural timelessness, organicism, and harmony. Vermont's tourist landscapes have always naturalized romantic, anti-modern, anti-urban discourses, and the state's tourist economy has always benefited from the enduring popularity of such discourses among some segments of American culture. The same can be said of other rural tourist communities nationwide; no history of rural tourism in the United States would be complete without recourse to the traditionalized rural perspectives that have done so much to motivate tourists and to shape the form and meaning of the places they see.

But no history of rural tourism would be complete without looking beyond these perspectives as well. As we have seen in the case of Vermont, social groups manipulated their surroundings according to a wide range of alternate discourses about tourism and rural life. In the process, they reproduced the validity and power of these discourses through the material landscape itself, both informing and challenging many of the traditionalized conceptions otherwise ascribed to the state. Sometimes that meant crafting tourist spaces with progressive, modern, or reformist goals in mind. Sometimes it meant transforming the rural according to the dictates of land-use planning or environmentalism. And sometimes it meant creating exclusionary spaces where broader social tensions moved to the fore.

I have tied this diversity together in this book by returning in each chapter to an examination of work-leisure relations. For over a century, I have argued, the construction of landscape and identity in Vermont has been guided consistently by the changing work-leisure relations that have emerged within any given tourist context. This "reworking of the rural landscape" has for decades yielded a consistent result: the spaces, meanings, and practices associated with rural work and rural leisure have continually overlapped with and informed one another, making it difficult to talk about one category independent of the other. Tourism obviously shapes the nature and meaning of work and its relationship to leisure in all manner of destinations, rural or otherwise. But that fact is particularly salient in a rural context, where, as we have seen here, working landscapes are so central to tourism's success, and where leisure-based land uses are so central to the ways in which those working landscapes are transformed and defined.

This overlap continues today in both subtle and overt ways. The popularity of "agritourism" in Vermont and elsewhere offers a good example. Agritourism — a contemporary term for a concept whose history in Vermont goes back more than a century — refers to the activities of the tens of thousands of travelers annually who visit farms and ranches nationwide. They come as others have before them, hoping to forge meaningful connections to rural life through an encounter with rural work.[1] Business organizations like the "Vermont Farms! Association" deploy a familiar logic to spread the message of agritourism: not only does agritourism promise to benefit visitors by putting them in contact with the restorative rhythms of rural life, it promises to benefit working farmers by supplementing and protecting their farm's agricultural heritage. "One of the major goals of the association," Vermont Farms! has noted in a recent profile, "is to sustain and further develop the working landscape that characterizes Vermont. Our farm visitors play a critical role in developing new direct marketing opportunities for Vermont farmers."[2]

Language like this reproduces the blurred boundaries between work and leisure that for decades have characterized Vermont's tourist experience. This mutual constitution of work and leisure is rural tourism's most common and enduring outcome, and it is one that gives voice to the shared experiences of rural tourist communities throughout the United States, both in the past and in the present. Growing numbers of Americans are now claiming a new stake in rural amenities at the start of the twenty-first century, whether as casual visitors, second home owners, or new residents who telecommute or who literally commute, often over great distances. These new interest groups typically bring with them a consumptive, leisure-based view of rural landscapes — one that is increasingly coming into conflict with preexisting working land uses geared toward the production of minerals, timber products, or even "offensive" agricultural practices such as hog farming or pesticide spraying.[3] But while that pattern is intensifying in rural areas from northern New England to the Rocky Mountains, it really is nothing new. Over the past century, Americans have learned through the tourist experience to see rural landscapes as sites for leisure and recreation while still wanting (and needing) them to be places of work. We expect rural landscapes to give us food and resources as well as scenery and fun — providing the former do not interfere with the latter. Consequently, many rural communities nationwide remain locked in a complicated and contentious intermediary position relative to the demands of insiders and outsiders, newcomers and old-timers, workers and visitors.

Each of the chapters in this book has explored this positioning through a combined topical and chronological approach that moved from one tourist activity to another across the decades. That sequencing, however, does not mean one particular set of work-leisure relations and the material and symbolic consequences with which they were associated necessarily supplanted the next. Nor does it mean that these sets apply only to the past and not to the present. Core questions about the nature of work and leisure and about the relative power of each to determine the future of rural places weave throughout the stories and decades explored here: To what degree should communities caught in changing economic circumstances embrace a tourist economy? Who gets to define the value of different types of work, different types of leisure? How should rural communities go about creating a fair, just, and respectful balance between the demands of each? Questions like these may be framed in different ways by different interest groups operating in different economic, social, and geographic contexts, but they persist nonetheless. They do so because tourism continues to make it difficult to separate rural work from the leisure-based contexts in which it is placed. Making sense of work, leisure, and the points of intersection between them is thus a challenge for historians and geographers of rural tourism as well as for those who make tourist policy in rural communities today.

Similarly, although our story ends with the immediate aftermath of Act 250, controversies over land use and tourism were by no means put to rest during the 1970s. Many of the same kinds of concerns and debates about tourism persist in Vermont, as does the tendency for groups to address those concerns and debates through formal policy — a legacy of legislative efforts to balance the rural during the 1960s and 1970s. Ongoing debates about snowmaking, for instance, attest to the enduring nature of tourist controversy. Since the early 1980s some of Vermont's larger ski areas have tried to tap into new water sources to boost their snowmaking capacities, arguing that their economic viability depends on their ability to expand, and that their ability to expand depends on snowmaking. Their efforts have continually been opposed by environmental groups and community members, however, many of which rely on environmental policy to protect critical animal habitats or to protect streams from diversion or draining. Killington's management was faced with such opposition in the mid-1980s, for instance, when they proposed spraying treated water from a nearby sewage facility through the resort's snowmaking system, thereby increasing their water supply while simultaneously completing the final step in the facility's treatment process — spray irrigation. Many Vermonters were outraged by the idea, and they leveled charges of arrogant elitism and environmental indifference at the ski industry — a message captured neatly by a popular bumper sticker that read "Killington: Where the Affluent Meet the Effluent." Killington's President and CEO, Preston Smith, responded to critics and appealed for public support in a 1986 magazine article entitled "The Mountains Are for Everyone." Smith argued that skiing was a critical part of the state's working economy, that the ski industry was a proven leader in environmental stewardship, and perhaps most importantly that his opponents had lost sight of a very basic point: all interest groups in Vermont, Smith argued, needed to figure out how to share the state's landscape with one another.[4]

Whether one approved of Killington's snowmaking proposal or not (and the ski area did eventually win the necessary permits, although it never went through with its plans), Smith's admonishments about sharing and cooperation captured the essence of rural tourism's core challenge. The title of his article, "The Mountains Are for Everyone," may have sounded wise and indisputable, but it had actually never been true: whether in the mountains or in the valleys, tourism has always turned landscape and identity into battlegrounds where the ideas and actions of some clashed with those of others. Rural tourist landscapes are political spaces born of the often competing demands of different social groups, each of which brings specific conceptions of rural land and life to bear on their experiences. Therefore, the reworking of rural Vermont has always been a political drama, where any claim to victory is measured by one's sense of control over the use and definition of the rural landscape itself. In this regard, the social and cultural politics of rural tourism transcend the context of tourism alone, spreading across the entire rural stage into the realm of what geographer Michael Woods has described as a new "politics of the rural." For Woods, decades of mounting pressures on rural resources have generated tensions across Western societies as traditional rural ways of life come under increasing pressure from new kinds of people and new kinds of land uses. The protests and debates that have emerged from this tend to focus not just on questions of formal, political governance but on the ability of one group or another to define the meaning of the category "rural." Control over the meaning and use of that category, Woods's thoughts suggest, is tantamount to control over the future of rural space itself.[5]

When we think about the history of rural tourism and the controversies it has engendered, we need to think about their connections to the kinds of broader debates about the future of rural America to which this "politics of the rural" point. Places imbued with traditionalized popular sentiment, such as Vermont, are on the front line of such debates. In Vermont, for instance, a number of recent debates over issues relating to taxes, education, environmental management, and even lifestyle reveal the extent to which the state's rural landscape and rural identity have become politicized, both in a general sense of the word and in more specific terms of governance. Two high-profile examples demonstrate this.

The first of these focuses on concerns about sprawl. The nonprofit National Trust for Historic Preservation has placed the entire state of Vermont on its annual list of "America's 11 Most Endangered Historic Places" twice in a dozen years, heightening concerns not only about rural Vermont but about rural landscapes nationwide. Vermont first made the group's list in

1993, when urban sprawl and "megamall developers" warned of an end to Vermont's "strong sense of community" and its "distinctive rural towns and countryside."[6] At that time, the trust and others were particularly concerned about attempts by Wal-Mart to open its first store in Vermont. Many from Vermont and elsewhere feared that Wal-Mart, among other national chains and "megamall developers," would destroy Vermont's idealized small-town, general-store way of life. But many Vermonters also saw such stores as welcome additions to their communities, and they saw efforts to block their construction as meddling by sentimental outsiders operating with an unrealistic image of the state in mind. The trust added Vermont to its list a second time in 2004, again citing threats posed by sprawl and "big-box stores" (to use a more contemporary phrase). Circumstances had changed for the worse in recent years, the trust noted, forcing them to list Vermont again in the hopes of calling attention to its urgent plight. In 1993 Vermont was the only state in the nation without a Wal-Mart. By 2004 the state had four and was facing an influx of still more "behemoth stores," all of which, one trust official warned, would "drastically alter the character of Vermont."[7]

A second example focuses on same-sex civil unions. In 2000 the Vermont legislature passed the state's controversial civil union law, legalizing same-sex unions and extending legal rights to same-sex partners. A vocal contingent in Vermont responded quickly, posting signs reading "Take Back Vermont" in their yards, on their cars, and on the sides of their houses and barns. For many in the opposition, the civil union law was merely the tip of an immense iceberg of discontent over rising taxes, strict land-use permits, and a liberalized educational system. Vermont's governor at the time, Howard Dean, and the state's Democratic-led legislature were turning their backs on "true" Vermonters, many argued, catering instead to a clientele of newcomers and potential newcomers, a growing percentage of which, some worried, would now be homosexual.[8] It was time, they felt, to take Vermont back in both a conceptual and a very real, physical sense. It was time for "true" Vermonters to regain control over the meaning and management of the state's rural landscape. Their opponents fought back with signs reading "Take Vermont Forward"—a progressive message they hoped would counteract what they saw as an inherently reactionary and hateful message of intolerance. Their image of Vermont was of a state that tolerated personal freedoms and change. Their Vermont was a state that made more room for the perspectives of non-natives of all kinds. Their Vermont was a truly special and unique place.

Issues like sprawl and civil unions may seem to have little to do with tourism. But in fact they share important connections to the kinds of stories about tourism and rural landscapes that we have explored in this book. In fact, they owe at least part of their meaning and resonance to tourism, for tourism throughout the twentieth century has exercised tremendous power over what people think of when they think of a properly ordered rural place. When opponents of sprawl or supporters of civil unions defined Vermont as a special or unique place, they echoed tourist promoters that have been espousing the same message for generations. And when supporters of Wal-Mart or opponents of civil unions rallied against outside influence in their state, they echoed longstanding arguments about the undue influence of tourists and about tourism's place in the state's economy and culture. Perhaps most importantly, though, concerns about sprawl, civil unions, and tourism all raise fundamental questions about the right to control the future of places that millions of Americans cherish. They all share a focus, that is, on the future of rural landscapes, both in Vermont and beyond. For this reason, they have all become part of the larger picture we see when we view the historical geography of rural tourism from the context of Vermont.

The view from Vermont is one that takes our gaze from a traditionally defined "middle landscape" to one of a different, more general, sort. It is a view that reminds us that the stories in this book, like those we might tell about rural tourist communities elsewhere, are about more than rural landscapes alone. The stories in this book are stories about landscapes more generally — stories about what landscapes are, about what they mean to us, and about what they do. No matter how Vermont's rural landscapes have been used, represented, or defined over the decades, they have always been true "landscapes" in a cultural- and historical-geographic sense of the term. Like all landscapes, they have been implicated in the production of dominant and competing discourses about society, culture, and place. Like all landscapes, they have reflected and reproduced such discourses, both in material and symbolic ways. And like all landscapes, they have been a constitutive part of the social relations and power dynamics that define American culture. We should think of rural tourist landscapes as quintessential American cultural landscapes not so much because of what they look like, then, but because of what they do. They are not merely quaint places to visit or esoteric topics to study. Rather, they play a key role in the processes by which American culture evolves. That role is what makes them landscapes. That role is why they matter.

It is also why it is useful to think of rural landscapes as middle landscapes in something other than a traditional sense of that phrase. Since the late nineteenth century, tourism has forced Vermont's visitors and residents to confront and to negotiate their relationships not only to the landscape but *to one another as well*. It continues to do so today: when we rework landscape and identity through the context of rural tourism — whether we do so as travelers or as rural residents — we rework our relationships with other people as well. The rural landscape is the ground on which and through which we do that; it is the ground on which and through which we make sense of ourselves, our relationships to our surroundings, and our relationships to others. The rural landscape is thus a middle landscape not strictly because of its positioning between nature and culture, or between urban and wild, but because of its positioning between people. Vermont's historical actors have always struggled to advance their own opinions about rural America by controlling the use, management, and representation of the state's working and leisure spaces — that is, by controlling the reworking of rural Vermont. In order to do so, they have struggled to gain control over more than physical space; they have struggled to gain control over one another as well.

That struggle — with all its points of contention and cooperation, with all its successes and shortcomings — is what we see unfolding everywhere before us when we contemplate the view from Vermont. It is what gives meaning and resonance to this story and this place, just as it does to the countless other stories and other places implicated in the ongoing development of American culture. Even when we strip aside the tourist rhetoric and the shades of sentimentality so commonly applied to Vermont, the state remains a truly special place to many people and for many reasons. But what makes its tourist history so special, at least from a geographic perspective, is what that history reminds us about ourselves and the worlds we create. If we allow our eyes to wander from Vermont's classically defined middle landscape of farms, forests, and villages toward a middle landscape of a different, more critical sort we find ourselves contemplating a view inscribed by a fundamental and inescapable message: No matter what kinds of landscapes we live in, visit, or study, we all take part in the construction of middle landscapes on which and through which our lives intersect with the lives of others. The story of Vermont should therefore remind us that we are all historical-geographic actors whose actions always have consequences, not only for the integrity of the physical world around us, but for the integrity of our relationships with one another. That is the larger message that Vermonters and their visitors have been recounting by their actions for over a century. That is the message we see in the view from Vermont.

Notes

Introduction: Tourism and the Reworking of Rural Vermont (pp. 1–16)

1. As John Fraser Hart points out, the United States Census Bureau's category "rural" is ill-defined and potentially misleading. On the pitfalls of rural census data, see "'Rural' and 'Farm' No Longer Mean the Same," in *The Changing American Countryside: Rural People and Places*, ed. Emery N. Castle (Lawrence: University Press of Kansas, 1995), 63–76.

2. Hal S. Barron, *Mixed Harvest: The Second Great Transformation in the Rural North, 1870–1930* (Chapel Hill: University of North Carolina Press, 1997).

3. Recent works on tourism and the production of group- and place-based identities include Jeffrey Sasha Davis, "Representing Place: 'Deserted Isles' and the Reproduction of Bikini Atoll," *Annals of the Association of American Geographers* 95 (September 2005): 607–25; Dydia DeLyser, *Ramona Memories: Tourism and the Shaping of Southern California* (Minneapolis: University of Minnesota Press, 2005); Thomas S. Bremmer, *Blessed with Tourists: The Borderlands of Religion and Tourism in San Antonio* (Chapel Hill: University of North Carolina Press, 2004), esp. 35–94; Stephen P. Hanna and Vincent J. Del Casino Jr., eds., *Mapping Tourism* (Minneapolis: University of Minnesota Press, 2003); Steven M. Schnell, "Creating Narratives of Place and Identity in 'Little Sweden, U.S.A.,'" *Geographical Review* 93 (January 2003): 1–29; Marguerite S. Shaffer, *See America First: Tourism and National Identity, 1880–1940* (Washington, D.C.: Smithsonian Institution Press, 2001); Bruce D'Arcus, "The 'Eager Gaze of the Tourist' Meets 'Our Grandfathers' Guns': Producing and Contesting the Land of Enchantment in Gallup, New Mexico," *Environment and Planning D: Society and Space* 18 (December 2000): 693–714; Steven Hoelscher, "The Photographic Construction of Tourist Space in Victorian America," *Geographical Review* 88 (October 1998): 548–70; Nuala C. Johnson, "Where Geography and History Meet: Heritage Tourism and the Big House in Ireland," *Annals of the Association of American Geographers* 86 (September 1996): 551–66.

4. "Rural tourism" is a broad term used here to refer to tourist travel and tourist activities that put visitors to Vermont in contact with the state's rural landscape, whether through outdoor recreation, visits to farms or summer homes, or simple drives in the countryside. For extended discussions about the scope and scale of rural tourism as they apply to Europe and North America, see Derek Hall, Lesley Roberts, and Morag Mitchell, eds., *New Directions in Rural Tourism* (Burlington, Vt.: Ashgate, 2003), and Lesley Roberts and Derek Hall, eds., *Rural Tourism and Recreation: Principles to Practice* (New York: CABI Publishing, 2001).

5. Michael Bunce, *The Countryside Ideal: Anglo-American Images of Landscape* (New York: Routledge, 1994), 2.

6. See Stephen Daniels, *Fields of Vision: Landscape Imagery and National Identity in England and the United States* (Princeton, N.J.: Princeton University Press, 1993); Peter J. Schmitt, *Back to Nature: The Arcadian Myth in Urban America* (1969; reprint, with a foreword by John Stilgoe, Baltimore: Johns Hopkins Uni-

versity Press, 1990); Leo Marx, *The Machine in the Garden: Technology and the Pastoral Ideal in America* (New York: Oxford University Press, 1964).

7. On the distrust of the city in American culture, see Morton White and Lucia White, *The Intellectual Versus the City: From Thomas Jefferson to Frank Lloyd Wright* (Cambridge, Mass.: Harvard University Press, 1962); Tom Slater, "Fear of the City, 1882–1967: Edward Hopper and the Discourse of Anti-Urbanism," *Social and Cultural Geography* 3 (June 2002): 135–54.

8. John Urry, *The Tourist Gaze: Leisure and Travel in Contemporary Societies* (London: Sage, 1990).

9. For example, see D. W. Meinig, "Symbolic Landscapes: Some Idealizations of American Communities," in *The Interpretation of Ordinary Landscapes: Geographical Essays*, ed. D. W. Meinig (New York: Oxford University Press, 1979), 164–92. On regional identity in Vermont and New England, including its links to landscape, see Joseph A. Conforti, *Imagining New England: Explorations of Regional Identity from the Pilgrims to the Mid-Twentieth Century* (Chapel Hill: University of North Carolina Press, 2001); Kent C. Ryden, *Landscape with Figures: Nature and Culture in New England* (Iowa City: University of Iowa Press, 2001); Jan Albers, *Hands on the Land: A History of the Vermont Landscape* (Cambridge, Mass.: MIT Press, 1999); William H. Truettner and Roger B. Stein, eds., *Picturing Old New England: Image and Memory* (Washington, D.C.: National Museum of American Art, Smithsonian Institution; New Haven, Conn.: Yale University Press, 1999); Joseph S. Wood, *The New England Village* (Baltimore: Johns Hopkins University Press, 1997); Stephen Nissenbaum, "New England as Region and Nation," in *All Over the Map: Rethinking American Regions*, ed. Edward L. Ayers, Patricia Nelson Limerick, Stephen Nissenbaum, and Peter S. Onuf (Baltimore: The Johns Hopkins University Press, 1996), 38–61; Dona Brown, *Inventing New England: Regional Tourism in the Nineteenth Century* (Washington, D.C.: Smithsonian Institution Press, 1995); Nancy Price Graff, ed., *Celebrating Vermont: Myths and Realities* (Middlebury, Vt.: The Christian A. Johnson Memorial Gallery, Middlebury College; Hanover, N.H.: University Press of New England, 1991). Also see Stephanie Foote, *Regional Fictions: Culture and Identity in Nineteenth-Century American Literature* (Madison: University of Wisconsin Press, 2001).

10. Brown, *Inventing New England*; Andrea Rebek, "The Selling of Vermont: From Agriculture to Tourism, 1860–1910," *Vermont History* 44 (Winter 1976): 14–27. Also see Conforti, *Imagining New England*, especially pages 203–309.

11. Roaldus Richmond, "The Green Mountain State," in *Vermont: A Profile of the Green Mountain State*, Vermont Writers' Project (New York: Fleming Publishing Company, 1941), no page number.

12. Vermont historians have struggled with this issue of Vermont's relevance for historical patterns that transcend the state's borders. For an engaging discussion, see Paul Searls, "America and the State that 'Stayed Behind': An Argument for the National Relevance of Vermont History," *Vermont History* 71 (Winter/Spring 2003): 75–87. Also see Michael Sherman, Gene Sessions, and P. Jeffrey Potash, *Freedom*

and Unity: A History of Vermont (Barre, Vt.: Vermont Historical Society, 2004), xiii–xix.

13. The classic statement on the reflective qualities of landscape remains Peirce F. Lewis's essay "Axioms for Reading the Landscape" in *The Interpretation of Ordinary Landscapes*, 11–32.

14. Richard H. Schein, "The Place of Landscape: A Conceptual Framework for Interpreting an American Scene," *Annals of the Association of American Geographers* 87 (December 1997): 663.

15. A rich vein of work in landscape studies within geography and other disciplines explores this relationship between landscape and social reproduction, between landscape and power. My thinking here has been influenced by key texts such as Denis E. Cosgrove, *Social Formation and Symbolic Landscape* (London: Croom Helm, 1984); James S. Duncan, *The City as Text: The Politics of Landscape Interpretation in the Kandyan Kingdom* (New York: Cambridge University Press, 1990); Barbara Bender, ed., *Landscape: Politics and Perspectives* (Oxford: Berg, 1993); W. J. T. Mitchell, ed., *Landscape and Power* (Chicago: University of Chicago Press, 1994); Don Mitchell, *Cultural Geography: A Critical Introduction* (Malden, Mass.: Blackwell, 2000). Also see James S. Duncan and Nancy G. Duncan, *Landscapes of Privilege: The Politics of the Aesthetic in an American Suburb* (New York: Routledge, 2004); Richard H. Schein, "Normative Dimensions of Landscape," in *Everyday America: Cultural Landscape Studies after J. B. Jackson*, ed. Chris Wilson and Paul Groth (Berkeley: University of California Press, 2003), 199–218; Nuala C. Johnson, *Ireland, the Great War, and the Geography of Remembrance* (New York: Cambridge University Press, 2003).

16. Dydia DeLyser, "Authenticity on the Ground: Engaging the Past in a California Ghost Town," *Annals of the Association of American Geographers* 89 (December 1999): 602–32. Also see Edward M. Bruner, *Culture on Tour: Ethnographies of Travel* (Chicago: University of Chicago Press, 2005); Steven D. Hoelscher, *Heritage on Stage: The Invention of Ethnic Place in America's Little Switzerland* (Madison: University of Wisconsin Press, 1998), 181–220; and Dean MacCannell, *The Tourist: A New Theory of the Leisure Class* (New York: Schocken Books, 1989, 1976).

17. Mitchell, *Cultural Geography*, 91–119; Don Mitchell, *The Lie of the Land: Migrant Workers and the California Landscape* (Minneapolis: University of Minnesota Press, 1996).

18. As a testimonial backlash to this fact, self-styled "savvy" tourists often seek out what Dean MacCannell has called "back regions" — everyday, lived-in places that, whether accurately or not, seem to provide more direct access to "authenticity" than the tourist-directed front experienced by most travelers. See MacCannell, *The Tourist*. Also see Karen E. Till, "Construction Sites and Showcases: Mapping 'The New Berlin' through Tourism Practices," in *Mapping Tourism*, 51–78. Also see DeLyser, *Ramona Memories*.

19. Raymond Williams, *The Country and the City* (New York: Oxford University Press, 1973), 120–26.

20. For similar arguments about the need for British rural geographers to look beyond the traditional "rural idyll" in their research agendas, see David Matless, *Landscape and Englishness* (London: Reaktion Books, 1998); Jo Little, "Otherness, Representation, and the Cultural Construction of Rurality," *Progress in Human Geography* 23 (September 1999): 437–42.

21. Hal K. Rothman, *Devil's Bargains: Tourism in the Twentieth-Century American West* (Lawrence: University Press of Kansas, 1998), 27.

22. Bruner, *Culture on Tour*, especially pp. 145–68. Also see David M. Wrobel, "Introduction: Tourists, Tourism, and the Toured Upon," in *Seeing and Being Seen: Tourism in the American West*, ed. David M. Wrobel and Patrick T. Long (Lawrence: University Press of Kansas, 2001), 1–34.

23. Here I follow works such as Paul Cloke, Marcus Doel, David Matless, Martin Phillips, Nigel Thrift, *Writing the Rural: Five Cultural Geographies* (London: Paul Chapman Publishing Ltd., 1994); Paul Cloke and Jo Little, eds., *Contested Countryside Cultures: Otherness, Marginalisation and Rurality* (New York: Routledge, 1997); Paul Milbourne, ed., *Revealing Rural "Others": Representation, Power, and Identity in the British Countryside* (London: Pinter, 1997); E. Melanie DuPuis and Peter Vandergeest, eds., *Creating the Countryside: The Politics of Rural and Environmental Discourse* (Philadelphia: Temple University Press, 1996); and Sarah Whatmore, Terry Marsden, and Philip Lowe, eds., *Gender and Rurality* (London: David Fulton Publishers, 1994).

24. Rothman, *Devil's Bargains*.

25. Cindy S. Aron, *Working at Play: A History of Vacations in the United States* (New York: Oxford University Press, 1999).

26. Urry, *The Tourist Gaze*, 2.

27. Paul S. Sutter, *Driven Wild: How the Fight Against Automobiles Launched the Modern Wilderness Movement* (Seattle: University of Washington Press, 2002), 52.

28. For related arguments on understanding nature through work, see Richard White, *The Organic Machine: The Remaking of the Columbia River* (New York: Hill and Wang, 1995). Also see Richard White, "'Are You an Environmentalist or Do You Work for a Living?': Work and Nature," in *Uncommon Ground: Rethinking the Human Place in Nature*, ed. William Cronon (New York: W. W. Norton and Company, 1996), 171–85; Ryden, *Landscape with Figures*, 67–95.

29. For more detailed discussions on landscapes of consumption and production, see Peter Walker and Louise Fortmann, "Whose Landscape?: A Political Ecology of the 'Exurban' Sierra," *Cultural Geographies* 10 (October 2003): 469–91, and Blake Harrison, "Shopping to Save: Green Consumerism and the Struggle for Northern Maine," *Cultural Geographies*, forthcoming.

30. On the need in geography to treat landscapes not merely as symbolic spaces but as socially mediated material spaces, and on the consequences for social justice offered by this perspective, see Tim Cresswell, "Landscape and the Obliteration of Practice," in *Handbook of Cultural Geography*, ed. Kay Anderson, Mona Domosh, Steve Pile, and Nigel Thrift (London: Sage Publications, 2003),

269–81, and Don Mitchell, "Cultural Landscapes: Just Landscapes or Landscapes of Justice?" *Progress in Human Geography* 27 (December 2003): 787–96.

Chapter 1. Two Worlds of Work: Nostalgia and Progress in
Turn-of-the-Century Vermont (pp. 17–49)

1. Vermont State Board of Agriculture (hereafter VSBA), *Resources and Attractions of Vermont, With a List of Desirable Homes for Sale* (Montpelier, Vt.: Press of the Watchman Publishing Co., 1892), 28.

2. *Sights in Barre* (Glens Falls, N.Y.: C. H. Possons, Publisher, 1887), 5.

3. I use the words "progress" and "progressive" in a general, conceptual sense, rather than in reference to the political, social, and economic reforms associated with America's turn-of-the-century Progressive Movement. Although aspects of Progressivism do surface in each of my first three chapters, I rely on the term "progressive" in this chapter to provide a contrast to the ideology of nostalgia so often associated with rural tourist landscapes.

4. I have argued this point earlier in Blake Harrison, "Rethinking the Rural: Nostalgia and Progress in Vermont's Tourist Industry," *Proceedings of the New England St. Lawrence Valley Geographical Society* 32 (2002): 31–43.

5. David L. Richards, *Poland Spring: A Tale of the Gilded Age, 1860–1900* (Hanover, N.H.: University Press of New England, 2005), 3–4.

6. Harold A. Meeks, "Stagnant, Smelly, and Successful: Vermont's Mineral Springs," *Vermont History* 47 (Winter 1979): 5–20.

7. J. Kevin Graffagnino, *Vermont in the Victorian Age: Continuity and Change in the Green Mountain State* (Bennington, Vt.: Vermont Heritage Press; Shelburne, Vt.: Shelburne Museum, 1985), 4.

8. "Spring Hotel, Newbury, Vt.," May 1, 1873, Broadside Collection, Vermont Historical Society, Barre, Vermont (hereafter VHS).

9. Eric Purchase, *Out of Nowhere: Disaster and Tourism in the White Mountains* (Baltimore: The Johns Hopkins University Press, 1999); Dona Brown, *Inventing New England: Regional Tourism in the Nineteenth Century* (Washington, D.C.: Smithsonian Institution Press, 1995), 41–74.

10. Marjorie Hope Nicolson, *Mountain Gloom and Mountain Glory: The Development of the Aesthetics of the Infinite* (1959; reprint, with a foreword by William Cronon, Seattle: University of Washington Press, 1997), esp. 271–323; Barbara Novak, *Nature and Culture: American Landscape Painting, 1825–1875* (London: Thames and Hudson, 1980).

11. Henry M. Burt, *Burt's Illustrated Guide of the Connecticut Valley* (Northampton, Mass.: New England Publishing Company, 1867), 245, 248.

12. "Two Trips to Lake Willoughby," *The Knickerbocker* 44 (September 1854), 263–70; Burt, *Burt's Illustrated Guide of the Connecticut Valley*, 144–48, 210–16;

William Cullen Bryant, ed., *Picturesque America; or, The Land We Live In*, vol. 2 (New York: D. Appleton and Company, 1874), 276–87.

13. Louise B. Roomet, "Vermont as a Resort Area in the Nineteenth Century," *Vermont History* 44 (Winter 1976): 1–13.

14. Christopher McGrory Klyza and Stephen C. Trombulak, *The Story of Vermont: A Natural and Cultural History* (Hanover, N.H.: University Press of New England, 1999), 63–74, 89.

15. Burt, *Burt's Illustrated Guide of the Connecticut Valley*, 154–55.

16. Novak, *Nature and Culture*, 157–200. Also see John Conron, *American Picturesque* (University Park: Pennsylvania State University Press, 2000), 17–32.

17. Leo Marx, *The Machine in the Garden: Technology and the Pastoral Ideal in America* (New York: Oxford University Press, 1964), 141.

18. For extended discussions, see T. J. Jackson Lears, *No Place of Grace: Antimodernism and the Transformation of Material Culture, 1880–1920* (Chicago: University of Chicago Press, 1981), and David Shi, *The Simple Life: Plain Thinking and High Living in American Culture* (New York: Oxford University Press, 1985).

19. Brown, *Inventing New England*. Also see Joseph A. Conforti, *Imagining New England: Explorations of Regional Identity from the Pilgrims to the Mid-Twentieth Century* (Chapel Hill: University of North Carolina Press, 2001), 203–62.

20. Dona Brown, ed., *A Tourist's New England: Travel Fiction, 1820–1920* (Hanover, N.H.: University Press of New England, 1999), 12.

21. "Mt. Mansfield Hotel, Stowe, Vermont," (no publisher, [1888]), 11, pamphlet collection, Wisconsin State Historical Society, Madison.

22. Brown, *Inventing New England*, 143.

23. VSBA, *Vermont, Its Opportunities for Investment in Agriculture, Manufacture, Minerals, Its Attractions for Summer Homes* (East Hardwick, Vt.: The Board, [1903]); Central Vermont Railroad, *Summer Homes Among the Green Hills of Vermont and Along the Shores of Lake Champlain* (St. Albans, Vt.: St. Albans Messenger Co., 1893); West River Valley Association, *The Call of the Country* (Brattleboro, Vt.: The Association, 1912). In addition to farm boarding enterprises, promotions like these also provided information on hotels and resorts.

24. Quoted in John K. Wright, "Stowe in Early Spring, 1919," in *Outsiders Inside Vermont: Travelers' Tales of 358 Years*, ed. T. D. Seymour Bassett (Brattleboro, Vt.: Stephen Green Press, 1967), 108.

25. Frank L. Greene, comp., *Vermont, the Green Mountain State: Past, Present, Prospective* (n.p.: Vermont Commission to the Jamestown Tercentennial Exposition, 1907), 78.

26. Central Vermont Railroad, *Summer Homes*, 3.

27. George Perry, "A Convenient and Profitable Home Market," in *Fifth Annual Report of the Vermont State Horticultural Society* (Bellows Falls, Vt.: P. H. Gobie Press, 1907), 84–90. Also see Brown, *Inventing New England*, 160.

28. Brown, *Inventing New England*, 157–67.

29. [Victor I. Spear], "Brattleboro Institute," in *Thirteenth Vermont Agricultural Report by the State Board of Agriculture, for the Year 1893* (Burlington, Vt.: Free Press Association, Printers and Binders, 1893), 56.

30. "Testimony of Mr. Milton Whitney," *Report of the Industrial Commission on Agriculture and Agricultural Labor* (Washington, D.C.: Government Printing Office, 1901), 870. Also see Harold Fisher Wilson, *The Hill Country of Northern New England: Its Social and Economic History, 1790–1930* (New York: Columbia University Press, 1936), 293–94.

31. Brown, *Inventing New England*, 144.

32. Lyle W. Dorsett, "Town Promotion in Nineteenth-Century Vermont," *New England Quarterly* 40 (June 1967): 275–79; Graffagnino, *Vermont in the Victorian Age*, ix; Michael Sherman, Gene Sessions, and P. Jeffrey Potash, *Freedom and Unity: A History of Vermont* (Barre, Vt.: Vermont Historical Society, 2004), 290–309.

33. Hamilton Child, comp., *Gazetteer and Business Directory of Windham County, Vermont, 1724–1884* (Syracuse, N.Y.: Published by the Compiler, 1884), 304.15–304.16; *Wood and Water: Mills in Searsburg, Vermont* ([Burlington, Vt.:] Consulting Archaeology Program, University of Vermont, 1980).

34. F.C. [*sic*], "Readsboro," *Deerfield Valley Times* [Wilmington, Vt.] (hereafter *DVT*), November 15, 1889, p. 1; Hamilton Child, comp., *Gazetteer and Business Directory of Bennington County, 1880–1881* (Syracuse, N.Y.: By the Compiler, 1880), 164–66; Bernard R. Carman, *Hoot Toot and Whistle: The Story of the Hoosac Tunnel and Wilmington Railroad* (Brattleboro, Vt.: Stephen Greene Press, 1963), 3–6.

35. George F. Wells, *The Status of Rural Vermont* (St Albans, Vt.: Cummings Printing Company for the Vermont State Agricultural Commission, 1903), 16.

36. *DVT*, November 22, 1889, p. 4.

37. *DVT*, July 19, 1888, p. 2.

38. "A Board of Trade Organized," *DVT*, November 8, 1889, p. 1; O. H. Jones, *Attractions of Wilmington and Vicinity* (Jacksonville, Vt.: F. L. Stetson and Printer, 1887); Frank Crosier, *General Views of Wilmington, Vermont* (Wilmington, Vt.: By the Author, [1889]); *DVT*, February 21, 1889, p. 3; "As the Birds See Us," *DVT*, May 15, 1891, local supplement, p. 1; *DVT*, August 29, 1890, p. 3.

39. C. M. Russell, ed., *Reunion of the Sons and Daughters of the Town of Wilmington* (Wilmington, Vt.: Deerfield Valley Times Press, 1890). On Old Home Week, also see Brown, *Inventing New England*, 138–42.

40. "The Deerfield Valley Railroad," *DVT*, November 6, 1891, p. 1.

41. William Gove, "Mountain Mills, Vermont and the Deerfield River Railroad," *Northeastern Logger and Timber Processor* 17 (May 1969): 17.

42. Victor I. Spear, *Report on Summer Travel for 1894* (Montpelier, Vt.: Press of the Watchman Publishing Co., 1894), 4.

43. "The Summer Resorts of Vermont," *The Vermonter* 10 (June 1905): 373.

44. Spear, *Report on Summer Travel for 1894*, 7.

45. *DVT*, August 30, 1889, p. 5.

46. "The Rayponda [sic] and Sylvan Lake Association," *DVT*, February 7, 1890, p. 5; "Raponda," *DVT*, July 4, 1890, p. 5; "Raponda's Boom," *DVT*, November 6, 1891, p. 2; "Mr. Former Resident," *DVT*, April 17, 1896, p. 7; "Hotel Raponda Burned," *DVT*, January 1, 1897, p. 1.

47. Land records and reports in the *DVT* show the Childs brothers as having purchased hundreds of acres of land that they later sold to the club. *DVT*, September 12, 1890, p. 5; *DVT*, October 17, 1890, p. 5; *DVT*, November 7, 1890, p. 5; Wilmington Town Office, Wilmington, Vermont, Deed Records, book 19, pp. 356–58, 364–71.

48. John F. Reiger, *American Sportsmen and the Origins of Conservation*, rev. ed. (Norman: University of Oklahoma Press, 1986).

49. Frankenstein [sic], "A Fine Fish and Game Park," reprinted from *Turf, Field, and Farm* in the *DVT*, April 15, 1892, p. 6; *Constitution, By-Laws, Rules and Regulations: Forest and Stream Club, Wilmington, Vermont, 1894* (Brattleboro, Vt.: Phoenix Job Printing Office, 1894).

50. "A Board of Trade Organized."

51. Hosea Mann Jr., "Will Wilmington and Whitingham Have a Railroad Completed and Running in 18 Months?" *DVT*, April 4, 1889, p. 2.

52. "Raponda," *DVT*, July 4, 1890, p. 5.

53. "Lake Raponda, Wilmington, VT," *DVT*, August 15, 1890, p. 1; "Raponda's Charms," *DVT*, August 14, 1891, p. 5.

54. *Constitution, By-Laws, Rules and Regulations*, 13.

55. "Forest and Stream Club," *DVT*, February 6, 1891, p. 5; "The Forest and Stream Club's Grand Opening," *DVT*, July 8, 1892, p. 5; *DVT*, October 21, 1892, p. 5.

56. "Forest and Stream Club" *DVT*, September 25, 1891, p. 5.

57. [Charles S. Forbes], "The Development of Vermont," *The Vermonter* 3 (January 1898): 125–28.

58. Richards, *Poland Spring*, 56.

59. Greene, *Vermont, The Green Mountain State: Past, Present, Prospective*, 74.

60. Barton Development Association, *Beautiful Barton, Vermont* (Rutland, Vt.: The Tuttle Company Printers, 1900), 7, 48.

61. Cindy S. Aron, *Working at Play: A History of Vacations in the United States* (New York: Oxford University Press, 1999). Also see Richards, *Poland Spring*.

62. "Lake Raponda Hotel, Wilmington, Vermont," (no publisher, [1901]), Wilmington Historical Society, Wilmington, Vermont.

63. William Irwin, *The New Niagara: Tourism, Technology, and the Landscape of Niagara Falls, 1776–1917* (University Park: The Pennsylvania University Press, 1996); David Nye, *American Technological Sublime* (Cambridge, Mass.: MIT Press, 1996); Patrick Vincent McGreevy, *Imagining Niagara: The Meaning and Making of Niagara Falls* (Amherst: University of Massachusetts Press, 1994); John F. Sears, *Sacred Places: American Tourist Attractions in the Nineteenth Century* (Amherst: University of Massachusetts Press, 1989), 182–208.

64. *Dorset, Vermont, As a Summer Home*, 3rd ed. (no publisher, 1904), 22.

65. "A Vacation in Vermont," *Harper's* 67 (November 1883): 823.

66. *DVT*, August 11, 1893, p. 5; John H. Walbridge, *Wilmington, Vermont* (Wilmington, Vt.: The [Deerfield Valley] Times Press, 1900), 15.

67. "New Industry," *DVT*, October 30, 1896, p. 1; "Raponda Reclining Chair," *DVT*, November 26, 1896, p. 4.

68. C. F. Ranney, "Newport and Lake Memphremagog," *The Vermonter* 3 (September 1897): 23–31; Bill Gove, "The Forest Industries of Lake Memphremagog," *Northern Logger and Timber Processor* 23 (March 1975): 18–19, 31–32, 34.

69. Also see Nye, *American Technological Sublime*, 110–13.

70. For a good example, see *Streets, Public Buildings, and General Views of Bellows Falls, Vt.* (n.p.: Published by F. J. Blake, 1885).

71. "To Saratoga," *DVT*, August 19, 1892, p. 5; "A Grand and Successful Excursion to Boston," *DVT*, October 14, 1892, p. 5.

72. *DVT*, January 10, 1890, p. 5; *DVT*, May 9, 1890, p. 5.

73. *DVT*, July 22, 1892, p. 5.

74. L. P. Tucker, "Beautiful for Situation," *DVT*, December 6, 1889, p. 1.

75. Richard W. Judd, *Common Lands, Common People: The Origins of Conservation in Northern New England* (Cambridge, Mass.: Harvard University Press, 1997).

76. *DVT*, January 10, 1890, p. 5.

77. "$75.00 Reward," *DVT*, July 31, 1891, p. 5.

78. A. H. Watson, "Forest and Stream Club," *DVT*, October 2, 1891, p. 5.

79. As Richard Judd has also argued, turn-of-the-century fish and game conservation in northern New England often emerged out of complex alliances between urban visitors and rural residents. Judd, *Common Lands, Common People*, 197–228.

80. *DVT*, April 15, 1892, p. 7.

81. Wilmington Town Office, Wilmington, Vermont, Deed Records, book 19, p. 358.

82. Philip G. Terrie, *Contested Terrain: A New History of Nature and People in the Adirondacks* (Blue Mountain Lake, N.Y.: The Adirondack Museum; Syracuse, N.Y.: Syracuse University Press, 1997), 115–24; Karl Jacoby, "Class and Environmental History: Lessons from the 'War in the Adirondacks,'" *Environmental History* 2 (July 1997): 324–42.

83. *DVT*, June 9, 1893, p. 5; "Illegal Fishing," *DVT*, June 26, 1896, p. 1.

84. "Make the Most of Our Advantages," *DVT*, July 19, 1889, p. 1.

85. "Emancipated from Inertia," reprinted from the *Springfield Republican* in the *DVT*, November 6, 1891, p. 1.

86. *DVT*, May 18, 1894, p. 5.

87. *DVT*, March 31, 1893, p. 5. Suggestions like this were a regular feature in the *DVT*.

88. Conforti, *Imagining New England*, 240–44. Also see Joseph S. Wood, *The New England Village* (Baltimore: The Johns Hopkins University Press, 1997), 135–60.

89. "Emancipated from Inertia," 1.

90. Colby Stoddard, "Barton and Its Development Association," *The Village: A Journal of Village Life* 1 (July 1907), 302. Also see Barton Development Association, *Beautiful Barton.*

91. Carman, *Hoot Toot and Whistle*, 14, 16; Gove, "Mountain Mills," 37–38.

Chapter 2. Abandonment and Resettlement: The Promises and Threats of Summer Homes (pp. 50–87)

1. James P. Taylor Collection, 1906–1949, Doc. T5, Folder 23, VHS.

2. For a good example, see Vrest Orton, ed., *And So Goes Vermont: A Picture Book of Vermont as It Is* (Weston, Vt.: Countryman Press; New York: Farrar and Rinehart, 1937). Also see Blake Harrison, "Tourism, Farm Abandonment, and the 'Typical' Vermonter," *Journal of Historical Geography* 31 (July 2005): 478–95.

3. On embodiment and rural geography, see Jo Little and Michael Leyshon, "Embodied Rural Geographies: Developing Research Agendas," *Progress in Human Geography* 27 (June 2003): 257–72.

4. Joseph A. Conforti, *Imagining New England: Explorations of Regional Identity from the Pilgrims to the Mid-Twentieth Century* (Chapel Hill: University of North Carolina Press, 2001), 203–309.

5. For extended explorations of exclusion as a geographic phenomenon, see David Sibley, *Geographies of Exclusion: Society and Difference in the West* (New York: Routledge, 1995), and James S. Duncan and Nancy G. Duncan, *Landscapes of Privilege: The Politics of the Aesthetic in an American Suburb* (New York: Routledge, 2004).

6. On regional writing and celebration of rural people see Stephanie Foote, *Regional Fictions: Culture and Identity in Nineteenth-Century American Literature* (Madison: University of Wisconsin Press, 2001), and Dona Brown, ed., *A Tourist's New England: Travel Fiction, 1820–1920* (Hanover, N.H.: University Press of New England, 1999).

7. Julian Agyeman and Rachel Spooner, "Ethnicity and the Rural Environment," in *Contested Countryside Cultures: Otherness, Marginalisation and Rurality*, ed. Paul Cloke and Jo Little (New York: Routledge, 1997), 197–217.

8. David Ward, *Poverty, Ethnicity, and the American City, 1840–1925: Changing Conceptions of the Slum and Ghetto* (New York: Cambridge University Press, 1989); Donald B. Cole, *Immigrant City: Lawrence, Massachusetts, 1845–1921* (Chapel Hill: University of North Carolina Press, 1963).

9. VSBA, *The Resources and Attractions of Vermont, With a List of Desirable Homes for Sale* (Montpelier, Vt.: Press of the Watchman Publishing Company, 1891), 8.

10. John A. Mead, "The Outlook for Vermont," *The Vermonter* 16 (December 1911): 416.

11. On anti-immigrant sentiments in Vermont, see John M. Lund, "Vermont Nativism: William Paul Dillingham and U.S. Immigration Legislation," *Vermont History* 63 (Winter 1995): 15–29.

12. Robert E. Shalhope, *Bennington and the Green Mountain Boys: The Emergence of Liberal Democracy in Vermont, 1760–1850* (Baltimore: Johns Hopkins University Press, 1996), 311–39.

13. For example, see Dorothy Canfield Fisher, "Vermont: Our Rich Little Poor State," *The Nation* 114 (May 31, 1922): 643–45. On the New England town meeting, see Frank Bryan, *Real Democracy: The New England Town Meeting and How It Works* (Chicago: University of Chicago Press, 2004).

14. Calvin Coolidge, "Address at Bennington," reprinted in *Vermont Prose: A Miscellany*, 2nd ed., ed. Arthur Wallace Peach and Harold Goddard Rugg (Brattleboro, Vt.: Stephen Daye Press, 1932), 247.

15. Walter Hard, "Vermont — A Way of Life," *The Survey* 68 (July 1, 1932): 301–3; Bernard DeVoto, "New England, There She Stands," *Harper's* 164 (March 1932): 405–15.

16. Peter Woolfson, "The Rural Franco-American in Vermont," *Vermont History* 50 (Summer 1982): 151–62; Ralph D. Vicero, "French-Canadian Settlement in Vermont Prior to the Civil War," *Professional Geographer* 23 (October 1971): 290–94.

17. United States Department of Commerce, Bureau of the Census, *Fifteenth Census of the United States: 1930, Population Volume III, Part 2* (Washington, D.C.: Government Printing Office, 1932), 1117, 1124. Also see J. L. Hypes, "Recent Immigrant Stocks in New England Agriculture," in *New England's Prospect: 1933*, ed. John K. Wright (New York: American Geographical Society, 1933), 189–205.

18. Clifton Johnson, *New England and Its Neighbors* (New York: Macmillan Company, 1924), 183.

19. Federal Writers Project, *Vermont: A Guide to the Green Mountain State* (Boston: Houghton Mifflin Company, 1937), 52.

20. On changing agricultural patterns and on out-migration from Vermont and New England in the late nineteenth century, see Harold Fisher Wilson, *The Hill Country of Northern New England: Its Social and Economic History, 1790–1930* (New York: Columbia University Press, 1936); Howard S. Russell, *A Long Deep Furrow: Three Centuries of Farming in New England* (1976; abridged, with a foreword by Mark Lapping, Hanover, N.H.: University Press of New England, 1982); Hal Barron, *Those Who Stayed Behind: Rural Society in Nineteenth-Century New England* (Cambridge: Cambridge University Press, 1984); and Michael M. Bell, "Did New England Go Downhill?" *Geographical Review* 79 (October 1989): 450–66.

21. Charles C. Nott, "A Good Farm for Nothing," *The Nation* 49 (November 21, 1889): 408.

22. Edward Hungerford, "Our Summer Migration: A Social Study," *Century* 20 (August 1891): 573.

23. Ibid.

24. Julian Ralph, "Our Tyrol and its Types," *Harper's* 106 (March 1903): 520.

25. Barron, *Those Who Stayed Behind.*

26. *DVT*, April 7, 1893, p. 5.

27. Examples from the nineteenth into the twentieth centuries include Peter Collier, ed., *Second Biennial Report of the Vermont State Board of Agriculture, Manufacturing and Mining, for the Years 1873–74* (Montpelier, Vt.: Freeman Steam Printing House and Bindery, 1874); Frederic Hathaway Chase, "Is Agriculture Declining in New England?" *New England Magazine* 2 (June 1890): 449–52; George F. Wells, comp., *The Status of Rural Vermont* (n.p.: Vermont State Agricultural Commission, 1903); and Guy Bailey and Orlando Martin, *Homeseekers' Guide to Vermont Farms* (St. Albans, Vt.: St. Albans Messenger Co., 1911).

28. Alonzo B. Valentine, *Report of the Commissioner of Agricultural and Manufacturing Interests of the State of Vermont, 1889–1890* (Rutland, Vt.: The Tuttle Company, Official State Printers, 1890), 15.

29. Valentine, *Report of the Commissioner*, 15–19; Dorothy Mayo Harvey, "The Swedes in Vermont," *Vermont History* 28 (January 1960): 39–58.

30. A. B. Valentine, "Swedish Immigration," *The Quill* 1 (September 1890): 27–32.

31. Johnson, *New England and Its Neighbors*, 183.

32. Florence W. Rich, "Vermont and Vermonters — Their Future," *The Vermonter* 27:7 (1922): 160.

33. Nancy L. Gallagher, *Breeding Better Vermonters: The Eugenics Project in the Green Mountain State* (Hanover, N.H.: University Press of New England, 1999); Kevin Dann, "From Degeneration to Regeneration: The Eugenics Survey of Vermont, 1925–1936," *Vermont History* 59 (Winter 1991): 5–29.

34. Vermont Commission on Country Life (hereafter VCCL), *Rural Vermont: A Program for the Future* (Burlington, Vt.: The Commission, 1931); Henry F. Perkins, "The Comprehensive Survey of Rural Vermont," in *New England's Prospect*, 206–12.

35. On the VSBA's early role in farm promotions, see VSBA, *The Resources and Attractions of Vermont* (1891), 65–67; Andrea Rebek, "The Selling of Vermont: From Agriculture to Tourism, 1860–1910," *Vermont History* 44 (Winter 1976): 14–27; and Dona Brown, *Inventing New England: Regional Tourism in the Nineteenth Century* (Washington, D.C.: Smithsonian Institution Press, 1995), 142–45.

36. West River Valley Association, *The Call of the Country* (Brattleboro, Vt.: The Association, 1912); Summer Homes in Vermont Corporation, *Your Summer Home in Vermont* (New York: The Corporation, 1927).

37. VSBA, *Resources and Attractions of Vermont, With a List of Desirable Homes for Sale* (Montpelier, Vt.: Press of the Watchman Publishing Co., 1892), 30. Also see VSBA, *Vermont, Its Opportunities for Investment in Agriculture, Manufacture, Minerals, Its Attractions for Summer Homes* (East Hardwick, Vt.: The Board, [1903]).

38. Albert Lewis, "The Abandoned Farms," *Forest and Stream* 38 (April 14, 1892): 347. Also see VSBA, *The Resources and Attractions of Vermont* (1891), 61.

39. An incomplete tally released by the Board of Agriculture in 1893 found that

nearly two hundred abandoned farms had been sold in the preceding year alone. They do not specify that these were sold for summer homes, but it is safe to assume that many of them were. There is no indication anywhere in the literature that this trend changed significantly in the coming decades. VSBA, *Good Homes in Vermont: A List of Desirable Farms for Sale* (Montpelier, Vt.: Press of the Watchman Publishing Co., 1893), 4–6.

40. West River Valley Association, *The Call of the Country*, 13.

41. Isabel C. Greene, "An Ideal Mountain Town," *The Vermonter* 13 (November–December 1908): 342.

42. For example, see Hungerford, "Our Summer Migration," 574–75.

43. "Beautiful Summer Homes," *DVT*, Holiday Edition, 1889, p. 4. Also see "No More Abandoned Farms," *The Nation* 69 (September 7, 1899): 184–85.

44. Nathaniel Coit Greene, "How You May Own a Paying Farm," *New England Magazine* 38 (July 1908): 537–41; E. Gordon Parker, "A Vacation on an Abandoned Farm," *Country Life in America* 20 (June 1, 1911): 49–51.

45. Greene, "How You May Own a Paying Farm," 538.

46. Edward A. Wright, "The Rural-Degeneracy Cry," *New England Magazine* 36 (April 1907): 152–58; Ripley Hitchcock, "The Summer Home: The Organization of Summer Communities," *Outing Magazine* 52 (July 1908): 498–501; Julius H. Ward, "The Revival of Our Country Towns" *New England Magazine* 1 (November 1889): 242–48.

47. Hungerford, "Our Summer Migration," 574, 576.

48. "No More Abandoned Farms," 184. Also see Clifton Johnson, "The Deserted Homes of New England," *The Cosmopolitan* 15 (May–October 1893): 215–22.

49. Wright, "The Rural-Degeneracy Cry," 154. Also see Ward, "The Revival of Our Country Towns."

50. Sinclair Lewis, "Back to Vermont," *Forum* 95 (April 1936): 255.

51. Lucy Margaret Clapp, "To Halifax — In Search of a Farm," *The Vermonter* 46 (February 1941): 35. Also see Allen Chamberlain, "The Ideal Abandoned Farm," *New England Magazine* 16 (June 1897): 473–78.

52. On the architecture of summer homes in New England, also see Bryant F. Tolles Jr., *Summer Cottages in the White Mountains: The Architecture of Leisure and Recreation, 1870–1930* (Hanover, N.H.: University Press of New England, 2000).

53. For an overview of tourism, leisure, and gender identity, see Cara Aitchison, Nicola E. Macleod, and Stephen J. Shaw, *Leisure and Tourism Landscapes: Social and Cultural Geographies* (New York: Routledge, 2000), 110–35.

54. Francine Watkins, "The Cultural Construction of Rurality: Gender Identities and the Rural Idyll," in *Thresholds in Feminist Geography: Difference, Methodology, Representation*, ed. John Paul Jones III, Heidi J. Nast, and Susan M. Roberts (Lanham, Md.: Rowman and Littlefield Publishers, 1997), 391.

55. Nathaniel Coit Greene, "A New Way of Enjoying Country Life," *New England Magazine* 38 (May 1908): 283–89. Also see Henry Hoyt Moore, "The Country Club and the Abandoned Farm," *Outlook* 111 (September 22, 1915): 203–10.

56. George Humphreys, interview by author, tape recording, West Dover, Vt., June 8, 1999.

57. Janet E. Schulte, "Summer Homes: A History of Family Summer Vacation Communities in Northern New England, 1880–1940" (Ph.D. diss., Brandeis University, 1994).

58. Florence E. Lemmon, "That Matter of Country House Names," *House Beautiful* 56 (July 1924): 32; Cromwell Childe, "What to Name the Country Home," *Woman's Home Companion* 38 (May 1911): 42.

59. Cullen Bryant Snell, "Our Search — Our Farm — Our Vacation," *The Vermonter* 22 (February–March 1917): 36.

60. Ibid.

61. Paul W. Thayer, "Evolving a Dormant Value," *The Vermonter* 38 (December 1933): 313.

62. Chamberlain, "The Ideal Abandoned Farm," 474. Also see Paul W. Thayer, "Selecting a Summer Home in Vermont," *The Vermonter* 31:11 (1926): 161–64.

63. Charles F. Speare, *We Found a Farm* (Brattleboro, Vt.: Stephen Daye Press, 1936), 19. Also see Thayer, "Evolving a Dormant Value," and M. S. Hinchman, "Renewing the Old Farm," *House and Garden* 48 (November 1925): 70, 116, 118.

64. For example, see Elisabeth Marbury, "When Is a Farm Not a Farm?" *House and Garden* 53 (May 1928): 126, 166, 196.

65. J. Horace McFarland, "The Surroundings of the Country Home," *Country Calendar* 1 (September 1905): 467.

66. Helen Dodd, "What Two Women Did with One Farmhouse," *Ladies' Home Journal* 29 (October 1912): 60.

67. *Report of the Department of Conservation and Development for the Term Ending June 30, 1940* (Springfield, Vt.: Springfield Printing Corporation, [1940]), 132. Also see H. H. Chadwick, *Vermont Bureau of Publicity: Its History, Expenditures and Activities* ([Montpelier, Vt.]: [Office of the Secretary of State], [1934]); Recreation Study Committee, Vermont State Planning Board, *Recreation Study Tentative Report,* 1939, James P. Taylor Collection, 1906–1949 (hereafter JPT), Doc. T9, Folder 14, VHS; "Press release for Evening Papers of Mar. 11, 1946," Publicity Office, Vermont Development Commission, in the binder "Old Publicity Service, Old Statistics," in unnumbered box "Vermont Travel Division [1879–1994]" (hereafter OPS,OS), Vermont State Archives, Office of the Secretary of State, Montpelier, Vermont (hereafter VSA).

68. On the history of the bureau, see Chadwick, *Vermont Bureau of Publicity.*

69. Tyler Resch, *Dorset: In the Shadow of the Marble Mountain* (West Kennebunk, Maine: Phoenix Publishing, 1989), 192–228.

70. Peter D. Watson, Wilhelmina Smith, Lewis Hill, Nancy Hill, Sally Fisher, Patricia Haslam, Rhoda Metraux, Dorothy Ling, Gail Sangree, *The History of Greensboro: The First Two Hundred Years* (Greensboro, Vt.: Greensboro Historical Society, 1990), 168–99; S. Whitney Landon, *Early Memories of Caspian Lake: How the Greensboro Summer Colony Began and What Made It Unique* (no pub-

lisher: [ca. 1975]); Peter D. Watson, *North Shore Summers: Memories of Greensboro in the 1920s* (Greensboro, Vt.: Greensboro Historical Society, 1981).

71. Lewis Hill, *Fetched Up Yankee: A New England Boyhood Remembered* (Chester, Conn.: Globe Pequot Press, 1990); Lewis Hill, *Yankee Summer: The Way We Were* (n.p.: 1st Book Library, 2000).

72. Hill, *Yankee Summer*, 179–81.

73. Constance Votey, *Growing Up with Aspenhurst*, 2nd ed. ([Greensboro, Vt.]: Printed by the Greensboro Historical Society, 1980), 4.

74. Estimates for annual income from mid-century repairs on summer homes in Vermont were averaging around $200,000 each year. "Press Release for Evening Papers of Mar. 11, 1946," OPS,OS, VSA.

75. Bernard DeVoto, "How to Live Among the Vermonters," *Harper's* 173 (August 1936): 336.

76. "The Native Sons," *House and Garden* 47 (April 1925): 62.

77. Kate Taylor Kemp, "A Summer Memory," *Outlook* 85 (April 20, 1907): 889.

78. Humphreys, interview, June 8, 1999.

79. Bruce Bliven, "Rock-ribbed," *New Republic* 33 (February 21, 1923): 346.

80. VCCL, *Rural Vermont*, 147; William Sidney Rossiter, "Three Sentinels of the North," *Atlantic Monthly* 132 (July 1923): 87–97.

81. For a related discussion, see Miguel De Oliver, "Historical Preservation and Identity: The Alamo and the Production of a Consumer Landscape," *Antipode* 28 (January 1996): 1–23.

82. R. Balfour Daniels, "Caspian Lake, Greensboro," *The Vermonter* 39 (June–July 1934): 169. Also see Watson et al., *The History of Greensboro*, 98.

83. H. W. Sachs, "Judaism in Vermont," *The Vermonter* 8 (February 1903): 227–28; Myron Samuelson, *The Story of the Jewish Community of Burlington, Vermont* (Burlington, Vt.: Published by the author, 1976).

84. Prejudice against Jewish travelers was by no means unique to Vermont. Also see James C. O'Connell, *Becoming Cape Cod: Creating a Seaside Resort* (Hanover, N.H.: University Press of New England, 2002), 76–77. For more on race, ethnicity, and tourism in America, see Cindy S. Aron, *Working at Play: A History of Vacations in the United States* (New York: Oxford University Press, 1999), 206–36.

85. N. C. Belth, *Barriers: Patterns of Discrimination Against Jews* (New York: Friendly House Publishers, 1958), 38–41. Also see Hal Goldman, "'A Desirable Class of People': The Leadership of the Green Mountain Club and Social Exclusivity, 1920–1936," *Vermont History* 65 (Summer/Fall 1997): 131–52.

86. *Vermont Cottages, Camps and Furnished Houses for Rent* (Montpelier, Vt.: Vermont Publicity Service, Department of Conservation and Development, 1938); for examples, see pages 20, 60.

87. *Lake Champlain Club, Malletts Bay, Vt.* (no publisher, [1920]). There are no page numbers in this text, but this quote can be found on the final page.

88. Surveys from the early 1930s, for instance, revealed that a third of all non-

property-owning visitors left the state with an interest in someday purchasing land. "Summary of Tourist Questionnaires, 1932," OPS,OS, VSA.

89. As quoted in Watson et al., *The History of Greensboro*, 44. Also see "Unenforceable Covenants Are in Many Deeds," *New York Times* (August 1, 1986), p. A9.

90. Hill, *Yankee Summer*, 120–21. On the Ku Klux Klan in Vermont, see Michael Sherman, Gene Sessions, and P. Jeffrey Potash, *Freedom and Unity: A History of Vermont* (Barre, Vt.: Vermont Historical Society, 2004), 404–7.

91. S. T. Hecht, "Vermont Pioneer from Flatbush," *Commentary* 19 (April 1955): 371. Although this was published in 1955, Hecht appears to have purchased his property sometime before World War II. The italics are original.

92. Hill, *Yankee Summer*, 45.

93. Votey, *Growing Up with Aspenhurst*, 4.

94. For example, see Snell, "Our Search — Our Farm — Our Vacation," 36.

95. Hill, *Yankee Summer*, 43–44.

96. George K. Holmes, "Movement from City and Town to Farms," in *Yearbook of the United States Department of Agriculture*, 1914 (Washington, D.C.: Government Printing Office, 1915): 268.

97. Bliven, "Rock-ribbed," 345.

98. As quoted in Jan Albers, *Hands on the Land: A History of the Vermont Landscape* (Cambridge, Mass.: MIT Press, 1999), 253. For similar recommendations, see VCCL, *Rural Vermont*, 130.

99. Zephine Humphrey, "The New Crop," *Outlook* 134 (July 11, 1923): 381.

100. Ibid.

101. Ibid.

102. VCCL, *Rural Vermont*, 372.

103. Ibid., 380.

104. Ida H. Washington, "Dorothy Canfield Fisher's *Tourists Accommodated* and Her Other Promotions of Vermont," *Vermont History* 65 (Summer/Fall 1997): 153–64.

105. Dorothy Canfield Fisher, "Vermonters," in *Vermont: A Guide to the Green Mountain State*, 3.

106. Fisher lobbied to create a state-level "inter-racial advisory committee" in Vermont, for instance, and she pressured the infamous Vermont Hotel Keepers' Association to condemn the anti-Semitic actions of its members. Dorothy Canfield Fisher to William Wills, June 8, 1944; Dorothy Canfield Fisher to William Wills, August 28, 1944, "Dorothy Canfield Fisher Collection," Box 22, Folder 19, University of Vermont Special Collections, Bailey/Howe Library, Burlington, Vermont (hereafter UVM). Fisher also spoke out against anti-Semitism in rural Vermont in the pages of her 1939 novel *Seasoned Timber*, and upon her death she won praise from the International Academy of Human Rights. Dorothy Canfield Fisher, *Seasoned Timber*, ed. Mark J. Madigan (Hanover, N.H.: University Press of New England, 1996, 1939). Also see David Baumgardt, "Dorothy Canfield Fisher: Friend of Jews in Life and Work," *Publication of the American Jewish Historical Society* 48 (June 1959): 245–55.

107. Dorothy Canfield, *Vermont Summer Homes* (Montpelier, Vt.: Vermont

Bureau of Publicity, 1932). The pages in this work are not numbered. Although the author typically went by her married name, Dorothy Canfield Fisher, she used the name Dorothy Canfield for this publication.

108. Ibid.

109. Ibid.

110. Ibid.

Chapter 3. Accessing the Rural Landscape: Driving, Hiking, and the Making of Unspoiled Vermont (pp. 88–131)

1. H. H. Chadwick, *Vermont's Tourist Business: A Study Covering Ten Years* (Burlington, Vt.: Free Press Printing Company, 1944), 11, 35–47; Lawrence W. Chidester, "The Importance of Recreation as a Land Use in New England," *Journal of Land and Public Utility Economics* 10 (May 1934): 202–9.

2. Bernard DeVoto, "How to Live Among the Vermonters," *Harper's* 173 (August 1936): 331–36; L. P. Thayer, "The Greater Vermont (Address Delivered before the Rotary Club, Rutland, Vermont, December 14, 1925)" (Rutland, Vt.: The Tuttle Company, 1925), 2; Dane Yorke, "The Florida of the North," *American Mercury* 20 (July 1930): 275–80.

3. Charles Edward Crane, *Let Me Show You Vermont* (New York: Alfred A. Knopf, 1937), 4–5.

4. Leon S. Gay, "Keeping Unspoiled Vermont Unspoiled," *The Vermonter* 40 (August 1935): 149–62.

5. Marguerite S. Shaffer, *See America First: Tourism and National Identity, 1880–1940* (Washington, D.C.: Smithsonian Institution Press, 2001); John A. Jakle, *The Tourist: Travel in Twentieth-Century North America* (Lincoln: University of Nebraska Press, 1985); Warren James Belasco, *Americans on the Road: From Autocamp to Motel, 1910–1945* (Baltimore: Johns Hopkins University Press, 1979).

6. Jakle, *The Tourist*, 121–33; Peter J. Hugill, "Good Roads and the Automobile in the United States, 1880–1929," *Geographical Review* 72 (July 1982): 327–49; Hal Barron, *Mixed Harvest: The Second Great Transformation in the Rural North, 1870–1930* (Chapel Hill: University of North Carolina Press, 1997), 19–42.

7. *The Vermont League for Good Roads* (Montpelier, Vt.: Watchman Publishing Company, 1892); *Proceedings of the First Annual Meeting of the Vermont Good Roads Association, May 5, 1904* (no publisher, [1904]); United States Department of Agriculture, Bureau of Public Roads, and the Vermont State Highway Department, *Report of a Survey of Transportation on the State Highways of Vermont* (no publisher, 1927), 9–14.

8. Chadwick, *Vermont's Tourist Business*, 13.

9. "The Automobile in Vermont," *The Vermonter* 10 (July 1905): 395; "The Coming of the Tourist," *The Vermonter* 13 (May 1908): 152.

10. "Summer Tourist Business (The Colonial Inn — South Woodstock)," Ver-

mont Travel Division/Development Department [1879–1950]: Statistics, Surveys, Reports, etc. (hereafter VTD/DD), Box 7, Folder "1929 Tourist Business Survey," VSA; "Green Hills: Vermont for Its Simplicity," *Boston Evening Transcript*, July 3, 1928, Travel-Shipping Section, p. 1.

11. Commissioner Bates, quoted in Arthur F. Stone, *The Vermont of Today: With Its Historic Background, Attractions, and People*, vol. 1 (New York: Lewis Historical Publishing Company, Inc., 1929), 704.

12. By the early 1930s 5,075 miles of Vermont's 15,031 miles of rural roads were surfaced, and 3,552 miles of these were overseen by the state. Of these state roads 3,168 miles were surfaced with a mix of gravel, sand, and clay, 122 with macadam, and 262 with concrete. Arthur W. Dean, "The Highways of New England," in *New England's Prospect: 1933*, ed. John K. Wright (New York: American Geographical Society, 1933), 364.

13. The records located in VTD/DD at the VSA contain the most comprehensive collection of these surveys and their summary reports.

14. *Report of Publicity Service, Department of Conservation and Development, State of Vermont, for Years Ending June 30, 1935 and 1936* (Burlington, Vt.: Free Press Printing Co., [1936]), 4; *Report of the Department of Conservation and Development for the Term Ending June 30, 1940* (Springfield, Vt.: Springfield Publishing, [1940]), 121.

15. Stephen Nissenbaum, "New England as Region and Nation," in *All Over the Map: Rethinking American Regions*, Edward L. Ayers, Patricia Nelson Limerick, Stephen Nissenbaum, and Peter S. Onuf (Baltimore: Johns Hopkins University Press, 1996), 38–61.

16. Robert L. Dorman, *Revolt of the Provinces: The Regionalist Movement in America, 1920–1945* (Chapel Hill: University of North Carolina Press, 1993).

17. See, for example, Shaffer, *See America First*.

18. VCCL, *Rural Vermont: A Program for the Future* (Burlington, Vt.: The Commission, 1932), 133.

19. Wallace Nutting, *Vermont Beautiful* (New York: Bonanza Books, 1922), 278.

20. William C. Lipke, "Changing Images of the Vermont Landscape," in *Vermont Landscape Images, 1776–1976*, ed. William C. Lipke and Phillip N. Grimes (Burlington, Vt.: Robert Hull Fleming Museum, 1976), 44. Also see Thomas Andrew Denenberg, *Wallace Nutting and the Invention of Old America* (New Haven, Conn.: Yale University Press, 2003).

21. Vermont Bureau of Publicity, "Unspoiled Vermont," [ca. 1933], uncatalogued pamphlet "Vermont — Description — Unspoiled Vermont," VHS.

22. "Real Vermont Villages," *The Vermonter* 42 (June 1937): 124–25. Also see Walter and Margaret Hard, *This Is Vermont* (Brattleboro, Vt.: Stephen Daye Press, 1936).

23. Herbert Wheaton Congdon, *The Covered Bridge: An Old American Landmark Whose Romance, Stability and Craftsmanship Are Typified by Structures Remaining in Vermont* (Brattleboro, Vt.: Stephen Daye Press, 1941), 15.

24. James P. Taylor, "A Champion of Saving, Safety, and Satisfaction," *Vermont Highways* (December 1930): 30.

25. K. R. B. Flint, "The Need for Zoning in Vermont," *The Vermonter* 34 (November 1929): 169–70. For more on zoning in Vermont, see JPT, Doc. T9, Folder 8, VHS; Andrew E. Nuquist and Edith W. Nuquist, *Vermont State Government and Administration: An Historical and Descriptive Study of the Living Past* (Burlington, Vt.: Government Research Center, University of Vermont, 1966), 508–10; John Nolen, Philip Sutler, Albert LaFleur, and Dana M. Doten, *Graphic Survey: A First Step in State Planning for Vermont* (no publisher, [1937]).

26. On Taylor's career, see Hal Goldman, "James Taylor's Progressive Vision: The Green Mountain Parkway," *Vermont History* 63 (Summer 1995): 158–79. Also see Arthur F. Stone, *The Vermont of Today: With Its Historic Background, Attractions and People*, vol. 2 (New York: Lewis Historical Publishing Company, Inc., 1929), 855–56.

27. "A Vermont Oracle," *The Vermonter* 35 (April 1930): 87

28. James P. Taylor, "The New Beauty Contest," *Vermont Highways* (September 1930): 20.

29. For examples, see JPT, Doc. T6, VHS, and JPT, Doc. T3, Folder 11, VHS.

30. James P. Taylor, "Hospitality de Lux: Pomp of Highways and Glory of Roadsides (An Address by James P. Taylor before the Vermont Federation of Women's Clubs, May 1929)," JPT, Doc. T6, Folder 6, VHS; Thayer, "The Greater Vermont."

31. For example, see James P. Taylor, "A Celebration Which Sells and Challenges," *Vermont Highways* (November 1930): 27–29.

32. James P. Taylor, "The Beginnings of the Wonderful Retouching of a Wonderful Picture," *Vermont Highways* (November 1931): 10. Also see Harold L. Tilton, "What Ho!!! Let's Mobilize Under a Grand Field Marshall," JPT, Doc. T3, Folder 11, VHS.

33. On depression era photography in Vermont, see Nancy Price Graff, *Looking Back at Vermont: Farm Security Administration Photographs, 1936–1942* (Middlebury, Vt.: Middlebury College Museum of Art; Hanover, N.H.: University Press of New England, 2002).

34. Mrs. W. A. Grover, "Along Vermont Roads," *The Vermonter* 44 (May 1939): 122.

35. Pearl Strachan, "Vermont Is America — Or Is It?" *Christian Science Monitor Weekly Magazine Section* (August 16, 1941): 9.

36. James P. Taylor to Wallace H. Gilpin, October 8, 1936, JPT, Doc. T3, Folder 11, VHS; James P. Taylor to Wallace H. Gilpin, August 18, 1936, JPT, Doc. T3, Folder 11, VHS.

37. Etta M. Wilson to James P. Taylor, June 28, 1939, JPT, Doc. T5, Folder 39, VHS.

38. Thomas D. Murphy, *New England Highways and Byways from a Motor Car: On Sunrise Highways* (Boston: L. C. Page and Company, 1924), 6–7.

39. Chadwick, *Vermont's Tourist Business*, 13. Exact numbers are hard to determine, in part due to terminology. According to public health and licensing

records, Vermont had 1,536 "tourist houses and cabin camps" in 1938—a figure that made no distinction between accommodations in private homes and those in roadside cabins. Vermont Development Commission, Vermont Department of Public Health, New England Council, *Vermont Hotels, Tourist Homes and Cabins, Restaurants, Boys and Girls Camps, Survey 1945*, VTD/DD, Box 7, Folder "VDC Recreation Surveys," VSA. Also see "Brief Summary of Work Carried on by Mrs. Pearl M. Brown for the Committee on Summer Homes and Tourists, Summer of 1929—in the State of Vermont," VTD/DD, Box 7, Folder "1929 Tourist Business Survey," VSA; Walter H. Crockett, "Vermont's Recreational Business," *The Vermonter* 34 (October 1929): 148; VCCL, *Rural Vermont*, 123.

40. Dorothy Canfield Fisher, introduction to *Tourists Accommodated: Some Scenes from Present-Day Summer Life in Vermont* (New York: Harcourt, Brace and Company, 1934), 1–10; Ida H. Washington, "Dorothy Canfield Fisher's *Tourists Accommodated* and Her Other Promotions of Vermont," *Vermont History* 65 (Summer/Fall 1997): 153–56. For quote, see *Tourists Accommodated*, 17.

41. "Conference on the Tourist Business in Vermont Homes," *Journal of Home Economics* 23 (January 1931): 41. Also see Lillian H. Johnson and Marianne Muse, *Cash Contribution to the Family Income Made by Vermont Farm Homemakers* (Burlington, Vt.: University of Vermont, Vermont Agricultural Experiment Station, 1933), 30–35 (hereafter UVM/VAES will be used to refer to University of Vermont, Vermont Agricultural Experiment Station).

42. Johnson and Muse, *Cash Contribution to the Family Income Made by Vermont Farm Homemakers*, 31.

43. "Diffusion of Spirit," *The Vermonter* 42 (August 1937): 157–58. Also see Bernice L. B. Graham, "Adventuring in Vermont," *Vermont Highways* (June 1931): 10–11, 14–16; Taylor, *Hospitality de Lux*.

44. On suspicion of automobile tourists in Vermont and beyond, see Belasco, *Americans on the Road*, 105–27; Hal Goldman, "'A Desirable Class of People': The Leadership of the Green Mountain Club and Social Exclusivity, 1920–1936," *Vermont History* 65 (Summer/Fall 1997): 131–52. On mobility as a source of social concern in American culture, see Tim Cresswell, *The Tramp in America* (London: Reaktion Books, 2001).

45. Fisher, *Tourists Accommodated*, 21.

46. Rose C. Field, "A Vermont Family," *The American Mercury* 9 (October 1926): 184.

47. Catherine Gudis, *Buyways: Billboards, Automobiles, and the American Landscape* (New York: Routledge, 2004), 165. Also see John A. Jakle and Keith A. Sculle, *Signs in America's Auto Age: Signatures of Landscape and Place* (Iowa City: University of Iowa Press, 2004), 135–66.

48. "Conservation: Vermont and Billboards," *Nature Magazine* 31 (March 1938): 165–66. Also see Rawson C. Myrick, "Outdoor Advertising in Vermont," *Vermont Highways* (April 1930): 16–17, 19; "Billboards/Outdoor Advertising," VSA.

49. On the VABR and billboard opposition in Vermont, see manuscript file

"Vermont Association for Billboard Restriction, 1938–1940," UVM; "Springfield Chamber of Commerce," 1938, JPT, Doc. T6, Folder 22, VHS; "Conservation: Vermont and Billboards."

50. See, for instance, Vermont Association for Billboard Research, "The Billboard Situation in Vermont, Primer No. 1," January 1938, "John Clement Papers," Doc. 152, Folder 17, VHS.

51. JPT, Doc. T6, Folder 22, VHS. For quote, see Recreation Study Committee, Vermont State Planning Board, *Recreation Study Tentative Report* (no publisher, 1939), 35, mss. copy, JPT, Doc. T9, Folder 14, VHS.

52. Dorothy Thompson Lewis to the Editor of the *Rutland Daily Herald*, October 11, 1937 (reprinted in *The Roadside Bulletin* 5 [January 1938]: 9), JPT, Doc. T3, Folder 11, VHS.

53. Sinclair Lewis, "Address Before the Rutland Rotary Club," reprinted in *Vermont Prose: A Miscellany*, 2nd ed., ed. Arthur Wallace Peach and Harold Goddard Rugg (Brattleboro, Vt.: Stephen Daye Press, 1932), 216.

54. F. J. Whalen to John M. Thomas, June 21, 1938, JPT, Doc. T6, Folder 22, VHS. Also see Munn Boardman to James Taylor, June 21, 1938, JPT, Doc. T6, Folder 22, VHS, and Justin B. Kelly to L. M. Tye, June 25, 1938, JPT, Doc. T6, Folder 22, VHS.

55. "How Vermont Solves Its Billboard Problem," *The Vermonter* 50 (January 1945): 11–13; Horace Brown, "Vermont's Big Boards Come Down," *The Vermonter* 50 (September 1945): 284–89. Billboard companies challenged the law in the Vermont Supreme Court and lost in a decision handed down in 1943 that ultimately increased control over billboards.

56. Roderick Nash, *Wilderness and the American Mind*, 3rd ed. (New Haven, Conn.: Yale University Press, 1982), 141–60; Laura Waterman and Guy Waterman, *Forest and Crag: A History of Hiking, Trail Blazing, and Adventure in the Northeast Mountains* (Boston: Appalachian Mountain Club, 1989).

57. This percentage, however, does not necessarily mean that all of this "farmland" was cleared and open. Some of it, for example, was in farm woodlots. Christopher McGrory Klyza and Stephen C. Trombulak, *The Story of Vermont: A Natural and Cultural History* (Hanover, N.H.: University Press of New England, 1999), 87–114. Their chart on page 89 offers a particularly useful summary.

58. Klyza and Trombulak, *The Story of Vermont*, 98–100; J. E. Riley, "The Forests a Great Natural Resource to Vermont," *The Vermonter* 27:3–4 (1922): 69–71; Joseph C. Kircher, "The Use of the Green Mountain National Forest," *Vermont Highways* (June 1931): 12–13; Otto G. Koenig, "Green Mountain National Forest in Vt.," *The Vermonter* 44 (December 1939): 264–66; W. W. Ashe, "A National Forest for Vermont," *The Vermonter* 36 (March 1931): 72–76.

59. Louis J. Paris, "The Green Mountain Club," *The Vermonter* 16 (May 1911): 170. Also see "The Green Mountain Club," *The Vermonter* 15 (April 1910): 114.

60. On the history of the GMC and the Long Trail, see Paris, "The Green Mountain Club," 151–70; Jane Curtis, Will Curtis, and Frank Lieberman, *Green*

Mountain Adventure, Vermont's Long Trail: An Illustrated History (Montpelier, Vt.: The Green Mountain Club, 1985); Waterman and Waterman, *Forest and Crag*, 356–73.

61. John B. Clark, "From Massachusetts to Canada," page 6 (a manuscript of a full trip on the trail in July and August 1928), Green Mountain Club Archives (hereafter GMC), Doc. 226, Folder 6, VHS.

62. Guy W. Bailey, "Foreword" to Vermont Bureau of Publicity, *Vermont: The Land of the Green Mountains* ([Essex Junction, Vt.]: Vermont Bureau of Publicity, Office of the Secretary of State, 1913), 4, 6.

63. This vision of broad economic development was not unlike that outlined soon after by Benton MacKaye for the Appalachian Trail. See Paul S. Sutter, *Driven Wild: How the Fight Against Automobiles Launched the Modern Wilderness Movement* (Seattle: University of Washington Press, 2002), 142–93; Mark Luccarelli, "Benton MacKaye's Appalachian Trail: Imagining and Engineering a Landscape," in *Technologies of Landscape: From Reaping to Recycling*, ed. David E. Nye (Amherst: University of Massachusetts Press, 1999), 207–17.

64. *The Long Trail* (Brandon, Vt.: The Brandon Inn, 1914), 5; Louis J. Paris, "The Loops of the Long Trail: With the Trail Possibilities of the Woodstock Quadrangle," *The Elm Tree Monthly and Spirit of the Age* (Woodstock, Vt.), n.s. 1 (December 1913): 40–43; "The Long Trail Brings Real Estate Development," *The Long Trail News* (hereafter *LTN*) 2 (August 1929): 2.

65. Curtis, Curtis, and Lieberman, *Green Mountain Adventure, Vermont's Long Trail*, 17–19.

66. Theron S. Dean to William S. Monroe, November 16, 1916, "Theron Dean Papers," Box 1, Folder 3, UVM.

67. Will S. Monroe, "The Monroe Sky-Line Trail," *The Vermonter* 21 (November/December 1916): 265–67; Curtis, Curtis, and Lieberman, *Green Mountain Adventure, Vermont's Long Trail*, 19–22.

68. Burlington Section, Green Mountain Club to William Monroe, October 27, 1916; Theron S. Dean to William S. Monroe, October 16, 1916, "Theron Dean Papers," Box 1, Folder 3, UVM.

69. See communications between Theron S. Dean and William S. Monroe, "Theron Dean Papers," Box 1, Folder 3, UVM.

70. See, for example, Roderic Marble Olzendam, *The Lure of Vermont's Silent Places: "The Green Mountains"* (Essex Junction, Vt.: The Vermont Bureau of Publicity, Office of the Secretary of State, [1918]), 17.

71. Theron Dean to William Monroe, January 3, 1917, "Theron Dean Papers," Box 1, Folder 4, UVM.

72. Nash, *Wilderness and the American Mind*, 150–56.

73. On women hikers and the contestation of traditional gender roles, see Karen M. Morin, "Peak Practices: Englishwomen's 'Heroic' Adventures in the Nineteenth-Century American West," *Annals of the Association of American Geographers* 89 (September 1999): 489–514.

74. "Equipment for Hiking on Long Trail, Especially for Women," *LTN* (July 1929): 4.

75. "'Three Musketeers' Make 300-Mile Trip Over the Long Trail in 30 Days," *Burlington Free Press* (hereafter *BFP*), September 5, 1927, p. 5; "Three Musketeers," typed manuscript, GMC, Doc. 229, Folder 49, VHS.

76. Walter H. Crockett to James P. Taylor, August 25, 1927, GMC, Doc. 229, Folder 48, VHS.

77. Edith M. Esterbrook, "The Long Trail Is Safe for Women Hikers," *The Vermonter* 34 (May 1929): 77–78.

78. "A Feminine Record," *LTN* (December 1927): 3.

79. Crane, *Let Me Show You Vermont,* 72.

80. Walter H. Crockett to James P. Taylor, August 25, 1927, GMC, Doc. 229, Folder 48, VHS.

81. Goldman, "'A Desirable Class of People,'" 134–35.

82. "Publicity," *LTN* (December 1927): 3; Curtis, Curtis, and Lieberman, *Green Mountain Adventure, Vermont's Long Trail,* 53–56.

83. "Three Musketeers," typed manuscript, GMC, Doc. 229, Folder 49, VHS.

84. For example, see Irving D. Appleby to Rawson C. Myrick, September 16, 1927, GMC, Doc. 229, Folder 48, VHS; Irving D. Appleby to James P. Taylor, September 23, 1927, GMC, Doc. 229, Folder 48, VHS.

85. "Appleby," *LTN* (December 1927): 1–2.

86. "Appleby Writes a Letter," *LTN* (April 1928): 2–3. The italics in this quote are original.

87. "Those Resolutions," *LTN* (June 1928): 2–3.

88. "Col. William J. Wilgus Explains Proposed Green Mt. Parkway," Vermont State Chamber of Commerce, "Informational Bulletins on State Problems, Bulletin No. 3, 28 August 1933," reference file "Green Mountain Parkway (1930s)," UVM.

89. On the NPS plan, see *The Green Mountain Parkway, Final Report by the Landscape Architects of the National Park Service,* bound manuscript, filed by name, UVM.

90. John Nolen, "Proposed Green Mountain Parkway Project in Vermont," August 1, 1934, p. 1, JPT, Doc. T3, Folder 3, VHS.

91. Richard Munson Judd, *The New Deal in Vermont: Its Impact and Aftermath* (New York: Garland Publishing, Inc., 1979), 82–87; Hannah Silverstein, "No Parking: Vermont Rejects the Green Mountain Parkway," *Vermont History* 63 (Summer 1995): 133–57; Goldman, "James Taylor's Progressive Vision."

92. Sutter, *Driven Wild,* 15–16.

93. "Special Meeting of the Trustees," *LTN* 7 (July 1933): 2.

94. *The Green Mountain Parkway, Final Report by the Landscape Architects of the National Park Service,* 38–39.

95. "To All Members of the Green Mountain Club," GMC, Doc. 225, Folder 27, VHS; "The Parkway," *LTN* 8 (September 1934): 2.

96. Clarence P. Cowles to Stanley C. Wilson, August 26, 1933, GMC, Doc.

225, Folder 18, VHS; Goldman, "'A Desirable Class of People,'" 145; Silverstein, "No Parking," 145–46.

97. "Thrills from on High," *BFP*, September 17, 1934, p. 6. Also see Nolen, "Proposed Green Mountain Parkway Project in Vermont," 2.

98. James P. Taylor to Harlan P. Kelsey, January 14, 1936, JPT, Doc. T3, Folder 10, VHS. Also see Goldman, "James Taylor's Progressive Vision."

99. *Rutland Daily Herald*, reprinted in "The Wilgus Dream," *LTN* 7 (September 1933): 3.

100. Nolen, "Proposed Green Mountain Parkway Project in Vermont," 3–5; Laurie Davidson Cox, "The Green Mountain Parkway," *Landscape Architecture* 25 (April 1935): 117–26.

101. "A Better Parkway Plan," *Rutland Daily Herald*, August 9, 1933, p. 8.

102. Roderic Olzendam to David W. Howe, August 16, 1934, JPT, Doc. T3, Folder 2, VHS. Also see Goldman, "'A Desirable Class of People,'" 143; "Do We Want a Mountain Parkway?" *LTN* 7 (September 1933): 3–4.

103. Greensboro Association, Inc., to Vermont Chamber of Commerce, August 27, 1934, JPT, Doc. T3, Folder 2, VHS; "The Parkway," *LTN*.

104. Harold J. Adams to James P. Taylor, September 26, 1933, JPT, Doc. T2, Folder 48, VHS.

105. James P. Taylor to William J. Wilgus, August 14, 1933, JPT, Doc. T2, Folder 47, VHS.

106. [David Howe], "Parkway Selfishness," GMC, Doc. 225, Folder 27, VHS.

107. Roderic Olzendam to David W. Howe.

108. Silverstein, "No Parking," 150–52; "Parkway Is Rejected by Voters of Vermont; No, 43,176; Yes, 31,101," *BFP*, March 4, 1936, pp. 1–2.

109. William Hazlett Upson, "The Green Mountain Parkway," in Vermont State Chamber of Commerce, "Informational Bulletins on State Problems, Bulletin No. 4, July 31, 1934," reference file "Green Mountain Parkway (1930s)," UVM.

110. "A Better Parkway Plan."

Chapter 4. The Four-Season State: Creating a New Seasonal Cycle (pp. 132–62)

1. Muriel Follett, *New England Year: A Journal of Vermont Farm Life* (Brattleboro, Vt.: Stephen Daye Press, 1939). On Vermont farm life and seasonal variations in work, see Scott E. Hastings Jr. and Geraldine S. Ames, *The Vermont Farm Year in 1890* (Woodstock, Vt.: Billings Farm and Museum, 1983); Scott E. Hastings Jr. and Elsie R. Hastings, *Up in the Morning Early: Vermont Farm Families in the Thirties* (Hanover, N.H.: University Press of New England, 1992); and "A Vermont Farm," *Fortune* 19 (February 1939): 48–53, 97–98, 100, 102, 104.

2. Charles F. Speare, *We Found a Farm* (Brattleboro, Vt.: Stephen Daye Press, 1936).

3. For statistics, see Harold A. Meeks, *Time and Change in Vermont: A Human*

Geography (Chester, Conn.: Globe Pequot Press, 1986), 159. Also see S. Axel Anderson and Florence M. Woodard, "Agricultural Vermont," *Economic Geography* 8 (January 1932): 12–42; Robert O. Sinclair and Malcom I. Bevins, *Small Farms in Vermont* (Burlington, Vt.: UVM/VAES, 1961); W. A. Anderson, *Population Change in Vermont, 1900 to 1950* (Burlington, Vt.: University of Vermont and State Agricultural College, Agricultural Experiment Station, 1955; hereafter UVMSAC/VAES will be used to refer to University of Vermont and State Agricultural College, Vermont Agricultural Experiment Station); Robert O. Sinclair, *Vermont's Dairy Industry* (Burlington, Vt.: UVMSAC/VAES, 1956).

4. Lewis Hill, *Fetched Up Yankee: A New England Boyhood Remembered* (Chester, Conn.: Globe Pequot Press, 1990), 126.

5. Wilson Follett, "Town Against Country," *Atlantic Monthly* 164 (September 1939), 367.

6. "New England Autumn," *Saturday Evening Post* 219 (October 12, 1946): 14.

7. Michael Kammen, *A Time to Every Purpose: The Four Seasons in American Culture* (Chapel Hill: University of North Carolina Press, 2004).

8. *Vermont Around the Calendar*, undated film produced and circulated by Vermont Publicity Service, Department of Conservation and Development, Montpelier, Vt., VSA.

9. Although the magazine was intended for native and non-native audiences alike, it more often than not felt as if it was directed primarily at readers from outside the state. "The aims of *Vermont Life*," its editors later described, "are simple: to so interest people with Vermont's charms that they may wish to visit and see more—perhaps eventually to live here." The magazine made great strides toward that goal; by the late 1950s 85 percent of its 41,000 subscribers were non-residents. *Vermont Life* 13 (Spring 1959): back cover. Also see *Seasons of Change: Fifty Years with Vermont Life, 1946–1996* (Montpelier, Vt.: Vermont Life, 1996).

10. Raymond Williams, *The Country and the City* (New York: Oxford University Press, 1973).

11. Scott E. Hastings Jr., "Introduction," in Hastings and Hastings, *Up in the Morning Early*, 19–25.

12. Helen and Scott Nearing, *The Maple Sugar Book: Together with Remarks on Pioneering as a Way of Living in the Twentieth Century* (New York: Schocken Books, 1970, 1950); Noel Perrin, *Amateur Sugar Maker* (Hanover, N.H.: University Press of New England, 1972); Noel Perrin, *First Person Rural: Essays of a Sometime Farmer* (New York: Penguin Books, 1980, 1978).

13. *A Vermont Cook Book by Vermont Cooks* (White River Junction, Vt.: Green Mountain Studios, Inc., 1947); Mary Pearl, *Vermont Maple Recipes* (Burlington, Vt.: Lane Press, 1952); Beatrice Vaughan, *Real, Old-Time Yankee Maple Cooking* (Brattleboro, Vt.: Stephen Greene Press, 1969).

14. Follett, *New England Year*, 51.

15. Muriel Follett, *A Drop in the Bucket: The Story of Maple Sugar Time on a Vermont Farm* (Brattleboro, Vt.: Stephen Daye Press, 1941), 7.

16. Rupert Blair, interview by Greg Sharrow, tape recording, Warren, Vermont, September 8, 1998, p. 15 of typed transcript, accession #TC92.2051, collection #VFC1992.0004, Vermont Folklife Center, Middlebury, Vermont (hereafter VFC).

17. Everett Palmer, quoted in Ron Strickland, ed., *Vermonter: Oral Histories from Downcountry to the Northeast Kingdom* (San Francisco: Chronicle Books, 1986), 120.

18. James P. Taylor, untitled talk in *Proceedings of the Twenty-Third Annual Meeting of the Vermont Maple Sugar Makers' Association*, comp. H. B. Chapin (St. Albans, Vt.: St. Albans Messenger Co. Print, 1916), 47–49.

19. George M. England and Enoch H. Tompkins, *Marketing Vermont's Maple Syrup* (Burlington, Vt.: UVMSAC/VAES, 1956), 16.

20. Also see C. Clare Hinrichs, "Consuming Images: Making and Marketing Vermont as a Distinctive Rural Place," in *Creating the Countryside: The Politics of Rural and Environmental Discourse*, ed. E. Melanie DuPuis and Peter Vandergeest (Philadelphia: Temple University Press, 1996), 268–72.

21. Perrin, *First Person Rural*, 104.

22. Advertisement, *New York Times*, April 5, 1931, p. 83. Also see Stephen Greene, "Sugar Weather in the Green Mountains," *National Geographic* 105 (April 1954): 482.

23. On links between food, purity, and place identity, see Mona Domosh, "Pickles and Purity: Discourses of Food, Empire and Work in Turn-of-the-Century USA," *Social and Cultural Geography* 4 (March 2003): 7–27.

24. W. H. Crockett, "Advertising Our Product," in *Proceedings of the Twenty-Eighth Annual Meeting of the Vermont Maple Sugar Makers' Association* (St. Albans, Vt.: St. Albans Messenger Co. Print, 1921), 23–28. For similar themes, also see England and Tompkins, *Marketing Vermont's Maple Syrup*, and Fred C. Webster and Christopher G. Barbieri, *Maple Marketing in Department Stores* (Burlington, Vt.: UVMSAC/VAES, 1964), 3–6, 23–26.

25. By 1954 Vermonters produced 721,000 gallons of syrup and 54,000 pounds of sugar, over half again as much syrup as they produced in 1909, although 160 times less than the total sugar production for that year, as even Vermonters turned toward cane sugar as a viable alternative. George M. England and Enoch H. Tompkins, *Marketing Vermont's Maple Syrup* (Burlington, Vt.: UVMSAC/VAES, 1956), 3–6. By the early 1960s the production of maple sugar for the market was negligible by comparison to syrup. For statistics, see the reference file "Maple Sugar Industry—History," UVM. Also see John P. Davis, "Vermont's Maple Sugar Industry," *The Vermont Review* 1 (March–April 1927), 139–41.

26. "Be a Part of a Vermont Farm, Harlow's Sugar House," reference file "Maple Sugar Industry—Marketing and Promotion," UVM.

27. For a more detailed discussion of this point, see Blake Harrison, "Shopping to Save: Green Consumerism and the Struggle for Northern Maine," *Cultural Geographies*, forthcoming.

28. Frederick M. Laing, Mary T. G. Lighthall, and James W. Marvin, *The Use of Plastic Tubing in Gathering Maple Sap* (Burlington, Vt.: UVMSAC/VAES, 1960).

29. Don Mitchell, *The Lie of the Land: Migrant Workers and the California Landscape* (Minneapolis: University of Minnesota Press, 1996).

30. For examples, see Vermont Development Commission, "Unspoiled Vermont: Beauty Corner of New England," no date, uncatalogued pamphlet "Vermont — Description — Unspoiled Vermont," VHS, and "The Story of Maple Time in Vermont" (Montpelier, Vt.: Vermont Sugar Makers' Association, Vermont Department of Agriculture, Vermont Development Department, [1971]), reference file "Maple Sugar Industry — Marketing and Promotion," UVM.

31. Frances Stockwell Lovell, "Land of the Quiet Hills," *Vermont Life* 8 (Autumn 1953): 6.

32. Philip DuVal, "We Spent Our Vacation on the Farm," *American Magazine* 152 (July 1951): 11.

33. Hap Gaylord, Hayden Gaylord, Eloise Gaylord, interview by Greg Sharrow, tape recording, Waitsfield, Vermont, June 10, 1991, p. 32 of typed transcript, accession #TC91.2008, Mad River Collection, collection #VFC1992.0004, VFC.

34. Arthur W. Dean, "The Highways of New England," in *New England's Prospect: 1933*, ed. John K. Wright (New York: American Geographical Society, 1933), 367; Leroy C. Flint, "Snow Rollers," *Rural Vermonter* 4 (January–February 1966): 22–23; Charles Edward Crane, *Winter in Vermont* (New York: Alfred A. Knopf, 1941), 127–30.

35. For a discussion of social meetings and committee activities, see Barbara Savage, "We Are Not Amused," *North American Review* 236 (July 1933): 41–47. Also see Follett, *New England Year*; Hill, *Fetched Up Yankee*, 78–90.

36. Blair, interview, September 8, 1998, pp. 27–29, VFC; Ruth Brooks, interview by Jane Beck, tape recording, Waitsfield Vermont, February 11, 1992, pp. 9–10 of typed transcript, accession #TC92.0013, collection #VFC1992.0004, VFC.

37. Crane, *Winter in Vermont*, 192.

38. On civic clubs and early winter sports in Vermont, see Dale S. Atwood, "Winter Sports at St. Johnsbury," *The Vermonter* 27:9 (1922): 217–22; Crane, *Winter in Vermont*, 192–98; *Vermont in the Victorian Age: Continuity and Change in the Green Mountain State, 1850–1900* (Bennington, Vt.: Vermont Heritage Press; Shelburne, Vt.: Shelburne Museum, 1985), 99–101; Betsy Beattie, "The Queen City Celebrates Winter: The Burlington Coasting Club and the Burlington Carnival of Winter Sports, 1886–1887," *Vermont History* 52 (Winter 1984): 5–16.

39. For more on winter carnivals, see Beattie, "The Queen City Celebrates Winter," and the Class of 1922, Stowe High School, "The Winter Carnival at Stowe," *The Vermonter* 26:2 (1921): 49–51. For the earliest statewide listings of early winter sports facilities, see Recreation Study Committee, Vermont State Planning Board, *Recreation Study Tentative Report* (no publisher, 1939), 51 (a copy of this report can be found in JPT, Doc. T9, Folder 14, VHS).

40. Frank Greene, comp., *Vermont, the Green Mountain State: Past, Present, Prospective* (n.p.: Vermont Commission to the Jamestown Tercentennial Exposition, 1907), 76.

41. E. John B. Allen, *From Skisport to Skiing: One Hundred Years of an American Sport, 1840–1940* (Amherst: University of Massachusetts Press, 1993), 29–62; Frederick F. Van de Water, "Brattleboro: Birth of the Winter Idea," *Vermont Life* 3 (Winter 1948–1949): 2–5; "America Cradles Scandinavia's Sport," *Literary Digest* 117 (February 10, 1934): 22; Melville Robertson, "Skiing—The Fastest of Sports," *Illustrated World* 24 (January 1916): 620–23.

42. Arthur F. Stone, *The Vermont of Today*, vol. 2 (New York: Lewis Historical Publishing Co., Inc., 1929), 747.

43. E. John B. Allen, "The Making of a Skier: Fred H. Harris, 1904–1911," *Vermont History* 53 (Winter 1985): 5–16; Jack Feth, "Hyah, Chubber?" *Yankee* 4 (February 1938): 4–8.

44. Fred M. Harris, "Skiing and Winter Sports in Vermont," *The Vermonter* 17 (November 1912): 677–81.

45. Walter H. Crockett to Dorothy Canfield Fisher, March 12, 1931, "Walter H. Crockett Papers/Writings, 1900–1931," Carton 1, Folder 25, UVM.

46. Charles Edward Crane, *Let Me Show You Vermont* (New York: Alfred A. Knopf, 1937), 44. Also see "Address of Governor Stanley C. Wilson of Vermont on Winter Attractions in Vermont," broadcast from Station WBZ [Boston], December 26, 1931, and issued by the Vermont Bureau of Publicity, and Walter H. Crockett, "Winter Touring in Vermont," *Vermont Highways* (December 1931): 7–9.

47. Crane, *Winter in Vermont*, 23.

48. Walter H. Crockett to Dorothy Canfield Fisher, March 12, 1931.

49. Crane, *Let Me Show You Vermont*, 52.

50. Crockett, "Winter Touring in Vermont," 9.

51. Chas. R. Cummings, "An Inspiring Revelation," *The Vermonter* 37 (February 1932): 31.

52. Annie Gilbert Coleman, *Ski Style: Sport and Culture in the Rockies* (Lawrence: University Press of Kansas, 2004), 59–72.

53. Cummings, "An Inspiring Revelation," 31.

54. John McDill, "Woodstock: Cradle of Winter Sports," *Vermont Life* 2 (Winter 1947–1948): 19; A. W. Coleman, "White Gold," *Vermont Life* 1 (Winter 1946–1947): 2–9. On the history of snow and winter weather in American culture, see Bernard Mergen, *Snow in America* (Washington, D.C.: Smithsonian Institution Press, 1997), and William B. Meyer, *Americans and Their Weather* (New York: Oxford University Press, 2000).

55. Roland Palmedo, "The Rediscovery of Snow," *Country Life in America* 67 (January 1935): 47–49, 88; Coleman, "White Gold," 8.

56. *The Green Mountain Parkway, Final Report by the Landscape Architects of the National Park Service*, 12, "Bound Manuscript," filed by name, UVM. Also see Margaret Godding, "Autumn through a Windshield," *Vermont Life* 6 (Autumn 1951): 32–39.

57. Examples include Thomas D. Murphy, *New England Highways and Byways from a Motor Car: On Sunrise Highways* (Boston: L. C. Page and Company,

1924), and Walter and Margaret Hard, *This Is Vermont* (Brattleboro, Vt.: Stephen Daye Press, 1936).

58. OPS, OS, VSA; reference file "Foliage," UVM.

59. Also see Kent C. Ryden, *Landscape with Figures: Nature and Culture in New England* (Iowa City: University of Iowa Press, 2001), 260–64.

60. C. F. Ranney, "Newport and Lake Memphremagog," *The Vermonter* 3 (September 1897): 23.

61. For accessible studies of the forest history of Vermont, see Christopher Mc-Grory Klyza and Stephen C. Trombulak, *The Story of Vermont: A Natural and Cultural History* (Hanover, N.H.: University Press of New England, 1999), and Tom Wessels, *Reading the Forested Landscape: A Natural History of New England* (Woodstock, Vt.: Countryman Press, 1997).

62. W. C. Prime, *Along New England Roads* (New York: Harper and Brothers, Franklin Square, 1892), 11.

63. Also see Godding, "Autumn through a Windshield," 36.

64. On the history of apples in Vermont and New England see Brian Donahue, *Reclaiming the Commons: Community Farms and Forests in a New England Town* (New Haven, Conn.: Yale University Press, 1999), 180–97; M. B. Cummings, "Apple Culture in Vermont," *The Vermont Review* 1 (March–April 1927): 136–39. Also see Michael Pollan, *The Botany of Desire: A Plant's-Eye View of the World* (New York: Random House, 2001), 3–58.

65. See, for example, Ray Allen, interview by Greg Sharrow, tape recording, South Hero, Vermont, February 8, 1990, accession #TC91.2004, VFC; Lewis Wood, interview by Jane Beck, tape recording, Middlebury, Vermont, October 3, 1978, accession #TC78.0005, Vermont Apple Project, VFC.

66. Gertrude L. Sylvester, "Harvest Dance," *Vermont Life* 5 (Autumn 1950): 34–35; Fred Copeland, "Apple Time in Yankee Land," *Vermont Life* 6 (Autumn 1951): 26–29.

67. Edward J. Tadejewski, *Roadside Marketing in Vermont* (Burlington, Vt.: UVMSAC/VAES, 1949). This survey also noted that 69 percent of sales at Vermont's roadside markets came from fruits and vegetables, 4 percent from eggs, 16 percent from maple products and honey, and 11 percent from crafts or other manufactured products. The most lucrative months for the roadside marketer were August through October, when produce was most plentiful.

68. *Windy Wood Farm Applegram: Harvest and Holiday, 1956–57* (Barre, Vt., 1956), uncatalogued pamphlet, VHS. Also see *Scott Farm Apple Book, Season 1926–1927* (Brattleboro, Vt., [1926]), uncatalogued pamphlet, VHS.

69. Mrs. Sidney Morse, interview by Jane Beck, tape recording, Calais, Vermont, October 17, 1978, accession #TC78.0008, Vermont Apple Project, VFC; Mildred Perkins, interview by Jane Beck, tape recording, Cornwall, Vermont, October 31, 1978, accession #TC78.0006, Vermont Apple Project, VFC.

70. Vrest Orton, *The American Cider Book: The Story of America's Natural Beverage* (New York: Farrar, Strauss and Giroux, 1973), 7–37.

71. Blair, interview, September 8, 1998, p. 17, VFC. Even with the benefit of tractors, mowers, and hayloaders, haying remained hot, dusty, labor-intensive work that often required the help of neighbors and hired hands. By the standards of many, cider was a good way to cap off a long day of haying. On the history of haying in Vermont, see Allen R. Yale Jr., *While the Sun Shines: Making Hay in Vermont, 1789–1990* (Montpelier, Vt.: Vermont Historical Society, 1991).

72. "Address of Governor Stanley C. Wilson of Vermont on Fall and Winter Touring in Vermont," broadcast from Station WBZ [Boston], December 26, 1931, and issued by the Vermont Bureau of Publicity.

73. *The Green Mountain Parkway, Final Report by the Landscape Architects of the National Park Service*, 11.

Chapter 5. New Paces for Old Places: Creating Vermont's
Mid-Century Ski Landscape (pp. 163–98)

1. Joe McCarthy and the Editors of Time-Life Books, *New England* (New York: Time Incorporated, 1967), 41.

2. Promotions such as these were fueled by sophisticated new surveys and demographic studies. Following the groundbreaking research done by the national Outdoor Recreation Resources Review Commission in 1958, Vermont's state officials and academic community stepped up their own studies of skiing, recreation marketing, and resort planning. See "Ski Vermont, The Beckoning Country," *BFP*, October 24, 1966, p. 24; "Ski Advertising Campaign," *BFP*, September 11, 1965, p. 2; "State Advertising Budget Gives Skiing Top Billing," *BFP*, May 2, 1969, p. 33; Vermont Development Department, *The Tourist and Recreation Industry in Vermont* (Montpelier, Vt.: The Department, 1963); Malcom I. Bevins, *The Outdoor Recreation Industry in Vermont* (Burlington, Vt.: Vermont Resources Research Center, UVM/VAES, 1964). Also see Outdoor Recreation Resources Review Commission, *Outdoor Recreation for America* (Washington, D.C.: Government Printing Office, 1962).

3. On the history of skiing in the United States, see E. John B. Allen, *From Skisport to Skiing: One Hundred Years of an American Sport, 1840–1940* (Amherst: University of Massachusetts Press, 1993), and Hal K. Rothman, *Devil's Bargains: Tourism in the Twentieth Century American West* (Lawrence: University Press of Kansas, 1998).

4. Craig O. Burt, quoted on page 29 of a manuscript entitled "How Skiing Began in Vermont" by Perry Merrill, "Perry Merrill Papers," unlabeled box, Folder "Multiple Use of Forest Lands, esp. Skiing," UVM.

5. For early ski statistics, see Vermont State Planning Board, *Report on Winter Sports Development* ([Montpelier, Vt.]: [The Board], 1938) (a copy of this report is available in VTD/DD, Box 7, Folder "Mount Mansfield Ski Lift, 1939–1940, 1941–1942," VSA); Recreation Study Committee, Vermont State Planning Board, *Recre-*

ation Study Tentative Report (no publisher, 1939), 47–51, mss. copy in JPT, Doc. T9, Folder 14, VHS; Albert S. Carlson, "Ski Geography of New England," *Economic Geography* 18 (July 1942): 307–20.

6. *The Snow Train* (Boston: Buck Printing Company, [1933]); Charles Edward Crane, *Winter in Vermont* (New York: Alfred A. Knopf, 1941), 226–31; Allen, *From Skisport to Skiing*, 104–9.

7. Perry Merrill, *Roosevelt's Forest Army: A History of the Civilian Conservation Corps, 1933–1942* (Montpelier, Vt.: published by the author, 1981), 183; Perry Merrill, *Vermont Skiing: A Brief History of Downhill and Cross Country Skiing* (Montpelier, Vt.: published by the author, 1987), 4–6; A. W. Coleman, "Vermont Ski Runs," *Appalachia* 20 (December 1934): 224–30.

8. Mort Lund, "Route 100," *Ski* 26 (December 1961): 72–75, 136; Hal Burton, "Ski 100," *Vermont Life* 17 (Winter 1962–1963): inside cover, 1, 4–11.

9. Merrill, *Vermont Skiing*, 7; Robert B. Williams to George T. Mazuzan, March 19, 1964, reference file "Skis and Skiing — Vermont — History," UVM; "Roads to Medical Center, 3 Ski Areas Pass Senate," *BFD*, June 19, 1965, p. 1; "Progress in Vermont's Ski Industry," *BFD*, October 14, 1966, p. 18.

10. Vermont State Planning Board, *Report on Winter Sports Development*, 7; William F. Corry, "Winter Highways," *Vermont Life* 1 (Winter 1946–1947): 28; Joseph G. Galway, "Snow Removal One of State's Biggest Jobs," *Eastern Skier* 3 (January 15, 1951): 4.

11. For more on ski technology and landscape design, see Blake Harrison, "The Technological Turn: Skiing and Landscape Change in Vermont, 1930–1970," *Vermont History* 71 (Summer/Fall 2003): 197–228. For similar sets of arguments about Colorado's ski landscapes, see Annie Gilbert Coleman, *Ski Style: Sport and Culture in the Rockies* (Lawrence: University Press of Kansas, 2004), 126–33.

12. On technological determinism, see Merritt Roe Smith and Leo Marx, eds., *Does Technology Drive History?: The Dilemma of Technological Determinism* (Cambridge, Mass.: MIT Press, 1994). Also see David Nye, ed., *Technologies of Landscape: From Reaping to Recycling* (Amherst: University of Massachusetts Press, 1999).

13. Ellen Lesser, *America's First Ski Tow, Commemorative Album* (South Pomfret, Vt.: Teago Publishing Company, 1983), 11–13; Crane, *Winter in Vermont*, 201–3; John McDill, "Woodstock: Cradle of Winter Sports," *Vermont Life* 2 (Winter 1947–1948): 14–19.

14. Research Division of the Vermont Development Commission, *Vermont Ski Facilities* ([Montpelier, Vt.]: [The Commission], 1948), 2 (a copy of this report is available in VTD/DD, Box 6, Folder "Ski: Vermont Ski Facilities, 1948," VSA).

15. Crane, *Winter in Vermont*, 210–12; Ted Cooke, "Ski Tows," *Appalachia* 20 (November 1935): 405–7; Merrill, *Vermont Skiing*, 7.

16. Robert Hagerman, *Mansfield: The Story of Vermont's Loftiest Mountain*, 2nd ed. (Canaan, N.H.: Phoenix Publishing, 1975), 67–71; George T. Mazuzan, "'Skiing Is Not Merely a Schport': The Development of Mount Mansfield as a Winter Recreation Area," *Vermont History* 40 (Winter 1972): 57–59; Edwin Bigelow,

Stowe, Vermont: Ski Capital of the East, 1763–1963 (Stowe, Vt.: Stowe Historical Society, 1964), 152–63; "World's Longest Chair Lift Opens at Mt. Mansfield," *Ski News* 3 (December 13, 1940): 5.

17. E. C. Nichols to Robert C. Lane, VTD/DD, Box 7, Folder "Skiing—General Data," VSA; "Millionth Rider on Chair Lift," *Eastern Ski Bulletin* 1 (March 2, 1953): 2.

18. Although they were popular with skiers during the 1960s, Vermont had only a few gondolas and one tramway. For complete chairlift statistics see "Ski Areas," OPS,OS, VSA. Also see Jan W. Sissener, "The Ski Lift Business in New England (Federal Reserve Bank of Boston Research Report No. 11)," ([Boston]: [The Bank], 1960); "New Slopes, Facilities Opened in State, Old Areas Improved," BFP, November 20, 1961, p. 2.

19. Central Planning Office, State of Vermont, *The Ski Industry in Vermont*, January 4, 1965 (a copy of this report is located in VTD/DD, Box 6, Folder "Ski: Misc. Reports," VSA); "Skiing Called Vermont's Fastest Growing Industry," BFP, April 15, 1961, p. 10.

20. William Gilman, "Ski Key: New England," *Travel* 105 (January 1956): 7. Also see "Lift Bonanza," *Ski* 22 (November 1957): 50–51, and "New Lifts in the East," *Ski* 21 (December 1956): 77–97.

21. The length of any ski lift dictates the vertical and lateral extent of the trails that emanate from it. Rope tows, for instance, were tiring and difficult to hold on to, and they required smooth, even terrain on which to move skiers up the mountain. As a result, trails accessed by rope tows were short and confined in their lateral reach as they spread to either side, winding down from the top of the tow to its base. By comparison, chairlifts and more technologically sophisticated surface lifts allowed for markedly longer ski trails and for a greater lateral spreading of trails to either side of the lift. Even more importantly, such lifts provided skiers with access to higher, more adventurous elevations. In 1948, for instance, ski areas organized around rope tows started at an average elevation of only 968 feet above sea level and traveled to an average of 1,478 feet. Ski areas serviced by more technologically sophisticated lifts started at an average elevation of 1,811 feet and traveled to an average of 3,045 feet. Research Division of the Vermont Development Commission, *Vermont Ski Facilities*, 2–3.

22. On ownership trends in the Vermont ski industry, see Rockwell Stephens, "Anatomy of the Ski Resort Biz," *Vermont Skiing* (Fall 1966): 22–26; Vermont Development Department, *The Tourist and Recreation Industry in Vermont*, 60–61. Also see Hal Burton, "Mr. Starr and His Pampered Mountain," *Saturday Evening Post* 226 (March 6, 1952): 32–33, 56, 60, 62.

23. Committee meetings of the Mount Mansfield Ski Club, April 26, 1945, Albert Gottlieb Papers, Folder "Ski Trails—Etc.," UVM.

24. A. W. Coleman, "The Development of Winter Sports Facilities," reprinted in Vermont State Planning Board, *Report on Winter Sports Development*, 10–12; Arthur C. Comey, "Ski Trail Standards," *Appalachia* 19 (June 1933): 427–31;

Charles N. Proctor, "Ski Trails and Their Design," *Appalachia* 20 (June 1934): 88–103; Robert W. Vincent, "Downhill Ski Trails," *The American School and University* 11 (1939): 244–48.

25. "New Big Bromley Area to Have Mile-Long Lift," *Ski News* 5 (December 11, 1942): 7; Arthur Zich, "A Brewer's Good Old Mountain Brew: Big Bromley," *Sports Illustrated* 17 (November 19, 1962): M7–M12; Pat Harty, "The Story of Bromley Mountain," *Vermont Life* 7 (Winter 1952–1953): 18–25.

26. Fred Pabst to H. H. Chadwick, November 8, 1943, VTD/DD, Box 7, Folder "Winter — General, 1941–42–43, Gas Shortage," VSA.

27. Vermont State Planning Board, *Report on Winter Sports Development*, 4.

28. Robert M. Coates, "My First Pair of Skis," *American Ski Annual and Skiing Journal* 39 (November 1955): 47. On early trail conditions and maintenance also see Federal Writers' Project, *Skiing in the East: The Best Trails and How to Get There* (New York: M. Barrows and Company, 1939), 13, and Crane, *Winter in Vermont*, 208–9.

29. Harry Ambrose, *The Horse's Mouth*, ed. Tom Bourne Jr. (Houston: A. & E. Products Co., Inc, 1997), 33.

30. Advertisements, *Ski Area Management* 1 (Fall 1962) and *Ski Area Management* 7 (Summer 1968).

31. "Walt Sprays 450 Tons of Ice over Bare Mohawk Slope, Breaks Even," *Eastern Skier* 2 (February 1, 1950): 1.

32. Photo with caption, *Boston Herald*, December 21, 1950, p. 34; Florence Zuckerbraun, "Snowmaking Is 'Practical,'" *Eastern Skier* 3 (January 15, 1951): 1–2; Harold F. Blaisdell, "Pipe the Snow!" *American Ski Annual and Skiing Journal* 36 (November 1952): 69–72. Also see Bernard Mergen, *Snow in America* (Washington, D.C.: Smithsonian Institution Press, 1997), 108–14.

33. "New Future for Skiers in the East and Midwest," *Ski* 22 (December 1957): 69–71; "Ski Areas to Make Snow," *Brattleboro Daily Reformer* (hereafter *BDR*), January 15, 1965, p. 1; "Snowmaking — It Pays on a Big Hill Too," *Skiing Area News* 1 (June 1966): 26–28; *Ski Vermont: America's First* (Montpelier, Vt.: Vermont Development Department, [1969]), 19.

34. *Ski Vermont*, 2.

35. Also see Louise Appleton, "Distillations of Something Larger: The Local Scale and American National Identity," *Cultural Geographies* 9 (October 2002): 421–47.

36. William C. Lipke, "From Pastoralism to Progressivism: Myth and Reality in Twentieth-Century Vermont," in *Celebrating Vermont: Myths and Realities*, ed. Nancy Price Graff (Middlebury, Vt.: The Christian A. Johnson Memorial Gallery, Middlebury College; Hanover, N.H.: University Press of New England, 1991), 72; Hank Powell, "White Christmas: The History," in *White Christmas: Movie Vocal Selections* (Milwaukee: Hal Leonard Corporation, 2000); Vrest Orton, "Norman Rockwell's Vermont," *Vermont Life* 1 (Summer 1947): 18–25.

37. Ralph N. Hill, Murray Hoyt, Walter R. Hard Jr., *Vermont: A Special World* (Montpelier, Vt.: Vermont Life Magazine, 1969).

38. Crane, *Winter in Vermont*, 214.

39. Richard White, *The Organic Machine: The Remaking of the Columbia River* (New York: Hill and Wang, 1995).

40. Coleman makes a similar point in chapter 4 of *Ski Style*.

41. For examples, see "Snow Valley Lift Speeded Up to Suit Steak-Eating Skiers," *Ski News* 8 (November 1945): 10; "New Sights on Slopes," *Life* 42 (February 18, 1957): 97–101.

42. "Million-Dollar Area Opened in the East," *National Skiing* 7 (December 15, 1954), 1, 15; "Sun Valley in Vermont," *Yankee* 19 (February 1955): 38–39; "The Scamp of the Ski Slope," *Life* 54 (February 8, 1963): 93–94; "Wizard of Palms, Pools, Mt. Snow Jobs," *Vermont Skiing* 4 (Fall 1967): 39–41; "The Jolly Green Giant Gets Even Bigger," *Ski Week* 1 (March 1, 1967): 6–7, 11.

43. "'Tsoo-gaar-boosch!' Skiing from Your Mountain Doorstep," *Newsweek* 64 (December 28, 1964): 48–51. Also see "Bolton Valley," *Vermont Life* 22 (Winter 1967): 22.

44. Coleman, *Ski Style*, 147–81. Also see Annie Gilbert Coleman, "The Unbearable Whiteness of Skiing," *Pacific Historical Review* 65 (November 1996): 590–92.

45. For additional commentaries on race and skiing, and on race in Vermont, see "Black Boom on the Ski Slopes," *Ebony* 29 (January 1974): 72–74; Laura B Randolph, "How Blacks Fare in the Whitest State," *Ebony* 43 (December 1987): 45–48, 50.

46. For the Hotel Tyrol and Alpenhorn Lodge, see *The National Survey Eastern Ski Atlas*, comp. and ed. The National Survey (Chester, Vt.: The National Survey, 1969), 14, 15; and *The National Survey Eastern Ski Atlas*, comp. and ed. by the publishers of the Eastern Ski Map (Chester, Vt.: The National Survey, 1964), 27. These are only a few of the hundreds of such advertisements one can encounter in the ski literature from the time.

47. Coleman, "The Unbearable Whiteness of Skiing," 590.

48. Gary Talese, "Social Climbing on the Slopes," *Saturday Evening Post* 236 (February 9, 1963): 30, 32–33.

49. "The Fashion for Skiing — Sugarbush," *Newsweek* 55 (March 14, 1960): 98–99; Ezra Bowen, "Some Spice for the Sugar," *Sports Illustrated* 19 (November 18, 1963): 37–38, 40, 43.

50. For extended discussions of gender, sex, and the ski landscape, see Coleman, *Ski Style*, 41–72, 184–88.

51. Annette Pritchard and Nigel J. Morgan, "Constructing Tourism Landscapes — Gender, Sexuality, and Space," *Tourism Geographies* 2 (May 2000): 125.

52. Talese, "Social Climbing on the Slopes," 33.

53. "Life Goes Skiing in Vermont," *Life* 22 (January 27, 1947): 107–10.

54. *Vermont Ski Story*, an undated film written by Geoffrey Orton and produced by the State of Vermont Development Department, Montpelier, VSA.

55. For more on ski instructors, their exoticism, and their appeal, see Coleman, *Ski Style*, 43–50; Allen, *From Skisport to Skiing*, 199–223.

56. Mike Erickson, "Sex and the Single Skier," *Vermont Skiing* 4 (Winter 1967–1968): 565.

57. "Airborne or Sitzbound — Skiing's for Fun," *Vermont Skiing* (Winter 1964–1965): 9.

58. Daniel Rochford, "New England Ski Trails," *National Geographic* 70 (November 1936): 646.

59. Coleman, *Ski Style*, 67.

60. Bradford Smith, "Border Flavor," *Vermont Life* 17 (Summer 1963): 27; Curtis W. Casewit, "Top U.S. Ski Spots," *Travel* 124 (December 1965): 29; Huston Horn, "New Sugar in an Old State," *Sports Illustrated* 19 (November 18, 1963): 30–34.

61. Also see Harrison, "The Technological Turn."

62. Burton, "Ski 100," 1. Also see Martin Luray, "The Vermont Experience," *Skiing* 24 (November 1971): 108–11, 155–56, 158–59.

63. John H. Ingersoll, "Lookout at Sugarbush," *House Beautiful* 109 (December 1967): 151. Also see Marshall Souers Jr., "Modern Planning, Mellow Charm," *Better Homes and Gardens* 31 (May 1953): 62–63; "Sugarbush: Where Myths Are Made," in the promotional booklet *Sugarbush: A Slice of the Good Life*, reference file "Warren, Vermont, Folder 2," UVM.

64. Casewit, "Top U.S. Ski Spots," 46.

65. John Fraser Hart, *The Changing Scale of American Agriculture* (Charlottesville: University of Virginia Press, 2003), 62–111.

66. On mid-century trends in Vermont agriculture see James G. Sykes, *Trends in Vermont Agriculture* (Burlington, Vt.: Vermont Resources Research Center, UVM/VAES, 1964), 7. Also see Harold A. Meeks, *Time and Change in Vermont: A Human Geography* (Chester, Conn.: Globe Pequot Press, 1986); M. I. Bevins and R. H. Tremblay, *Dairy Farming in Vermont* (Burlington, Vt.: UVM/VAES, 1967); Robert O. Sinclair, *Vermont's Dairy Industry* (Burlington, Vt.: UVMSAC/AES, 1956); and Robert O. Sinclair and Malcom I. Bevins, *Small Farms in Vermont* (Burlington, Vt.: UVM/VAES, 1961).

67. William Gilman, "Farm-Built Ski Tow," *Popular Mechanics* 100 (October 1953): 104–5; "Fight Winter? Why Not Join It?" *Farm Journal* 85 (February 1961): 66–67.

68. For a comprehensive look at mid-century ski employment, see Vermont Development Department, *The Tourist and Recreation Industry in Vermont*, 67–76.

69. Research Division of the Vermont Development Commission, *Vermont Ski Facilities*, 6.

70. Vermont Development Department, *The Tourist and Recreation Industry in Vermont*, 72–73.

71. Hap Gaylord, Hayden Gaylord, Eloise Gaylord, interview by Greg Sharrow, tape recording, Waitsfield, Vermont, June 10, 1991, pp. 11–19 of typed manuscript, accession #TC91.2008, Mad River Collection, collection #VFC1992.0004, VFC. Also see Joe Sherman, *Fast Lane on a Dirt Road: Vermont Transformed, 1946–1990* (Woodstock, Vt.: Countryman Press, 1991), 67–69.

72. Joe Heaney, "Skiing Business Puts New Life into Three Vermont Towns," *BFP*, July 25, 1963, p. 16.

73. Hugh Moffett, "The Ruckus in Irasburg," *Life* 66 (April 4, 1969): 62–64, 67–72, 74; J. Kevin Graffagnino, Samuel B. Hand, and Gene Sessions, eds., *Vermont Voices, 1609 through the 1990s: A Documentary History of the Green Mountain State* (Montpelier, Vt.: Vermont Historical Society, 1999), 353–55.

74. For examples, see "Unhappy with Ski Industry," *BFP*, January 5, 1967, p. 16; . . . *So Goes Vermont*, prod. John Karol, 23 minutes, Environmental Planning Information Center, a social project of the Vermont Natural Resources Council, 1971, 1992, videocassette.

75. Vermont Development Department, *The Tourist and Recreation Industry in Vermont*, 70.

76. David Sellers, "Mad River Madness?" *Vermont Environmental Report* 1 (August 1972): 2.

77. "The Fashion for Skiing — Sugarbush," *Newsweek* 55 (March 14, 1960): 99. Also see Sherman, *Fast Lane on a Dirt Road*, 65–66.

78. "The Vermont Way," *Ski News* 10 (January 1, 1948): 8.

79. Bevins, *The Outdoor Recreation Industry in Vermont*, 36, 42–45.

80. Gaylord, Gaylord, Gaylord, interview, June 10, 1991, p. 38 of typed manuscript, VFC.

81. "The Vermont Way," 8.

82. Typed manuscript, no author, page 1, VTD/DD, Box 7, Folder "Mount Mansfield Ski Lift, 1939–1940, 1941–1942," VSA.

83. Crane, *Winter in Vermont*, 213–14.

84. P. Tenney Kudgett, "Them Skiers," *Rural Vermonter* (January–February 1966): 7.

85. Henry Wheeler, quoted from *Tales from the Mountain: Mount Snow's First 40 Years*, prod. and dir. Allen and Sally Seymour, 30 min., Vermont Studios, Inc., 1994, videocassette.

86. Pat Orvis, "Atom Bomb to Be Used in Vermont? Schoenknecht Has Asked Permission," *BFP*, March 15, 1963, p. 4; Don Guy, "Atom Blast May Make New Trail," *Keene Evening Sentinel* (Keene, N.H.), January 11, 1963, p. 9.

87. A. R. Costello to West Dover Selectmen, February 15, 1962, Mount Snow scrapbook, Dover Town Office, West Dover, Vermont.

88. Mudgett, "Them Skiers," 6, 7.

89. Jack and Jessamine Larrow, interview by Jane Beck, tape recording, Waitsfield, Vermont, March 10, 1994, pp. 25–27 of typed manuscript, accession #TC94.0017, Mad River Valley Collection, collection #VFC1992.0004, VFC.

90. George Humphreys, interview by the author, tape recording, West Dover, Vt., June 18, 1999.

91. Katherine Toll and Kenneth A. Henderson, "Ski Tows, Right or Wrong," *Appalachia* 23 (December 1941): 480–84; Roland Palmedo, "Too Many Lifts?" *Ski* 22 (December 1957): 40–45; M. M. Martin, "What's Happened to Skiing?" *Yankee* 12 (March 1948): 35, 51.

92. A. W. Coleman, "Skiing in Transition," *American Ski Annual and Skiing*

Journal 38 (February 1954): 38. Also see Albert E. Sigal, "This Business of Skiing," *American Ski Annual and Skiing Journal* 37 (June 1953): 38.

93. John Henry Auren, "Sugarbush Reconsidered," *Ski* 26 (January 1962): 62–65, 78.

94. Heinz Herrmann, "The Case against the Ski Tow," *Colorado Quarterly* 2 (Spring 1954): 448–49.

95. "Cross-Country Craze," *Life* 72 (January 14, 1972): 65; Woody Woodhall, "Ski the Quiet Way in Vermont," *BFP*, January 7, 1972, p. 8A. Also see "Vermont Step by Step," *Skiing* 22 (February 1970): 62–69.

96. Miriam Chapin, "Vermont: Where Are All Those Yankees?" *Harper's* 215 (December 1957): 50, 54. Also see Hal Burton, "Vermont: Last Stand of the Yankees," *Saturday Evening Post* 234 (July 22, 1961): 22–23, 43–45.

Chapter 6. Balancing the Rural: Planning, Legislation, and the Search for Control (pp. 199–235)

1. Hal K. Rothman, *Devil's Bargains: Tourism in the Twentieth-Century American West* (Lawrence: University Press of Kansas, 1998).

2. Central Planning Office [hereafter CPO], *The Preservation of Roadside Scenery through the Police Power* (Montpelier, Vt.: The Office, 1966), 1.

3. Richard W. Judd and Christopher S. Beach, *Natural States: The Environmental Imagination in Maine, Oregon, and the Nation* (Washington, D.C.: Resources for the Future, 2003).

4. Sidney Plotkin, *Keep Out: The Struggle for Land Use Control* (Berkeley: University of California Press, 1987), 9.

5. For statistics and quote, see Vermont Development Department, *The Tourist and Recreation Industry in Vermont* (Montpelier, Vt.: The Department, 1963), 5.

6. Harold A. Meeks, *Time and Change in Vermont: A Human Geography* (Chester, Conn.: Globe Pequot Press, 1986), 207, 298–318; Benjamin L. Huffman, *Getting Around Vermont: A Study of Twenty Years of Highway Building in Vermont with Respect to Economics, Automotive Travel, Community Patterns, and the Future* (Burlington, Vt.: University of Vermont, 1974).

7. Leonard U. Wilson, "Land Use, Planning, and Environmental Protection," in *Vermont State Government since 1965*, ed. Michael Sherman (Burlington, Vt.: The Center for Research on Vermont, The Snelling Center for Government, University of Vermont, 1995), 454–56; *History of Planning in Vermont* (Montpelier, Vt.: Vermont Department of Housing and Community Affairs, 1999), 3–6. Also see Joe Sherman, *Fast Lane on a Dirt Road: Vermont Transformed, 1945–1990* (Woodstock, Vt.: Countryman Press, Inc., 1991), 49–63.

8. CPO, *State Planning in Vermont* (Montpelier, Vt.: The Office, 1964), 1–2.

9. CPO, *Vermont Scenery Preservation* (Montpelier, Vt.: The Office, 1966), 3.

10. CPO, *The Preservation of Roadside Scenery through the Police Power* (Montpelier, Vt.: The Office, 1966), 9.

11. Philip H. Hoff Papers (hereafter PHH), Box 23, Folder 4–5–5, UVM; PHH, Box 56, Folder 9–3–73, UVM.

12. Peter Blake, *God's Own Junkyard: The Planned Deterioration of America's Landscape* (New York: Holt, Rinehart, and Winston, 1964).

13. Lewis L. Gould, *Lady Bird Johnson and the Environment* (Lawrence: University Press of Kansas, 1988), 136–68.

14. CPO, *The Preservation of Roadside Scenery through the Police Power*, 1.

15. "No. 70," *Acts and Resolves Passed by the General Assembly of the State of Vermont at the Forty-Ninth Biennial Session, 1967* ([Montpelier, Vt.]: Published by Authority of the General Assembly, 1967), 656–57. This special session also yielded a broader act "to preserve and to enhance scenic values in the state of Vermont and to create a scenery preservation council." (pp. 643–56). Although not specifically road-based, this act drew similar links between tourism, scenery, and economic development.

16. Catherine Gudis, *Buyways: Billboards, Automobiles, and the American Landscape* (New York: Routledge, 2004), 221–26.

17. "State Dooms Scores of Billboards," *BFP*, August 27, 1965, p. 3; "2,000 Vermonters to Move Signs, Billboards," *BFP*, January 13, 1966, p. 3; "Senate Passes Vt. Billboard Control Bill," *BFP*, March 9, 1966, p. 1; "State Tourist Groups Ask Exemption from Billboard Rules," *BFP*, January 26, 1967, p. 1; "Two Ski Area Representatives Point Up Value of Road Signs," *BFP*, February 2, 1967, p. 16.

18. *Report of the Committee to Study Outdoor Advertising* (n.p.: Legislative Council of the State of Vermont, 1967); "Riehle Would Ban Billboards in State," *BFP*, July 25, 1967, p. 1.

19. "Natural Resources Committee Meeting of January 25, 1967," reference file "Billboards/Outdoor Advertising," VSA. For more opinions on billboards, see "Readers Express Opinions on the Billboard Issue," *BFP*, August 2, 1967, p. 16; "Views on Billboard Issue," *BFP*, August 7, 1967, p. 12; PHH, Box 66, Folders 11–1-16 and 11–1-18, UVM.

20. "No. 333," *Acts and Resolves Passed by the General Assembly of the State of Vermont at the Forty-Ninth Biennial Session, 1967*, 346–56.

21. CPO, *State Planning in Vermont* (Montpelier, Vt.: The Office, 1964), 1.

22. "Lodging for Skiers Expands Greatly," *BFP*, February 6, 1963, 36; Vermont Development Department, *The Tourist and Recreation Industry in Vermont*, 5, 7.

23. Blake Harrison, "Tracks across Vermont: *Vermont Life* and the Landscape of Downhill Skiing, 1946–1970," *Journal of Sport History* 28 (Summer 2001): 253–70.

24. George A. Donovan, *Vermont Vacation Home Survey, 1968* (Montpelier, Vermont Development Department, 1968), 3. This report defined vacation homes as privately owned, fixed structures not used as permanent dwellings. Although Vermonters owned many thousands of Vermont's vacation homes (many of which were hunting camps), visitors from New York, Connecticut, and Massachusetts

owned the majority of them. Over 50 percent of Vermont's vacation homes in 1968 were concentrated in only 20 percent of the state's towns, typically those located near water or skiing. Also see Vermont Agency of Environmental Conservation, *Vermont Vacation Home Inventory, 1973* (Montpelier, Vt.: The Agency, 1974); Kathy Frazer and Terry Donovan, *Vacation Home Survey of Eight Vermont Towns* (n.p.: New England Board of Higher Education, 1972); Robert O. Sinclair and Stephen B. Meyer, *Nonresident Ownership of Property in Vermont* (Burlington, Vt.: UVM/VAES, 1972); Department of the Interior, Bureau of Outdoor Recreation, *Northern New England Vacation Home Study, 1966* (Washington, D.C.: Government Printing Office, 1967).

25. Vermont Agency of Environmental Conservation, *Vermont Vacation Home Inventory, 1973*, 36.

26. "Five Major Land Developments, Southern Vermont, 1969," Records of Deane C. Davis (hereafter DCD), Box 13B, Folder 30, VSA; Virginia Page, "Southern Vermont's Own Beauty Creates Problems of Development," *BDR*, October 1, 1969, 13.

27. Simmons Associates, Inc., and Goldweitz Company, Inc., "Cavanagh Leasing Corporation," 1969, p. 7, DCD, Box 13B, Folder 29, VSA. Also see Pete Horton, "It's Dover Hills," *BDR*, November 20, 1969, p. 4.

28. For audio and slide images from this sales video, see . . . *So Goes Vermont*, prod. John Karol, 23 minutes, Environmental Planning Information Center, a social project of the Vermont Natural Resources Council, 1971, 1992, videocassette.

29. Peter M. Miller, "Bull Market Boom in Ski Land," *Vermont Skiing* (Winter 1964–1965): 21–27; "Everybody Wants a Home on the Hills," *Vermont Skiing* 4 (Winter 1967–1968): 13–15; Richard M. Klein, "Green Mountains, Green Money," *Natural History* 79 (March 1970): 10–12, 14, 18–20, 22, 24, 26.

30. Lovilla H. Bromley, quoted in "Quotes from Brokers on Ski Real Estate," *Vermont Skiing* (Winter 1965): 48.

31. "Land Developers Follow Ski Boom," *BDR*, February 2, 1968, p. 4.

32. Horton, "It's Dover Hills."

33. Amory C. Smith and Frederic O. Sargent, *The Rural Land Market in Vermont* (Burlington, Vt.: Vermont Resources Research Center, UVM/VAES, 1964); Robert O. Sinclair, *Trends in Rural Land Prices in Vermont* (Burlington, Vt.: UVM/VAES, Bulletin 659, 1969).

34. Malcom I. Bevins, *Agriculture and Recreation — Competitive or Compatible* (no publisher, 1968), mss. copy located in VTD/DD, Box 6, no folder, VSA.

35. Examples include Margaret D. Smith, "More Non-Residents than Cows," *Yankee* 31 (August 1967): 64–65, 127–31; Klein, "Green Mountains, Green Money."

36. Press Release, July 11, 1969, DCD, Box 13B, Folder 30, VSA.

37. Simmons Associates, Inc., and Goldweitz Company, Inc., "Cavanagh Leasing Corporation."

38. Clifford Jarvis quoted in "Cry, Vermont," *Time* 94 (September 16, 1969), 50.

39. Jack and Jessamine Larrow, interview by Jane Beck, tape recording, Waits-

field, Vermont, March 10, 1994, page 13 of typed manuscript, ascension # TC94.
0017, Mad River Valley Collection, collection #1992–0004, VFC.

40. For example, see Stephen Hedger, *Downhill in Warren: The Effects of the Ski Industry and Land Development on Warren, Vermont* (Montpelier, Vt.: Vermont Public Interest Research Group, Inc., 1972); Edwin L. Johnson, Stanley Judkins, Stephen M. Hedger, and Frederick P. Jagels, *Effect of Second Home Development on Ludlow, Vermont, 1973* (Windsor, Vt.: Envico, 1973).

41. Vermont Agency of Environmental Conservation, *Vermont Vacation Home Inventory, 1973*, 3, 6.

42. Page, "Southern Vermont's Own Beauty Creates Problems of Development," 13.

43. Samuel Hays, *Beauty, Health, and Permanence: Environmental Politics in the United States, 1955–1985* (New York: Cambridge University Press, 1987).

44. On the place of ecology in environmental thought, see Barry Commoner, *The Closing Circle: Nature, Man, and Technology* (New York: Alfred A. Knopf, 1971), especially 14–48. Also see Donald Worster, *Nature's Economy: A History of Ecological Ideas* (New York: Cambridge University Press, 1977).

45. Adam Rome, *The Bulldozer in the Countryside: Suburban Sprawl and the Rise of American Environmentalism* (New York: Cambridge University Press, 2001), 87–118.

46. "Developer's Zoning Plea Weighed in Wilmington," *BDR*, April 23, 1968, pp. 1, 6; "Variance Is Refused for Chimney Hill," *BDR*, May 7, 1968, pp. 1, 5; "Chimney Hill Developer Approved," *BDR*, July 5, 1968, pp. 1–2. Also see Richard Wien, "The Collapse of a Development," *Country Journal* 2 (June 1975): 48–55.

47. Grady Holloway, "Dover Hills Plan Attacked by Many," *BDR*, November 5, 1969, pp. 1, 9.

48. "Destruction and Devastation: That's View on Dover Hills," *BDR*, November 21, 1969, p.1. Also see Grady Holloway, "The Dover Hills Problem: What Now?" *BDR*, November 22, 1969, pp. 1, 9; Victor Harrison, "Stratton Mt. Plans Questioned Sharply," *BDR*, June 25, 1969, pp. 1, 5

49. "Jeffords on Stowe: 'Sewage Capital,'" *BDR*, November 7, 1969, p. 3.

50. Vermont Planning Council, *Vision and Choice: Vermont's Future, The State Framework Plan* ([Montpelier, Vt.]: The Council, 1968), 13.

51. "Environmental Control Board Formed in State," *BFP*, May 15, 1969, p. 4.

52. "Gov. Davis and Entourage Will Spend Day in Town," *BDR*, May 27, 1969, p. 1; "Gov. Davis Lends Ear to Problems." *BDR*, May 28, 1969, pp. 1, 5; "Davis Notes Urgency of Area Problem," *BDR*, May 29, 1969, p. 1; "Relief for Wilmington to Be Aim of Governor," *BDR*, May 29, 1969, p. 1.

53. Telegram, Governor Deane C. Davis to Edward B. Hinman, June 24, 1969, DCD, Box 13B, Folder 30, VSA. Also see Harrison, "Stratton Mt. Plans Questioned Sharply"; "Developers Set to Meet with Governor Davis," *BDR*, July 1, 1969, p. 5.

54. Press Release, July 11, 1969, DCD, Box 13B, Folder 30, VSA; untitled announcement from the International Paper Company, July 17, 1969, DCD, Box 13B, Folder 30, VSA; "Stratton-Winhall Development Halted Pending Mutual Review," *BFP*, July 19, 1969, p.1.

55. Deane C. Davis with Nancy Price Graff, *Deane C. Davis: An Autobiography* (Shelburne, Vt.: New England Press, 1991), 250; Bob Babcock Jr., "Development Aggressiveness," *BFP*, September 9, 1969, p. 13.

56. "Health Board Passes Subdivision Rules," *BFP*, September 19, 1969, p. 15.

57. *State of Vermont, Governor's Commission on Environmental Control, Reports to Governor, January 19, 1970, May 18, 1970* (no publisher, [1970]), 1–3.

58. For more on the passage and specifications of Act 250, see Robert M. Sanford and Mark B. Lapping, "The Beckoning Country: Act 200, Act 250 and Regional Planning in Vermont," in *Big Places, Big Plans*, ed. Mark B. Lapping and Owen J. Furuseth (Burlington, Vt.: Ashgate, 2004), 5–27; Art Gibb and Sam Lloyd, "The Evolution of Act 250," in *Vermont Environmental Board: Twenty-fifth Anniversary Report, 1970–1995* (Montpelier, Vt.: The Board, 1995), 4–6; Phyllis Myers, *So Goes Vermont: An Account of the Development, Passage, and Implementation of State Land Use Legislation in Vermont* (Washington, D.C.: The Conservation Foundation, 1974).

59. For more on the embrace of simple living during the 1960s and 1970s, see David E. Shi, *The Simple Life: Plain Living and High Thinking in American Culture* (New York: Oxford University Press, 1985), 248–76.

60. On hippies and communes in Vermont, see Sherman, *Fast Land on a Dirt Road*, 82–98; Robert Houriet, *Getting Back Together* (New York: Coward, McCann, and Geoghegan, 1971); Raymond Mungo, *Total Loss Farm: A Year in the Life* (New York: Dutton., 1970).

61. Becker Research Corp., *An Environmental/Economic Profile of Vermont by Vermonters* (Montpelier, Vt.: Vermont Natural Resources Council, 1972). There are no page numbers in this report, but this quote was taken from the summary statement. The report was based on 561 forty-five-minute interviews conducted throughout the state.

62. "Green Up Campaign Hopes to Halt Litterbug Urge," *BFP*, March 5, 1970, p. 17; "Massive Vermont Green Up Underway This Morning," *BFP*, April 18, 1970, p.1; "'Bend and Pick Up' Greens Up Vermont," *BFP*, April 20, 1970, p. 13.

63. Peter Franchot, *Bottles and Cans: The Story of the Vermont Deposit Law*, ed. Susan Bartlett (Washington, D.C.: National Wildlife Federation, [1978]).

64. Leo O'Connor, "Pros, Cons of Vermont Development Examined," *BFP*, November 30, 1972, p. 23.

65. Becker Research Corp., *An Environmental/Economic Profile of Vermont by Vermonters*.

66. For an extended discussion of this point, see John McClaughry, "The New Feudalism," *Environmental Law* 5 (Spring 1975): 675–702.

67. As quoted in . . . *So Goes Vermont*, prod. John Karol.

68. See Judd and Beach, *Natural States*. Also see Plotkin, *Keep Out*.

69. Becker Research Corp., *An Environmental/Economic Profile of Vermont by Vermonters*.

70. Norman Runnion, "Ahmeek — The Story of a Development," *BDR*, October 23, 1969, p. 5.

71. Steve Chontos, *The Death of Dover, Vermont* (New York: Vantage Press, 1974).

72. Vermont State Planning Office, *Vermont's Land Use Plan and Act 250* (Montpelier, Vt.: The Office, 1974) — a copy of this report is available at the Dover Free Library, Dover, Vermont; Vermont State Planning Office, *Vermont Interim Land Capability Plan* ([Montpelier, Vt.]: [The Office], 1971); Stephen Carlson, "State Interim Land Plan Gets OK," *BFP*, December 30, 1971, p. 1; Stephen Carlson, "Land Capability Plan OK'd," *BFP*, March 9, 1972, p.1. This phase of the post–Act 250 agenda did not require legislative approval.

73. "Vermonters to Receive Land Use Plans," *BFP*, November 20, 1972, p. 7; Wilson, "Land Use, Planning, and Environmental Protection," 458–59; Keith Wallace, "No to Mandatory State Zoning," *BFP*, February 1, 1973, p. 17; "Northeast Kingdom to Fight Land Use Plan," *BFP*, September 29, 1973, p. 17.

74. Neil A. Davis, "Land Plan Backed, Opposed at Hearing," *BFP*, December 5, 1972, p. 10.

75. "Landowner Outrage Spreads," *BFP*, November 24, 1972, p.1.

76. Chontos, *The Death of Dover*, 56, 60, 117.

77. For examples, see Rothman, *Devil's Bargains*, and Annie Gilbert Coleman, *Ski Style: Sport and Culture in the Rockies* (Lawrence: University Press of Kansas, 2004), 193–206.

78. "Ski Area Head Upset by Dr. Aiken's Pollution Comments," *BFP*, July 1, 1971, p. 22.

79. L. Dana Gatlin, "An Environmental Look at Verdant Vermont," *Ski Area News* 7 (Fall 1972): 15–19, 44, 46. Also see "Ski Operators Told to Watch the Image," *BFP*, January 8, 1970, p. 23; John Hitchcock, "Vermont's Ecology Act One Year Later: The Meaning of Greening," *Ski Area Management* 10 (October 1971): 22–24; I. William Berry, "Vermont: The Beckoning Country," *Ski* 35 (October 1970): 78–82, 143, 145–46, 148; I. William Berry, "Vermont's Last Stand," *Ski Area Management* 9 (Fall 1970): 37–39.

80. Calvin Trillin, "U.S. Journal: Vermont, 250 and Beyond," *New Yorker* 50 (November 1974): 130.

81. "Environment-Economy Balance Sought," *BFP*, August 7, 1971, p. 8.

82. "Moulton Favors Development over Environment," *BFP*, April 30, 1970, p. 8.

83. Stephen Carlson, "Development, Environment Equal Goals, Says Governor," *BFP*, April 24, 1970, p. 13.

84. Becker Research Corp., *An Environmental/Economic Profile of Vermont by Vermonters*.

85. Robert Burley Associates, *People on the Land: Settlement Patterns for Vermont* ([Montpelier, Vt.]: Vermont State Planning Office, 1973), 3.

86. Neil Davis, "Board Adopts Land Capability Plan," *BFP*, January 3, 1973, p. 19; "Land Capability Plan Gets Signature of Gov. Davis," *BFP*, January 4, 1973, p. 8; Bruce Talbot, "Land Plan Passes House Vote, 122 to 26." *BFP*, March 21, 1973, p. 1; Stephen Carlson, "Property Tax Relief Bill, Land Plan Gain Passage: Bottle Lobbyists' Amendment Ousted," *BFP*, April 16, 1973, pp. 1, 3; Wilson, "Land Use, Planning, and Environmental Protection," 458–60; Neil Davis, "Land Plan Termed 'Dead,'" *BFP*, March 20, 1974, p. 1.

87. "Remarks made by Elbert G. Moulton at a Hearing on the Scenic Corridors Bill before the Natural Resources Committee — Tuesday, February 7, 1967," PHH, Box 55, Folder 9-2-74, UVM.

88. Lorna Lecker, "Development in State Not Slowed by Act 250," *BFP*, September 20, 1972, p. 4.

Epilogue: The View from Vermont (pp. 236–44)

1. For more on agritourism, see Norma Polovitz Nickerson, Rita J. Black, Stephen F. McCool, "Agritourism: Motivations Behind Farm/Ranch Business Diversification," *Journal of Travel Research* 40 (August 2001): 19–26; Derek Hall and Greg Richards, eds., *Tourism and Sustainable Community Development* (New York: Routledge, 2003); Lesley Roberts and Derek Hall, eds., *Rural Tourism and Recreation: Principles to Practice* (New York: CABI Publishing, 2001).

2. Vermont Farms! Association, "Vermont Farms! Working Farms Open to the Public," 2005 brochure, page 1 (author's collection).

3. For representative discussions, see W. B. Beyers and P. B. Nelson, "Contemporary Development Forces in the Non-Metropolitan West: New Insights from Rapidly Growing Communities," *Journal of Rural Studies* 16 (October 2000): 459–74; Peter Walker and Louise Fortmann, "Whose Landscape: A Political Ecology of the 'Exurban' Sierra," *Cultural Geographies* 10 (October 2003): 469–91; Kenneth B. Beesley, Hugh Millward, Brian Ilbery, and Lisa Harrington, eds., *The New Countryside: Geographic Perspectives on Rural Change* (Brandon, Manitoba: Brandon University, Halifax, Nova Scotia: St. Mary's University, 2003).

4. "Impure as the Driven Snow," *Time* 126 (October 21, 1985): 45; Preston Smith, "The Mountains Are for Everyone," *Vermont Affairs* 1 (Winter 1986): 20–24. Also see J. Kevin Graffagnino, Samuel B. Hand, and Gene Sessions, eds., *Vermont Voices, 1609 through the 1990s: A Documentary History of the Green Mountain State* (Montpelier, Vt.: Vermont Historical Society, 1999), 348, 366–70.

5. Michael Woods, "Deconstructing Rural Protest: The Emergence of a New Social Movement," *Journal of Rural Studies* 19 (July 2003): 309–25.

6. "America's 11 Most Endangered Historic Places, 1993," *News Release, Na-*

tional Trust for Historic Preservation, 1993. The trust has released its "11 Most Endangered Historic Places" list each year since 1988.

7. Pam Belluck, "Preservationists Call Vermont Endangered, by Wal-Mart,"
New York Times, May 25, 2004, p. A18.

8. Carey Goldberg, "Vermont Residents Split Over Civil Unions Law," *New
York Times,* September 3, 2000, p. 14.

Selected Bibliography

Manuscript Sources

New England Ski Museum, Franconia Notch, New Hampshire

Uncatalogued Resources, misc.

University of Vermont Special Collections,
Bailey/Howe Library, Burlington, Vermont

Walter H. Crockett Papers/Writings, 1900–1931
Theron Dean Papers
Dorothy Canfield Fisher Collection
Albert Gottlieb Papers
The Green Mountain Parkway, Final Report by the Landscape Architects of the National Park Service, bound manuscript
Philip H. Hoff Papers
Manuscript Files, misc.
Perry Merrill Papers
Reference Files, misc.
Governor Stanley C. Wilson Radio Addresses from the early 1930s, unpublished bound volume

The Vermont Folklife Center, Middlebury, Vermont

Mad River Collection
The Vermont Apple Project

Vermont Historical Society, Barre, Vermont

Broadside Collection, misc.
John Clement Papers
Green Mountain Club Archives
James P. Taylor Collection, 1906–1949
Uncatalogued Pamphlet Collection, misc.

Vermont State Archives, Office of the Secretary
of State, Montpelier, Vermont

Publicity Films, misc.
Records of Deane C. Davis
Vermont Travel Division [1879–1994]
Vermont Travel Division/Development Department [1879–1950]: Statistics, Surveys, Reports, etc.

Periodicals

The American Mercury
American Magazine
American Ski Annual and Skiing Journal
Appalachia
Atlantic Monthly
Better Homes and Gardens
Boston Evening Transcript
Boston Herald
Brattleboro Daily Reformer (Brattleboro, Vt.)
Burlington Free Press (Burlington, Vt.)
Century
Commentary
Cosmopolitan
Country Calendar
Country Journal
Country Life in America
Deerfield Valley Times (Wilmington, Vt.)
Eastern Ski Bulletin
Eastern Skier
Elm Tree Monthly and Spirit of the Age
Farm Journal

Forest and Stream
Fortune
Forum
Harper's
House and Garden
House Beautiful
Illustrated World
Journal of Home Economics
Knickerbocker
Ladies' Home Journal
Life
Literary Digest
Long Trail News
Nation
National Geographic
Natural History
Nature Magazine
New England Magazine
New Republic
New York Times
New Yorker
Newsweek
North American Review
Outing Magazine
Outlook
Popular Mechanics
Quill
Rural Vermonter
Rutland Daily Herald (Rutland, Vt.)
Saturday Evening Post
Ski
Ski Area Management
Ski Area News
Ski News
Ski Week
Skiing
Skiing Area News
Sports Illustrated
Survey
Time
Travel
Travel/Holiday
Vermont Affairs

Vermont Environmental Report
Vermont Highways
Vermont Life
Vermont Review
Vermont Skiing
Vermonter
Women's Home Companion
Yankee

Books, Articles, Reports, and Films

Agyeman, Julian, and Rachel Spooner. "Ethnicity and the Rural Environment." In *Contested Countryside Cultures: Otherness, Marginalisation and Rurality*, ed. Paul Cloke and Jo Little, 197–217. New York: Routledge, 1997.

Aitchison, Cara, Nicola E. Macleod, and Stephen J. Shaw. *Leisure and Tourism Landscapes: Social and Cultural Geographies.* New York: Routledge, 2000.

Albers, Jan. *Hands on the Land: A History of the Vermont Landscape.* Cambridge, Mass.: MIT Press, 1999.

Allen, E. John B. "The Making of a Skier: Fred H. Harris, 1904–1911." *Vermont History* 53 (Winter 1985): 5–16.

———. *From Skisport to Skiing: One Hundred Years of an American Sport, 1840–1940.* Amherst: University of Massachusetts Press, 1993.

Ambrose, Harry. *The Horse's Mouth: The Story of the Birth of Modern Skiing at Woodstock, Vermont, and the Saga of Woodstock High School's Vermont Interscholastic Championship Teams of 1940, 1941, and 1942.* 2nd ed. Ed. Tom Bourne Jr. Houston: A&E Products Co., Inc, 1997.

Anderson, Kay, Mona Domosh, Steve Pile, and Nigel Thrift, eds. *Handbook of Cultural Geography.* London: Sage, 2003.

Anderson, S. Axel, and Florence M. Woodard. "Agricultural Vermont." *Economic Geography* 8 (January 1932): 12–42.

Anderson, W. A. *Population Change in Vermont, 1900 to 1950.* Burlington, Vt.: University of Vermont and State Agricultural College, Agricultural Experiment Station, 1955.

Appleton, Louise. "Distillations of Something Larger: The Local Scale and American National Identity." *Cultural Geographies* 9 (October 2002): 421–47.

Aron, Cindy S. *Working at Play: A History of Vacations in the United States.* New York: Oxford University Press, 1999.

Ayers, Edward L., Patricia Nelson Limerick, Stephen Nissenbaum, Peter S. Onuf. *All Over the Map: Rethinking American Regions.* Baltimore: Johns Hopkins University Press, 1996.

Bailey, Guy, and Orlando Martin. *Homeseekers' Guide to Vermont Farms.* St. Albans, Vt.: St. Albans Messenger Co., 1911.

Bard, Albert S. "Vermont Billboard Decision." *American Planning and Civic Annual* (1943): 46–49.

Barron, Hal S. *Those Who Stayed Behind: Rural Society in Nineteenth-Century New England.* New York: Cambridge University Press, 1984.

——. *Mixed Harvest: The Second Great Transformation in the Rural North, 1870–1930.* Chapel Hill: University of North Carolina Press, 1997.

Bassett, T. D. Seymour, ed. *Outsiders Inside Vermont: Travelers' Tales of 358 Years.* Brattleboro, Vt.: Stephen Green Press, 1967.

Baumgardt, David. "Dorothy Canfield Fisher: Friend of Jews in Life and Work." *Publication of the American Jewish Historical Society* 48 (June 1959): 245–55.

Beattie, Betsy. "The Queen City Celebrates Winter: The Burlington Coasting Club and the Burlington Carnival of Winter Sports, 1886–1887." *Vermont History* 52 (Winter 1984): 5–16.

Becker Research Corp. *An Environmental/Economic Profile of Vermont by Vermonters.* Montpelier, Vt.: Vermont Natural Resources Council, 1972.

Beesley, Kenneth B., Hugh Millward, Brian Ilbery, and Lisa Harrington, eds. *The New Countryside: Geographic Perspectives on Rural Change.* Brandon, Manitoba: Brandon University; Halifax, Nova Scotia: St. Mary's University, 2003.

Belasco, Warren James. *Americans on the Road: From Autocamp to Motel, 1910–1945.* Baltimore: Johns Hopkins University Press, 1979.

Bell, Michael M. "Did New England Go Downhill?" *Geographical Review* 79 (October 1989): 450–66.

Belth, N. C. *Barriers: Patterns of Discrimination Against Jews.* New York: Friendly House Publishers, 1958.

Bender, Barbara, ed. *Landscape: Politics and Perspectives.* Oxford: Berg, 1993.

Bevins, M. I., and R. H. Tremblay. *Dairy Farming in Vermont.* Burlington, Vt.: University of Vermont, Vermont Agricultural Experiment Station, 1967.

——. *The Outdoor Recreation Industry in Vermont.* Burlington, Vt.: Vermont Resources Research Center, Vermont Agricultural Experiment Station, University of Vermont, 1964.

Bevins, Malcom I. *Agriculture and Recreation — Competitive or Compatible.* No publisher, 1968.

Beyers, W. B., and P. B. Nelson. "Contemporary Development Forces in the Non-Metropolitan West: New Insights from Rapidly Growing Communities." *Journal of Rural Studies* 16 (October 2000): 459–74.

Bigelow, Edwin. *Stowe, Vermont: Ski Capital of the East, 1763–1963.* Stowe, Vt.: Stowe Historical Society, 1964.

Blake, Peter. *God's Own Junkyard: The Planned Deterioration of America's Landscape.* New York: Holt, Rinehart, and Winston, 1964.

Bonnet, R. D'Arcy. "National Forest Planning: Landscape and Recreational Policies." *Landscape Architecture* 26 (April 1936): 114–18.

Bremmer, Thomas S. *Blessed with Tourists: The Borderlands of Religion and Tourism in San Antonio.* Chapel Hill: University of North Carolina Press, 2004.

Brown, Dona. *Inventing New England: Regional Tourism in the Nineteenth Century.* Washington, D.C.: Smithsonian Institution Press, 1995.

———. *A Tourist's New England: Travel Fiction, 1820–1920.* Hanover, N.H.: University Press of New England, 1999.

Bruner, Edward M. *Culture on Tour: Ethnographies of Travel.* Chicago: University of Chicago Press, 2005.

Bryan, Frank. *Real Democracy: The New England Town Meeting and How It Works.* Chicago: University of Chicago Press, 2004.

Bryant, William Cullen, ed. *Picturesque America; or, the Land We Live In.* Vol. 2. New York: D. Appleton and Company, 1874.

Bunce, Michael. *The Countryside Ideal: Anglo-American Images of Landscape.* New York: Routledge, 1994.

Burt, Henry M. *Burt's Illustrated Guide of the Connecticut Valley, Containing Descriptions of Mount Holyoke, Mount Mansfield, White Mountains, Lake Memphremagog, Lake Willoughby, Montreal, Quebec, Etc.* Northhampton, Mass.: New England Publishing Company, 1867.

Carlson, Albert S. "Ski Geography of New England." *Economic Geography* 18 (July 1942): 307–20.

Carman, Bernard R. *Hoot Toot and Whistle: The Story of the Hoosac Tunnel and Wilmington Railroad.* Brattleboro, Vt.: Stephen Greene Press, 1963.

Castle, Emery N., ed. *The Changing American Countryside: Rural People and Places.* Lawrence: University Press of Kansas, 1995.

Central Planning Office. *The Preservation of Roadside Scenery through the Police Power.* Montpelier, Vt.: The Office, 1966.

———. *Vermont Scenery Preservation.* Montpelier, Vt.: The Office, 1966.

Chadwick, H. H. *Vermont Bureau of Publicity: Its History, Expenditures, and Activities.* [Montpelier, Vt.]: [Office of the Secretary of State], [1934].

Chapin, H. B., comp. *Proceedings of the Twenty-Third Annual Meeting of the Vermont Maple Sugar Makers' Association.* St. Albans, Vt.: St. Albans Messenger Co. Print, 1916.

Chidester, Lawrence W. "The Importance of Recreation as a Land Use in New England." *Journal of Land and Public Utility Economics* 10 (May 1934): 202–9.

Child, Hamilton, comp. *Gazetteer and Business Directory of Bennington County, 1880–1881.* Syracuse, N.Y.: By the Compiler, 1880.

———. *Gazetteer and Business Directory of Windham County, Vermont, 1724–1884.* Syracuse, N.Y.: By the Compiler, 1884.

Chontos, Steve. *The Death of Dover, Vermont.* New York: Vantage Press, 1974.

Cloke, Paul, and Jo Little, eds. *Contested Countryside Cultures: Otherness, Marginalisation and Rurality.* New York: Routledge, 1997.

Cloke, Paul, Marcus Doel, David Matless, Martin Phillips, Nigel Thrift. *Writing the Rural: Five Cultural Geographies.* London: Paul Chapman Publishing Ltd., 1994.

Cole, Donald B. *Immigrant City: Lawrence, Massachusetts, 1845–1921.* Chapel Hill: University of North Carolina Press, 1963.

Coleman, Annie Gilbert. "The Unbearable Whiteness of Skiing." *Pacific Historical Review* 65 (November 1996): 583–614.

——. *Ski Style: Sport and Culture in the Rockies.* Lawrence: University Press of Kansas, 2004.

Collier, Peter, ed. *Second Biennial Report of the Vermont State Board of Agriculture, Manufacturing and Mining, for the Years 1873–74.* Montpelier, Vt.: Freeman Steam Printing House and Bindery, 1874.

Commoner, Barry. *The Closing Circle: Nature, Man, and Technology.* New York: Alfred A. Knopf, 1971.

Conforti, Joseph A. *Imagining New England: Explorations of Regional Identity from the Pilgrims to the Mid-Twentieth Century.* Chapel Hill: University of North Carolina Press, 2001.

Congdon, Herbert Wheaton. *The Covered Bridge: An Old American Landmark Whose Romance, Stability and Craftsmanship Are Typified by Structures Remaining in Vermont.* Brattleboro, Vt.: Stephen Daye Press, 1941.

Conron, John. *American Picturesque.* University Park: Pennsylvania State University Press, 2000.

Constitution, By-Laws, Rules and Regulations: Forest and Stream Club, Wilmington, Vermont, 1894. Brattleboro, Vt.: Phoenix Job Printing Office, 1894.

Coolidge, Calvin. "Address at Bennington." In *Vermont Prose: A Miscellany,* 2nd ed., ed. Arthur Wallace Peach and Harold Goddard Rugg, 246–47. Brattleboro, Vt.: Stephen Daye Press, 1932.

Cosgrove, Denis E. *Social Formation and Symbolic Landscape.* London: Croom Helm, 1984.

Cox, Laurie Davidson. "The Green Mountain Parkway." *Landscape Architecture* 25 (April 1935): 117–26.

Crane, Charles Edward. *Let Me Show You Vermont.* New York: Alfred A. Knopf, 1937.

——. *Winter in Vermont.* New York: Alfred A. Knopf, 1941.

Cresswell, Tim. *The Tramp in America.* London: Reaktion Books, 2001.

——. "Landscape and the Obliteration of Practice." In *Handbook of Cultural Geography,* ed. Kay Anderson, Mona Domosh, Steve Pile, and Nigel Thrift, 269–81. London: Sage Publications, 2003.

Crockett, W. H. "Advertising our Product." In *Proceedings of the Twenty-Eighth Annual Meeting of the Vermont Maple Sugar Makers' Association,* 23–28. St. Albans, Vt.: St. Albans Messenger Co. Print, 1921.

Cronon, William, ed. *Uncommon Ground: Rethinking the Human Place in Nature.* New York: W. W. Norton and Company, 1996.

Curtis, Jane, Will Curtis, and Frank Lieberman. *Green Mountain Adventure, Vermont's Long Trail: An Illustrated History.* Montpelier, Vt.: Green Mountain Club, 1985.

Dann, Kevin. "From Degeneration to Regeneration: The Eugenics Survey of Vermont, 1925–1936." *Vermont History* 59 (Winter 1991): 5–29.

Daniels, Stephen. *Fields of Vision: Landscape Imagery and National Identity in England and the United States.* Princeton, N.J.: Princeton University Press, 1993.

D'Arcus, Bruce. "The 'Eager Gaze of the Tourist' Meets 'Our Grandfathers' Guns': Producing and Contesting the Land of Enchantment in Gallup, New Mexico." *Environment and Planning D: Society and Space* 18 (December 2000): 693–714.

Davis, Deane C., with Nancy Price Graff. *Deane C. Davis: An Autobiography.* Shelburne, Vt.: New England Press, 1991.

Davis, Jeffrey Sasha. "Representing Place: 'Deserted Isles' and the Reproduction of Bikini Atoll." *Annals of the Association of American Geographers* 95 (September 2005): 607–25.

DeLyser, Dydia. "Authenticity on the Ground: Engaging the Past in a California Ghost Town." *Annals of the Association of American Geographers* 89 (December 1999): 602–32.

———. *Ramona Memories: Tourism and the Shaping of Southern California.* Minneapolis: University of Minnesota Press, 2005.

DeOliver, Miguel. "Historical Preservation and Identity: The Alamo and the Production of a Consumer Landscape." *Antipode* 28 (January 1996): 1–23.

Dean, Arthur W. "The Highways of New England." In *New England's Prospect: 1933,* ed. John K. Wright, 362–71. New York: American Geographical Society, 1933.

Denenberg, Thomas Andrew. *Wallace Nutting and the Invention of Old America.* New Haven, Conn.: Yale University Press, 2003.

Department of the Interior, Bureau of Outdoor Recreation. *Northern New England Vacation Home Study, 1966.* Washington, D.C.: U.S. Government Printing Office, 1967.

Domosh, Mona. "Pickles and Purity: Discourses of Food, Empire and Work in Turn-of-the-Century USA." *Social and Cultural Geography* 4 (March 2003): 7–27.

Donahue, Brian. *Reclaiming the Commons: Community Farms and Forests in a New England Town.* New Haven, Conn.: Yale University Press, 1999.

Donovan, George A. *Vermont Vacation Home Survey, 1968.* Montpelier, Vt.: Vermont Development Department, 1968.

Dorman, Robert L. *Revolt of the Provinces: The Regionalist Movement in America, 1920–1945.* Chapel Hill: University of North Carolina Press, 1993.

Dorsett, Lyle W. "Town Promotion in Nineteenth-Century Vermont." *New England Quarterly* 40 (June 1967): 275–79.

DuPuis, E. Melanie, and Peter Vandergeest, eds. *Creating the Countryside: The Politics of Rural and Environmental Discourse.* Philadelphia: Temple University Press, 1996.

Duncan, James S. *The City as Text: The Politics of Landscape Interpretation in the Kandyan Kingdom.* Cambridge: Cambridge University Press, 1990.

Duncan, James S., and Nancy G. Duncan. *Landscapes of Privilege: The Politics of the Aesthetic in an American Suburb.* New York: Routledge, 2004.

England, George M., and Enoch H. Tompkins. *Marketing Vermont's Maple Syrup.*

Burlington, Vt.: Vermont Agricultural Experiment Station, University of Vermont and State Agricultural College, 1956.

Federal Writers' Project. *Vermont: A Guide to the Green Mountain State.* Boston: Houghton Mifflin Company, 1937.

———. *Skiing in the East: The Best Trails and How to Get There.* New York: M. Barrows and Company, 1939.

Fifth Annual Report of the Vermont State Horticultural Society. Bellows Falls, Vt.: P. H. Gobie Press, 1907.

[Fisher], Dorothy Canfield. *Vermont Summer Homes.* Montpelier, Vt.: Vermont Bureau of Publicity, 1932.

Fisher, Dorothy Canfield. *Tourists Accommodated: Some Scenes from Present-Day Summer Life in Vermont.* New York: Harcourt, Brace and Company, 1934.

———. "Vermonters." In *Vermont: A Guide to the Green Mountain State,* Workers of the Federal Writers' Project of the Works Progress Administration for the State of Vermont, 3–9. Boston: Houghton Mifflin Company, 1937.

———. *Seasoned Timber.* Ed. Mark J. Madigan. Hanover, N.H.: University Press of New England, 1996.

Follett, Muriel. *New England Year: A Journal of Vermont Farm Life.* Brattleboro, Vt.: Stephen Daye Press, 1939.

———. *A Drop in the Bucket: The Story of Maple Sugar Time on a Vermont Farm.* Brattleboro, Vt.: Stephen Daye Press, 1941.

Foote, Stephanie. *Regional Fictions: Culture and Identity in Nineteenth-Century American Literature.* Madison: University of Wisconsin Press, 2001.

Franchot, Peter. *Bottles and Cans: The Story of the Vermont Deposit Law.* Ed. Susan Bartlett. Washington, D.C.: National Wildlife Federation, [1978].

Frazer, Kathy, and Terry Donovan. *Vacation Home Survey of Eight Vermont Towns.* N.p.: New England Board of Higher Education, 1972.

Gallagher, Nancy. *Breeding Better Vermonters: The Eugenics Project in the Green Mountain State.* Hanover, N.H.: University Press of New England, 1999.

Goldman, Hal. "James Taylor's Progressive Vision: The Green Mountain Parkway." *Vermont History* 63 (Summer 1995): 158–79.

———. "'A Desirable Class of People': The Leadership of the Green Mountain Club and Social Exclusivity, 1920–1936." *Vermont History* 65 (Summer/Fall 1997): 131–52.

Gould, Lewis L. *Lady Bird Johnson and the Environment.* Lawrence: University Press of Kansas, 1988.

Gove, Bill. "The Forest Industries of Lake Memphremagog." *Northern Logger and Timber Processor* 23 (March 1975): 18–19, 31–32, 34.

Gove, William. "Mountain Mills, Vermont, and the Deerfield River Railroad." *Northeastern Logger and Timber Processor* 17 (May 1969): 18–20, 36–38.

Graff, Nancy Price, ed. *Celebrating Vermont: Myths and Realities.* Middlebury, Vt.: Christian A. Johnson Memorial Gallery, Middlebury College; Hanover, N.H.: University Press of New England, 1991.

———. *Looking Back at Vermont: Farm Security Administration Photographs, 1936–1942.* Middlebury, Vt.: Middlebury College Museum of Art; Hanover, N.H.: University Press of New England, 2002.

Graffagnino, J. Kevin. *Vermont in the Victorian Age: Continuity and Change in the Green Mountain State.* Bennington, Vt.: Vermont Heritage Press; Shelburne, Vt.: Shelburne Museum, 1985.

Graffagnino, J. Kevin, Samuel B. Hand, and Gene Sessions, eds. *Vermont Voices, 1609 through the 1990s: A Documentary History of the Green Mountain State.* Montpelier, Vt.: Vermont Historical Society, 1999.

Greene, Frank, comp. *Vermont, the Green Mountain State: Past, Present, Prospective.* N.p.: Vermont Commission to the Jamestown Tercentennial Exposition, 1907.

Gudis, Catherine. *Buyways: Billboards, Automobiles, and the American Landscape.* New York: Routledge, 2004.

Hagerman, Robert L. *Mansfield: The Story of Vermont's Loftiest Mountain.* Essex Junction, Vt.: Essex Publishing Company, 1971.

Hall, Derek, and Greg Richards, eds. *Tourism and Sustainable Community Development.* New York: Routlege, 2003.

Hall, Derek, Lesley Roberts, and Morag Mitchell, eds. *New Directions in Rural Tourism.* Burlington, Vt.: Ashgate, 2003.

Hanna, Stephen P., and Vincent J. Del Casino Jr., eds. *Mapping Tourism.* Minneapolis: University of Minnesota Press, 2003.

Hard, Walter, and Margaret Hard. *This Is Vermont.* Brattleboro, Vt.: Stephen Daye Press, 1936.

Harrison, Blake. "Tracks Across Vermont: *Vermont Life* and the Landscape of Downhill Skiing, 1946–1970." *Journal of Sport History* 28 (Summer 2001): 253–70.

———. "Rethinking the Rural: Nostalgia and Progress in Vermont's Tourist Industry." *Proceedings of the New England St. Lawrence Valley Geographical Society* 32 (2002): 31–43.

———. "The Technological Turn: Skiing and Landscape Change in Vermont, 1930–1970." *Vermont History* 71 (Summer/Fall 2003): 197–228.

———. "Tourism, Farm Abandonment, and the 'Typical' Vermonter." *Journal of Historical Geography* 31 (July 2005): 478–95.

———. "Shopping to Save: Green Consumerism and the Struggle for Northern Maine." *Cultural Geographies*, forthcoming.

Hart, John Fraser. "'Rural' and 'Farm' No Longer Mean the Same." In *The Changing American Countryside: Rural People and Places,* ed. Emery N. Castle, 63–76. Lawrence: University Press of Kansas, 1995.

———. *The Changing Scale of American Agriculture.* Charlottesville: University of Virginia Press, 2003.

Harvey, Dorothy Mayo. "The Swedes in Vermont." *Vermont History* 28 (January 1960): 39–58.

Hastings, Scott E. Jr., and Elsie R. Hastings. *Up in the Morning Early: Vermont Farm Families in the Thirties.* Hanover, N.H.: University Press of New England, 1992.

Hastings, Scott E. Jr., and Geraldine S. Ames. *The Vermont Farm Year in 1890.* Woodstock, Vt.: Billings Farm and Museum, 1983.

Hays, Samuel P. *Beauty, Health, and Permanence: Environmental Politics in the United States, 1955–1985.* New York: Cambridge University Press, 1987.

Hedger, Stephen. *Downhill in Warren: The Effects of the Ski Industry and Land Development on Warren, Vermont.* Montpelier, Vt.: Vermont Public Interest Research Group, Inc., 1972.

Hill, Lewis. *Fetched Up Yankee: A New England Boyhood Remembered.* Chester, Conn.: Globe Pequot Press, 1990.

——. *Yankee Summer: The Way We Were.* N.p.: 1st Book Library, 2000.

Hill, Ralph N., Murray Hoyt, and Walter R. Hard Jr. *Vermont: A Special World.* Montpelier, Vt.: Vermont Life Magazine, 1969.

Hinrichs, C. Clare. "Consuming Images: Making and Marketing Vermont as a Distinctive Rural Place." In *Creating the Countryside: The Politics of Rural and Environmental Discourse,* ed. E. Melanie DuPuis and Peter Vandergeest, 259–78. Philadelphia: Temple University Press, 1996.

History of Planning in Vermont. Montpelier, Vt.: Vermont Department of Housing and Community Affairs, 1999.

Hoelscher, Steven D. *Heritage on Stage: The Invention of Ethnic Place in America's Little Switzerland.* Madison: University of Wisconsin Press, 1998.

——. "The Photographic Construction of Tourist Space in Victorian America." *Geographical Review* 88 (October 1998): 548–70.

Holmes, George K. "Movement from City and Town to Farms." In *Yearbook of the United States Department of Agriculture, 1914,* 257–74. Washington, D.C.: Government Printing Office, 1915.

Houriet, Robert. *Getting Back Together.* London: Abacus, 1973.

Huffman, Benjamin L. *Getting Around Vermont: A Study of Twenty Years of Highway Building in Vermont with Respect to Economics, Automotive Travel, Community Patterns, and the Future.* Burlington, Vt.: The Environmental Program, University of Vermont, 1974.

Hugill, Peter J. "Good Roads and the Automobile in the United States, 1880–1929." *Geographical Review* 72 (July 1982): 327–49.

Hypes, J. L. "Recent Immigrant Stocks in New England Agriculture." In *New England's Prospect: 1933,* ed. John K. Wright, 189–205. New York: American Geographical Society, 1933.

Irwin, William. *The New Niagara: Tourism, Technology, and the Landscape of Niagara Falls, 1776–1917.* University Park: Pennsylvania University Press, 1996.

Jacoby, Karl. "Class and Environmental History: Lessons from the 'War in the Adirondacks.'" *Environmental History* 2 (July 1997): 324–42.

Jakle, John. *The Tourist: Travel in Twentieth-Century North America.* Lincoln: University of Nebraska Press, 1985.

Jakle, John, and Keith A. Sculle. *Signs in America's Auto Age: Signatures of Landscape and Place.* Iowa City: University of Iowa Press, 2004.

Johnson, Clifton. *New England and Its Neighbors.* New York: Macmillan Company, 1924.

Johnson, Edwin L., Stanley Judkins, Stephen M. Hedger, and Frederick P. Jagels. *Effect of Second Home Development on Ludlow, Vermont, 1973.* Windsor, Vt.: Envico, 1973.

Johnson, Lillian H., and Marianne Muse. *Cash Contribution to the Family Income Made by Vermont Farm Homemakers.* Burlington, Vt.: University of Vermont, Vermont Agricultural Experiment Station, 1933.

Johnson, Nuala C. "Where Geography and History Meet: Heritage Tourism and the Big House in Ireland." *Annals of the Association of American Geographers* 86 (September 1996): 551–66.

———. *Ireland, the Great War, and the Geography of Remembrance.* New York: Cambridge University Press, 2003.

Jones, John Paul, III, Heidi J. Nast, and Susan M. Roberts, eds. *Thresholds in Feminist Geography: Difference, Methodology, Representation.* Lanham, Md.: Rowman and Littlefield Publishers, 1997.

Jones, O. H. *Attractions of Wilmington and Vicinity.* Jacksonville, Vt.: F. L. Stetson, 1887.

Judd, Richard. *Common Lands, Common People: The Origins of Conservation in Northern New England.* Cambridge, Mass.: Harvard University Press, 1997.

Judd, Richard, and Christopher S. Beach. *Natural States: The Environmental Imagination in Maine, Oregon, and the Nation.* Washington, D.C.: Resources for the Future, 2003.

Judd, Richard Munson. *The New Deal in Vermont: Its Impact and Aftermath.* New York: Garland Publishing, Inc., 1979.

Kammen, Michael. *A Time to Every Purpose: The Four Seasons in American Culture.* Chapel Hill: University of North Carolina Press, 2004.

Klyza, Christopher McGrory, and Stephen C. Trombulak. *The Story of Vermont: A Natural and Cultural History.* Hanover, N.H.: University Press of New England, 1999.

Laing, Frederick M., Mary T. G. Lighthall, and James W. Marvin. *The Use of Plastic Tubing in Gathering Maple Sap.* Burlington, Vt.: University of Vermont and State Agricultural College, Vermont Agricultural Experiment Station, 1960.

Lapping, Mark B., and Owen J. Furuseth, eds. *Big Places, Big Plans.* Burlington, Vt.: Ashgate, 2004.

Lears, T. J. Jackson. *No Place of Grace: Antimodernism and the Transformation of Material Culture, 1880–1920.* Chicago: University of Chicago Press, 1981.

Lesser, Ellen. *America's First Ski Tow, Commemorative Album.* South Pomfret, Vt.: Teago Publishing Company, 1983.

Lewis, Peirce F. "Axioms for Reading the Landscape." In *The Interpretation of Ordinary Landscapes: Geographical Essays,* ed. Donald Meinig, 11–32. New York: Oxford University Press, 1979.

Lewis, Sinclair. "Address Before the Rutland Rotary Club." Reprinted in *Vermont*

Prose: A Miscellany, 2nd ed., ed. Arthur Wallace Peach and Harold Goddard Rugg, 215–18. Brattleboro, Vt.: Stephen Daye Press, 1932.

Lipke, William C. "Changing Images of the Vermont Landscape." In *Vermont Landscape Images, 1776–1976*, ed. William C. Lipke and Philip N. Grimes, 33–48. Burlington, Vt.: Robert Hull Fleming Museum, 1976.

——. "From Pastoralism to Progressivism: Myth and Reality in Twentieth-Century Vermont." In *Celebrating Vermont: Myths and Realities*, ed. Nancy Price Graff, 61–88. Middlebury, Vt.: Christian A. Johnson Memorial Gallery, Middlebury College, 1991.

Lipke, William C., and Philip N. Grimes, eds. *Vermont Landscape Images, 1776–1976*. Burlington, Vt.: Robert Hull Fleming Museum, 1976.

Little, Jo. "Otherness, Representation and the Cultural Construction of Rurality." *Progress in Human Geography* 23 (September 1999): 437–42.

Little, Jo, and Michael Leyshon. "Embodied Rural Geographies: Developing Research Agendas." *Progress in Human Geography* 27 (June 2003): 257–72.

Luccarelli, Mark. "Benton MacKaye's Appalachian Trail: Imagining and Engineering a Landscape." In *Technologies of Landscape: From Reaping to Recycling*, ed. David E. Nye, 207–17. Amherst: University of Massachusetts Press, 1999.

Lund, John M. "Vermont Nativism: William Paul Dillingham and U.S. Immigration Legislation." *Vermont History* 63 (Winter 1995): 15–29.

MacCannell, Dean. *The Tourist: A New Theory of the Leisure Class*. New York: Schocken Books, 1989, 1976.

Marx, Leo. *The Machine in the Garden: Technology and the Pastoral Ideal in America*. New York: Oxford University Press, 1964.

Matless, David. *Landscape and Englishness*. London: Reaktion Books, 1998.

Mazuzan, George T. "'Skiing Is Not Merely a Schport': The Development of Mount Mansfield as a Winter Recreation Area." *Vermont History* 40 (Winter 1972): 47–63.

McCarthy, Joe, and the Editors of Time-Life Books. *New England*. New York: Time Incorporated, 1967.

McClaughry, John. "The New Feudalism." *Environmental Law* 5 (Spring 1975): 675–702.

McGreevy, Patrick Vincent. *Imagining Niagara: The Meaning and Making of Niagara Falls*. Amherst: University of Massachusetts Press, 1994.

Meeks, Harold. "Stagnant, Smelly, and Successful: Vermont's Mineral Springs." *Vermont History* 47 (Winter 1979): 5–20.

——. *Time and Change in Vermont: A Human Geography*. Chester, Conn.: Globe Pequot Press, 1986.

Meinig, D. W., ed. *The Interpretation of Ordinary Landscapes: Geographical Essays*. New York: Oxford University Press, 1979.

——. "Symbolic Landscapes: Some Idealizations of American Communities." In *The Interpretation of Ordinary Landscapes: Geographical Essays*, ed. D. W. Meinig, 164–92. New York: Oxford University Press, 1979.

Mergen, Bernard. *Snow in America*. Washington, D.C.: Smithsonian Institution Press, 1997.

Merrill, Perry. *Roosevelt's Forest Army: A History of the Civilian Conservation Corps, 1933–1942*. Montpelier, Vt.: Published by the Author, 1981.

——. *Vermont Skiing: A Brief History of Downhill and Cross Country Skiing*. Montpelier, Vt.: Published by the Author, 1987.

Meyer, William B. *Americans and Their Weather*. New York: Oxford University Press, 2000.

Milbourne, Paul, ed. *Revealing Rural "Others": Representation, Power and Identity in the British Countryside*. London: Pinter, 1997.

Mitchell, Don. *The Lie of the Land: Migrant Workers and the California Landscape*. Minneapolis: University of Minnesota Press, 1996.

——. *Cultural Geography: A Critical Introduction*. Malden, Mass.: Blackwell Publishers, 2000.

——. "Cultural Landscapes: Just Landscapes or Landscapes of Justice?" *Progress in Human Geography* 27 (December 2003): 787–96.

Mitchell, W. J. T., ed. *Landscape and Power*. 2nd ed. Chicago: University of Chicago Press, 2002.

Morin, Karen M. "Peak Practices: Englishwomen's 'Heroic' Adventures in the Nineteenth-Century American West." *Annals of the Association of American Geographers* 89 (September 1999): 489–514.

Mungo, Raymond. *Total Loss Farm: A Year in the Life*. New York: E. P. Dutton & Co., Inc., 1970.

Murphy, Thomas D. *New England Highways and Byways from a Motor Car: On Sunrise Highways*. Boston: L. C. Page and Company, 1924.

Myers, Phyllis. *So Goes Vermont: An Account of the Development, Passage, and Implementation of State Land-Use Legislation in Vermont*. Washington, D.C.: The Conservation Foundation, 1974.

Nash, Roderick. *Wilderness and the American Mind*. 3rd ed. New Haven, Conn.: Yale University Press, 1982.

National Survey, The, comp., ed. *The National Survey Eastern Ski Atlas*. Chester, Vt.: The National Survey, 1969.

Nearing, Helen, and Scott Nearing. *The Maple Sugar Book: Together with Remarks on Pioneering as a Way of Living in the Twentieth Century*. New York: Schocken Books, 1970, 1950.

Nickerson, Norma Polovitz, Rita J. Black, Stephen F. McCool. "Agritourism: Motivations Behind Farm/Ranch Business Diversification." *Journal of Travel Research* 40 (August 2001): 19–26.

Nicolson, Marjorie Hope. *Mountain Gloom and Mountain Glory: The Development of the Aesthetics of the Infinite*. 1959. Reprint, with a foreword by William Cronon, Seattle: University of Washington Press, 1997.

Nissenbaum, Stephen. "New England as Region and Nation." In *All Over the Map: Rethinking American Regions*, Edward L. Ayers, Patricia Nelson Limerick, Stephen Nissenbaum, and Peter S. Onuf, 38–61. Baltimore: Johns Hopkins University Press, 1996.

Nolen, John, Philip Shutler, Albert LaFleur, and Dana M. Doten. *Graphic Survey: A First Step in State Planning for Vermont (A Report Submitted to the Vermont State Planning Board and National Resource Board)*. No publisher, [1937].

Novak, Barbara. *Nature and Culture: American Landscape Painting, 1825–1875*. London: Thames and Hudson, 1980.

Nuquist, Andrew E., and Edith W. Nuquist. *Vermont State Government and Administration: An Historical and Descriptive Study of the Living Past*. Burlington, Vt.: Government Research Center, University of Vermont, 1966.

Nutting, Wallace. *Vermont Beautiful*. New York: Bonanza Books, 1922.

Nye, David E. *American Technological Sublime*. Cambridge, Mass.: MIT Press, 1996.

Nye, David E., ed. *Technologies of Landscape: From Reaping to Recycling*. Amherst: University of Massachusetts Press, 1999.

O'Connell, James C. *Becoming Cape Cod: Creating a Seaside Resort*. Hanover, N.H.: University Press of New England, 2002.

Olzendam, Roderic Marble. *The Lure of Vermont's Silent Places: "The Green Mountains."* Essex Junction, Vt.: [Vermont Publicity Bureau, Office of the Secretary of State], [1918].

Orton, Vrest, ed. *And So Goes Vermont: A Picture Book of Vermont as It Is*. Weston, Vt: Countryman Press; New York: Farrar and Rinehart, 1937.

——. *The American Cider Book: The Story of America's Natural Beverage*. New York: Farrar, Strauss and Giroux, 1973.

Passenger Department, Central Vermont Railroad. *Summer Homes Among the Green Hills of Vermont and Along the Shores of Lake Champlain*. St. Albans, Vt.: St. Albans Messenger Co., 1893.

Peach, Arthur Wallace, and Harold Goddard Rugg, eds. *Vermont Prose: A Miscellany*. 2nd ed. Brattleboro, Vt.: Stephen Daye Press, 1932.

Pearl, Mary. *Vermont Maple Recipes*. Burlington, Vt.: Lane Press, 1952.

Perkins, Henry F. "The Comprehensive Survey of Rural Vermont." In *New England's Prospect: 1933*, ed. John K. Wright, 206–12. New York: American Geographical Society, 1933.

Perrin, Noel. *Amateur Sugar Maker*. Hanover, N.H.: University Press of New England, 1972.

——. *First Person Rural: Essays of a Sometime Farmer*. New York: Penguin Books, 1980, 1978.

Perry, George. "A Convenient and Profitable Home Market." In *Fifth Annual Report of the Vermont State Horticultural Society*, 84–90. Bellows Falls, Vt.: P. H. Gobie Press, 1907.

Plotkin, Sidney. *Keep Out: The Struggle for Land Use Control*. Berkeley: University of California Press, 1987.

Pollan, Michael. *The Botany of Desire: A Plant's-Eye View of the World*. New York: Random House, 2001.

Prime, William Cowper. *Along New England Roads*. New York: Harper and Brothers, 1892.

Pritchard, Annette, and Nigel J. Morgan. "Constructing Tourism Landscapes — Gender, Sexuality and Space." *Tourism Geographies* 2 (May 2000): 115–39.

Proceedings of the Twenty-Eighth Annual Meeting of the Vermont Maple Sugar Makers' Association. St. Albans, Vt.: St. Albans Messenger Co. Print, 1921.

Publishers of the Eastern Ski Map, comp., ed. *The National Survey Eastern Ski Atlas*. Chester, Vt.: The National Survey, 1964.

Purchase, Eric. *Out of Nowhere: Disaster and Tourism in the White Mountains*. Baltimore: Johns Hopkins University Press, 1999.

Rebek, Andrea. "The Selling of Vermont: From Agriculture to Tourism, 1860–1910." *Vermont History* 44 (Winter 1976): 14–27.

Recreation Study Committee, Vermont State Planning Board. *Recreation Study Tentative Report*. No publisher, 1939.

Reiger, John F. *American Sportsmen and the Origins of Conservation*. Rev. ed. Norman: University of Oklahoma Press, 1986.

Report of the Department of Conservation and Development for the Term Ending June 30, 1940. Springfield, Vt.: Springfield Printing Corporation, [1940].

Report of the Industrial Commission on Agriculture and Agricultural Labor. Washington, D.C.: Government Printing Office, 1901.

Report of Publicity Service, Department of Conservation and Development, State of Vermont, for Years Ending June 30, 1935 and 1936. Burlington, Vt.: Free Press Printing Co., [1936].

Resch, Tyler. *Dorset: In the Shadow of the Marble Mountain*. West Kennebunk, Maine: Phoenix Publishing, 1989.

Research Division of the Vermont Development Commission. *Vermont Ski Facilities*. [Montpelier, Vt.]: [The Commission], 1948.

Richards, David L. *Poland Spring: A Tale of the Gilded Age, 1860–1900*. Hanover, N.H.: University Press of New England, 2005.

Richmond, Roaldus. "The Green Mountain State." In *Vermont: A Profile of the Green Mountain State*, Vermont Writers' Project, no page numbers. New York: Fleming Publishing Company, 1941.

Robert Burley Associates. *People on the Land: Settlement Patterns for Vermont*. Waitsfield, Vt.: The Associates, 1973.

Roberts, Lesley, and Derek Hall, eds. *Rural Tourism and Recreation: Principles to Practice*. New York: CABI Publishing, 2001.

Rome, Adam. *The Bulldozer in the Countryside: Suburban Sprawl and the Rise of American Environmentalism*. New York: Cambridge University Press, 2001.

Roomet, Louise B. "Vermont as a Resort Area in the Nineteenth Century." *Vermont History* 44 (Winter 1976): 1–13.

Rothman, Hal K. *Devil's Bargains: Tourism in the Twentieth-Century American West*. Lawrence: University Press of Kansas, 1998.

Russell, C. M., ed. *Reunion of the Sons and Daughters of the Town of Wilmington.* Wilmington, Vt.: Deerfield Valley Times Press, 1890.

Russell, Howard S. *A Long Deep Furrow: Three Centuries of Farming in New England.* 1976. Abridged, with a foreword by Mark Lapping, Hanover, N.H.: University Press of New England, 1982.

Ryden, Kent C. *Landscape with Figures: Nature and Culture in New England.* Iowa City: University of Iowa Press, 2001.

Samuelson, Myron. *The Story of the Jewish Community of Burlington, Vermont.* Burlington, Vt.: Published by the Author, 1976.

Sanford, Robert M., and Mark B. Lapping. "The Beckoning Country: Act 200, Act 250 and Regional Planning in Vermont." In *Big Places, Big Plans,* ed. Mark B Lapping and Owen J. Furuseth, 5–27. Burlington, Vt.: Ashgate, 2004.

Schein, Richard H. "The Place of Landscape: A Conceptual Framework for Interpreting an American Scene." *Annals of the Association of American Geographers* 87 (December 1997): 660–80.

———. "Normative Dimensions of Landscape." In *Everyday America: Cultural Landscape Studies after J. B. Jackson,* ed. Chris Wilson and Paul Groth, 199–218. Berkeley: University of California Press, 2003.

Schmitt, Peter. *Back to Nature: The Arcadian Myth in Urban America.* 1969. Reprint with a foreword by John Stilgoe, Baltimore: Johns Hopkins University Press, 1990.

Schnell, Steven M. "Creating Narratives of Place and Identity in 'Little Sweden, U.S.A.'" *Geographical Review* 93 (January 2003): 1–29.

Schulte, Janet E. "Summer Homes: A History of Family Summer Vacation Communities in Northern New England, 1880–1940." Ph.D. diss., Brandeis University, 1994.

Searls, Paul. "America and the State That 'Stayed Behind': An Argument for the National Relevance of Vermont History." *Vermont History* 71 (Winter/Spring 2003): 75–87.

Sears, John F. *Sacred Places: American Tourist Attractions in the Nineteenth Century.* Amherst: University of Massachusetts Press, 1989.

Seasons of Change: Fifty Years with Vermont Life, 1946–1996. Montpelier, Vt.: Vermont Life, 1996.

Shaffer, Marguerite S. *See America First: Tourism and National Identity, 1880–1940.* Washington, D.C.: Smithsonian Institution Press, 2001.

Shalhope, Robert E. *Bennington and the Green Mountain Boys: The Emergence of Liberal Democracy in Vermont, 1760–1850.* Baltimore: Johns Hopkins University Press, 1996.

Sherman, Joe. *Fast Track on a Dirt Road: Vermont Transformed, 1945–1990.* Woodstock, Vt.: Countryman Press, Inc., 1991.

Sherman, Michael. *Vermont State Government since 1965.* Burlington, Vt.: The Center for Research on Vermont, The Snelling Center for Government, University of Vermont, 1995.

Sherman, Michael, Gene Sessions, and P. Jeffrey Potash. *Freedom and Unity: A History of Vermont.* Barre, Vt.: Vermont Historical Society, 2004.

Shi, David. *The Simple Life: Plain Living and High Thinking in American Culture.* New York: Oxford University Press, 1985.

Sibley, David. *Geographies of Exclusion: Society and Difference in the West.* New York: Routledge, 1995.

Sights in Barre. Glens Falls, N.Y.: C. H. Possons, Publisher, 1887.

Silverstein, Hannah. "No Parking: Vermont Rejects the Green Mountain Parkway." *Vermont History* 63 (Summer 1995): 133–57.

Sinclair, Robert O. *Vermont's Dairy Industry.* Burlington, Vt.: University of Vermont and State Agricultural College, Vermont Agricultural Experiment Station, 1956.

———. *Trends in Rural Land Prices in Vermont.* Burlington, Vt.: Agricultural Experiment Station, University of Vermont, 1969.

Sinclair, Robert O., and Malcom I. Bevins. *Small Farms in Vermont.* Burlington, Vt.: University of Vermont, Vermont Agricultural Experiment Station, 1961.

Sinclair, Robert O., and Stephen B. Meyer. *Nonresident Ownership of Property in Vermont.* Burlington, Vt.: Agricultural Experiment Station, University of Vermont, 1972.

Slater, Tom. "Fear of the City, 1882–1967: Edward Hopper and the Discourse of Anti-Urbanism." *Social and Cultural Geography* 3 (June 2002): 135–54.

Slayton, Tom. "Five Decades, Six Editors, One Magazine: A History of *Vermont Life.*" In *Seasons of Change: Fifty Years with Vermont Life, 1946–1996,* 14–19. Montpelier, Vt.: Vermont Life, 1996.

Smith, Amory C., and Frederic O. Sargent. *The Rural Land Market in Vermont.* Burlington, Vt.: Vermont Resources Research Center, Vermont Agricultural Experiment Station, University of Vermont, 1964.

Smith, Henry Nash. *Virgin Land: The American West as Symbol and Myth.* Cambridge, Mass.: Harvard University Press, 1950.

Smith, Merritt Roe, and Leo Marx, eds. *Does Technology Drive History?: The Dilemma of Technological Determinism.* Cambridge, Mass.: MIT Press, 1994.

. . . *So Goes Vermont.* Videocassette. Prod. John Karol. 23 min. Environmental Planning Information Center, a social project of the Vermont Natural Resources Council. 1971, 1992.

Spear, Victor. "Brattleboro Institute." In *Thirteenth Vermont Agricultural Report by the State Board of Agriculture, for the Year 1893,* 50–57. Burlington, Vt.: Free Press Association, Printers and Binders, 1893.

———. *Report on Summer Travel for 1894.* Montpelier, Vt.: Press of the Watchman Publishing Co., 1894.

Speare, Charles F. *We Found a Farm.* Brattleboro, Vt.: Stephen Daye Press, 1936.

State of Vermont, Governor's Commission on Environmental Control, Reports to Governor, January 19, 1970, May 18, 1970. No publisher, [1970].

Stone, Arthur F. *The Vermont of Today.* Vol. 1. New York: Lewis Historical Publishing Company, 1929.

——. *The Vermont of Today*. Vol. 2. New York: Lewis Historical Publishing Company, Inc., 1929.

Summer Homes in Vermont Corporation. *Your Summer Home in Vermont*. New York: The Corporation, 1927.

Sutter, Paul S. *Driven Wild: How the Fight Against Automobiles Launched the Modern Wilderness Movement*. Seattle: University of Washington Press, 2002.

Sykes, James G. *Trends in Vermont Agriculture*. Burlington, Vt.: Vermont Resources Research Center, Vermont Agricultural Experiment Station, University of Vermont, 1964.

Tadejewski, Edward J. *Roadside Marketing in Vermont*. Burlington, Vt.: University of Vermont and State Agricultural College, Agricultural Experiment Station, 1949.

Tales from the Mountain: Mount Snow's First 40 Years. Videocassette. Prod. and dir. Allen and Sally Seymour. 30 min. Vermont Studios, Inc., 1994.

Taylor, Fred H. *Variation in Sugar Content of Maple Sap*. Burlington, Vt.: University of Vermont and State Agricultural College, Agricultural Experiment Station, 1956.

Terrie, Philip G. *Contested Terrain: A New History of Nature and People in the Adirondacks*. Blue Mountain Lake, N.Y.: Adirondack Museum; Syracuse, N.Y.: Syracuse University Press, 1997.

Terry, Stephen C. *A History of the Hoff Years, 1963–1969*. Ed. Sally Johnson and Sam Hand. No publisher, 1990.

Thirteenth Vermont Agricultural Report by the State Board of Agriculture, for the Year 1893. Burlington, Vt.: Free Press Association, Printers and Binders, 1893.

Thompson, John. *The Tourist and Recreation Industry in Vermont*. Montpelier, Vt.: The Vermont Development Department, 1963.

Till, Karen E. "Construction Sites and Showcases: Mapping 'The New Berlin' through Tourism Practices." In *Mapping Tourism*, ed. Stephen P. Hanna and Vincent J. Del Casino Jr., 51–78. Minneapolis: University of Minnesota Press, 2003.

Tolles, Bryant F. Jr. *Summer Cottages in the White Mountains: The Architecture of Leisure and Recreation, 1870–1930*. Hanover, N.H.: University Press of New England, 2000.

Truettner, William H., and Roger B. Stein, eds. *Picturing Old New England: Image and Memory*. Washington, D.C.: National Museum of American Art, Smithsonian Institution; New Haven, Conn.: Yale University Press, 1999.

United States Department of Agriculture, Bureau of Public Roads, and the Vermont State Highway Department. *Report of a Survey of Transportation on the State Highways of Vermont*. No publisher, 1927.

United States Department of Commerce, Bureau of the Census. *Fifteenth Census of the United States: 1930, Population Volume III, Part 2*. Washington, D.C. Government Printing Office, 1932.

Urry, John. *The Tourist Gaze: Leisure and Travel in Contemporary Societies*. London: Sage, 1990.

Valentine, Alonzo B. *Report of the Commissioner of Agricultural and Manufactur-*

ing Interests of the State of Vermont, 1889–1890. Rutland, Vt.: Tuttle Company, Official State Printers, 1890.

Vaughan, Beatrice. *Real, Old-Time Yankee Maple Cooking*. Brattleboro, Vt.: Stephen Greene Press, 1969.

Vermont Agency of Environmental Conservation. *Vermont Vacation Home Inventory, 1973*. Montpelier, Vt.: The Agency, 1974.

Vermont Bureau of Publicity. *Vermont: The Land of the Green Mountains*. [Essex Junction, Vt.]: Vermont Bureau of Publicity, Office of the Secretary of State, 1913.

Vermont Central Planning Office. *State Planning in Vermont*. Montpelier, Vt.: The Office, 1964.

Vermont Commission on Country Life. *Rural Vermont: A Program for the Future*. Burlington, Vt.: The Commission, 1931.

Vermont Cook Book by Vermont Cooks, A. White River Junction, Vt.: Green Mountain Studios, Inc., 1947.

Vermont Cottages, Camps and Furnished Houses for Rent. Montpelier, Vt.: Vermont Publicity Service, Department of Conservation and Development, 1938.

Vermont Environmental Board. *Vermont Environmental Board: Twenty-Fifth Anniversary Report, 1970–1995*. Montpelier, Vt.: The Board, 1995.

Vermont Maple Sugar and Syrup. Montpelier, Vt.: The Vermont Development Commission and the Vermont Department of Agriculture, 1957.

Vermont Planning Council. *Vision and Choice: Vermont's Future; The State Framework Plan*. Montpelier, Vt.: The Council, 1968.

Vermont State Board of Agriculture. *The Resources and Attractions of Vermont. With a List of Desirable Homes for Sale*. Montpelier, Vt.: Press of the Watchman Publishing Company, 1891.

———. *Resources and Attractions of Vermont. With a List of Desirable Homes for Sale*. Montpelier, Vt.: Press of the Watchman Publishing Co., 1892.

———. *Good Homes in Vermont: A List of Desirable Farms for Sale*. Montpelier, Vt.: Press of the Watchman Publishing Co. 1893.

———. *Vermont, Its Opportunities for Investment in Agriculture, Manufacture, Minerals, Its Attractions for Summer Homes*. East Hardwick, Vt.: The Board, [1903].

Vermont State Planning Board. *Report on Winter Sports Development*. [Montpelier, Vt.]: [The Board], 1938).

Vermont State Planning Office. *Interim Land Capability Plan*. [Montpelier, Vt.]: [The Office], 1971.

———. *Vermont's Land Use Plan and Act 250*. Montpelier, Vt.: The Office, 1974.

Vermont Writers' Project. *Vermont: A Profile of the Green Mountain State*. New York: Fleming Publishing Company, 1941.

Vicero, Ralph D. "French-Canadian Settlement in Vermont Prior to the Civil War." *Professional Geographer* 23 (October 1971): 290–94.

Vincent, Robert W. "Downhill Ski Trails." *The American School and University* 11 (1939): 244–48.

Votey, Constance. *Growing Up with Aspenhurst.* 2nd ed. [Greensboro, Vt.]: Printed by the Greensboro Historical Society, 1980.

Walbridge, John H. *Wilmington, Vermont.* Wilmington, Vt.: The [Deerfield Valley] Times Press, 1900.

Walker, Peter, and Louise Fortmann. "Whose Landscape: A Political Ecology of the 'Exurban' Sierra." *Cultural Geographies* 10 (October 2003): 469–91.

Ward, David. *Poverty, Ethnicity, and the American City, 1840–1925: Changing Conceptions of the Slum and Ghetto.* New York: Cambridge University Press, 1989.

Washington, Ida H. "Dorothy Canfield Fisher's *Tourists Accommodated* and Her Other Promotions of Vermont." *Vermont History* 65 (Summer/Fall 1997): 151–64.

Waterman, Laura, and Guy Waterman. *Forest and Crag: A History of Hiking, Trail Blazing, and Adventure in the Northeast Mountains.* Boston: Appalachian Mountain Club, 1989.

Watkins, Francine. "The Cultural Construction of Rurality: Gender Identities and the Rural Idyll." In *Thresholds in Feminist Geography: Difference, Methodology, Representation,* ed. John Paul Jones III, Heidi J. Nast, and Susan M. Roberts, 383–92. Lanham, Md.: Rowman and Littlefield Publishers, 1997.

Watson, Peter D., Wilhelmina Smith, Lewis Hill, Nancy Hill, Sally Fisher, Patricia Haslam, Rhoda Metraux, Dorothy Ling, and Gail Sangree. *History of Greensboro: The First Two Hundred Years.* Greensboro, Vt.: Greensboro Historical Society, 1990.

Webster, Fred C., and Christopher G. Barbieri. *Maple Marketing in Department Stores.* Burlington, Vt.: Vermont Agricultural Experiment Station, University of Vermont and State Agricultural College, 1964.

Wells, George F. *The Status of Rural Vermont.* St. Albans, Vt.: Cummings Printing Company for the Vermont State Agricultural Commission, 1903.

Wessels, Tom. *Reading the Forested Landscape: A Natural History of New England.* Woodstock, Vt.: Countryman Press, 1997.

West River Valley Association. *The Call of the Country.* Brattleboro, Vt.: The Association, 1912.

Whatmore, Sarah, Terry Marsden, and Philip Lowe, eds. *Gender and Rurality.* London: David Fulton Publishers, 1994.

White, Morton, and Lucia White. *The Intellectual Versus the City: From Thomas Jefferson to Frank Lloyd Wright.* Cambridge, Mass.: Harvard University Press, 1962.

White, Richard. *The Organic Machine: The Remaking of the Columbia River.* New York: Hill and Wang, 1995.

———. "'Are You an Environmentalist or Do You Work for a Living?': Work and Nature." In *Uncommon Ground: Rethinking the Human Place in Nature,* ed. William Cronon, 171–85. New York: W. W. Norton and Company, 1996.

Williams, Raymond. *The Country and the City.* New York: Oxford University Press, 1973.

Wilson, Harold Fisher. *The Hill Country of Northern New England: Its Social and Economic History, 1790–1930.* New York: Columbia University Press, 1936.

Wilson, Leonard U. "Land Use, Planning, and Environmental Protection." In *Vermont State Government since 1965*, ed. Michael Sherman, 453–66. Burlington, Vt.: The Center for Research on Vermont, The Snelling Center for Government, University of Vermont, 1995.

Wood, Joseph S. *The New England Village*. Baltimore: Johns Hopkins University Press, 1997.

Wood and Water: Mills in Searsburg, Vermont. [Burlington, Vt.]: Consulting Archaeology Program, University of Vermont, 1980.

Woods, Michael. "Deconstructing Rural Protest: The Emergence of a New Social Movement." *Journal of Rural Studies* 19 (July 2003): 309–25.

Woolfson, Peter. "The Rural Franco-American in Vermont." *Vermont History* 50 (Summer 1982): 151–62.

Worster, Donald. *Nature's Economy: A History of Ecological Ideas*. New York: Cambridge University Press, 1977.

Wright, John K., ed. *New England's Prospect: 1933*. New York: American Geographical Society, 1933.

———. "Stowe in Early Spring, 1919." In *Outsiders Inside Vermont: Travelers' Tales of 358 Years*, ed. T. D. Seymour Bassett, 107–10. Brattleboro, Vt.: Stephen Green Press, 1967.

Wrobel, David M. and Patrick T. Long, eds. *Seeing and Being Seen: Tourism in the American West*. Lawrence: University Press of Kansas, 2001.

———. "Introduction: Tourists, Tourism, and the Toured Upon." In *Seeing and Being Seen: Tourism in the American West*, ed. David M. Wrobel and Patrick T. Long, 1–34. Lawrence: University Press of Kansas, 2001.

Yale, Allen R. Jr. *While the Sun Shines: Making Hay in Vermont, 1789–1990*. Montpelier, Vt.: Vermont Historical Society, 1991.

Yearbook of the United States Department of Agriculture, 1914. Washington, D.C.: Government Printing Office, 1915.

Index

Page numbers in *italics* represent figures.